COMPUTABILITY AND LOGIC

COMPUTABILITY AND LOGIC

Second edition

GEORGE BOOLOS
Professor of Philosophy
Massachusetts Institute of Technology

RICHARD JEFFREY
Professor of Philosophy
Princeton University

CAMBRIDGE UNIVERSITY PRESS
CAMBRIDGE
LONDON NEW YORK NEW ROCHELLE
MELBOURNE SYDNEY

Published by the Press Syndicate of the University of Cambridge
The Pitt Building, Trumpington Street, Cambridge CB2 1RP
32 East 57th Street, New York, NY 10022, USA
296 Beaconsfield Parade, Middle Park, Melbourne 3206, Australia

First published 1974
Second edition 1980

Typeset in Great Britain by William Clowes & Sons Limited
Printed in Great Britain at the University Press, Cambridge

British Library cataloguing in publication data

Boolos, George

Computability and logic—2nd ed.

1. Logic, Symbolic and mathematical
I. Title II. Jeffrey, Richard Carl
511'.3 BC135 80-40432
ISBN 0 521 23479 4 hard covers
ISBN 0 521 29967 5 paperback
(ISBN 0 521 20402 X 1st edition)

FOR
REBECCA
AND
EDITH

Logical dependence of the chapters
('↓' means 'presupposes')

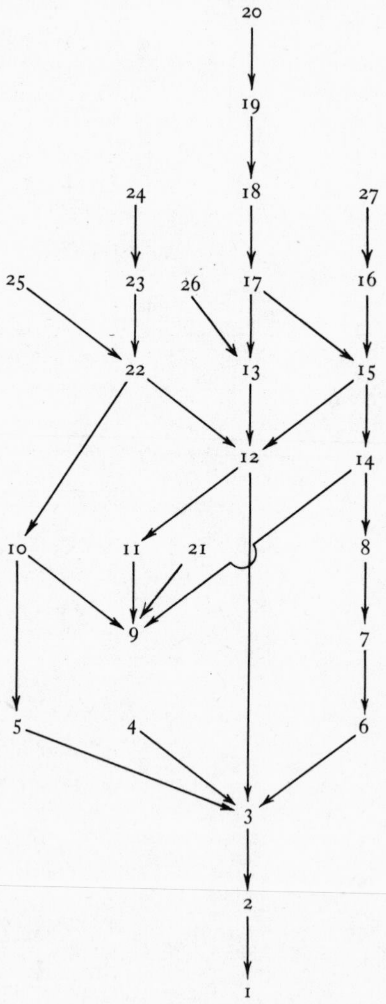

Contents

Preface

Computability and Logic is intended for the student in philosophy or pure or applied mathematics who has mastered the material ordinarily covered in a first course in logic and who wishes to advance his or her acquaintance with the subject. The aim of the book is to present the principal fundamental theoretical results *about* logic, and to cover certain other metalogical results whose proofs are not easily obtainable elsewhere. We have tried to make the exposition as readable as was compatible with the presentation of complete proofs, to use the most elegant proofs we knew of, to employ standard notation, and to reduce *hair* (as it is technically known).

A brief overview: We first introduce the notion of enumerability (countability, denumerability), and demonstrate the existence of non-enumerable sets. We then introduce the notion of computability by means of Turing machines, whose halting problem is shown to be unsolvable. We introduce two other notions of computability and show their equivalence to the Turing machine characterization. The undecidability of first-order (elementary) logic is proved directly from the unsolvability of the halting problem, without appeal to 'arithmetization'. The soundness and the completeness of a certain system of deduction are then shown. The Skolem–Löwenheim theorem, in the form 'every interpretation has an elementarily equivalent subinterpretation with an enumerable domain', is a by-product of the proofs of soundness and completeness. We next show that all recursive functions are representable in Q (Robinson's arithmetic) and then derive Church's theorem on the undecidability of first-order logic (again), Tarski's theorem on the indefinability of truth, and Gödel's first incompleteness theorem, in the form 'There is no complete, consistent, axiomatizable extension of Q'. We next prove versions of Gödel's second incompleteness theorem and Löb's theorem, discuss the existence and nature of non-standard models of arithmetic, and point out the (sharp) differences between second- and first-order logic. We show that, in contrast to Tarski's indefinability theorem, arithmetical truth can be defined in second-order arithmetic, and certain 'approximations' to arithmetical truth can be defined in (first-order) arithmetic.

We then prove Addison's theorem, which states that the class of sets (of natural numbers) that are definable in arithmetic is not itself definable in arithmetic. We then demonstrate Presburger's theorem, that arithmetic becomes decidable if multiplication is dropped. The Craig interpolation lemma is proved and the Robinson consistency theorem and the Beth definability theorem are derived from Craig's 'lemma'. Finally, we show the decidability of monadic logic with identity and the undecidability of logic without identity and with only a single two-place predicate letter.

R.J. was the principal author of Chapters 1 through 9; G.B., of Chapters 13 through 25. 10, 11, and 12 are joint.

We wish to thank Simon Kochen and Saul Kripke for two suggestions which greatly improved Chapter 14. Kochen suggested that the recursive functions be shown to be the functions obtainable by composition and minimization from the identity functions, plus, times, and the characteristic function of the identity relation; the proof that all recursive functions are representable in Q is then much simplified. The idea of avoiding the Chinese remainder theorem by using concatenation over a *variable* prime base to define a simple β-function is due to Kripke. We are also grateful to Paul Benacerraf, Burton Dreben, Hartry Field, Clark Glymour, Warren Goldfarb, David Lewis, Paul Mellema, Hilary Putnam, W. V. O. Quine, T. M. Scanlon, James Thomson, and Peter Tovey for comments, corrections, and suggestions. We are also indebted to the books of Church, Enderton, Kleene, Kreisel and Krivine, Shoenfield, and Tarski, Mostowski, and Robinson.

GEORGE BOOLOS
RICHARD JEFFREY

August 1973

Preface to the second edition

Two new chapters have been written for this second edition, a third has been completely revised, several new sections and exercises have been added, and we have made numerous typographical corrections. We are grateful to Leroy Meyers, David Auerbach, David Fair, Osvaldo Jaeggli, Stephen Leeds, Michael Levin, Michael Pendlebury, and many others for pointing typographical errors out to us, and we are also indebted to Douglas Edwards for discovering the flaw in the old version of Chapter 8. The new chapters, 26 and 27, deal with the famous combinatorial theorems of F. P. Ramsey and with the connections between modal logic and the concepts of provability and consistency.

The response to the first edition of *Computability and Logic* has delighted us, and we are greatly pleased to have the opportunity to improve our book.

GEORGE BOOLOS
RICHARD JEFFREY

February 1980

1
Enumerability

An *enumerable* set is one whose members can be enumerated: arranged in a single list with a first entry, a second entry, etc., so that every member of the set appears sooner or later in the list. Examples: the set P of positive integers is enumerated by the list

$$1, 2, 3, 4, \ldots$$

and the set E of even positive integers is enumerated by the list

$$2, 4, 6, 8, \ldots$$

Of course, the entries in these lists are not integers, but names of integers. In listing the members of a set (say, the members of the United States Senate) you manipulate names, not the things named. (You don't have the Senators form a queue; rather, you arrange their *names* in a list, perhaps alphabetically.)

By courtesy, we regard as enumerable the empty set, $\varnothing$, which has no members. (*The* empty set: there is only one. The terminology is a bit misleading: it suggests a comparison of empty sets with empty containers. But sets are more aptly compared with contents, and it should be considered that all empty containers have the same, null content.)

A list that enumerates a set may be finite or unending. An infinite set which is enumerable is said to be *enumerably infinite* or *denumerable*. Let us get clear about what things count as infinite lists, and what things do not. The positive integers can be arranged in a single infinite list as indicated above, but the following is not acceptable as a list of the positive integers:

$$1, 3, 5, 7, \ldots, 2, 4, 6, 8, \ldots$$

Here, all the odd positive integers are listed, and then all the even ones. This will not do. In an acceptable list, each item must appear sooner or later as the nth entry, for some *finite* n. But in the unacceptable arrangement above, none of the even positive integers are represented in this way. Rather, they appear (so to speak) as entries number $\infty + 1, \infty + 2$, etc.

To make this point perfectly clear we might define an enumeration of a set not as a listing, but as an arrangement in which each member of the set is *associated with* one of the positive integers 1, 2, 3, … Actually, a list *is* such an arrangement. The thing named by the first entry in the list is associated with the positive integer 1, the thing named by the second entry is associated with the positive integer 2, and in general, the thing named by the nth entry is associated with the positive integer n.

In enumerating a set by listing its members, it is perfectly all right if a member of the set shows up more than once in the list. The requirement is rather that each member show up *at least once*. It does not matter if the list is redundant; all we require is that it be complete. Indeed, a redundant list can always be thinned out to get an irredundant list. (Examine each entry in turn, comparing it with the finite segment of the list that precedes it. Erase the entry if it already appears in that finite segment.)

In mathematical parlance, an infinite list determines a *function* (call it 'f') which takes positive integers as *arguments* and takes members of the set as *values*. The value of the function f for the argument n is denoted by '$f(n)$'. This value is simply the thing denoted by the nth entry in the list. Thus, the list

$$2, 4, 6, 8, \ldots$$

which enumerates the set E of even positive integers determines the function f for which we have

$$f(1) = 2, \quad f(2) = 4, \quad f(3) = 6, \quad f(4) = 8, \ldots$$

And conversely, the function f determines the list, except for notation. (The same list would look like this, in binary notation: 10, 100, 110, 1000, ...) Thus, we might have defined the function f first, by saying that for any positive integer n, the value of f is $f(n) = 2n$; and then we could have described the list by saying that for each positive integer n, its nth entry is the decimal representation of the number $f(n)$, i.e., of the number $2n$.

Then we may speak of sets as being enumerated by functions, as well as by lists. Instead of enumerating the odd positive integers by the list 1, 3, 5, 7, ..., we may enumerate them by the function which assigns to each positive integer n the value $2n-1$. And instead of enumerating the set P of all positive integers by the list 1, 2, 3, 4, ..., we may enumerate P by the function which assigns to each positive integer n the value n itself. (This is the *identity function*. If we call it 'i', we have $i(n) = n$ for each positive integer n.)

If one function enumerates a nonempty set, so does some other; and so, in fact, do infinitely many others. Thus, the set of positive integers is enumerated not only by the function i, but also by the function (call it 'g') which is determined by the following list:

$$2, 1, 4, 3, 6, 5, \ldots$$

This list is obtained from the list 1, 2, 3, 4, 5, 6, ... by interchanging entries in pairs: 1 with 2, 3 with 4, 5 with 6, and so on. This list is a strange but perfectly acceptable enumeration of the set P: every positive integer shows up in it, sooner or later. The corresponding function, g, can be defined as follows:

$$g(n) = \begin{cases} n+1 & \text{if } n \text{ is odd,} \\ n-1 & \text{if } n \text{ is even.} \end{cases}$$

This definition is not as neat as the definitions $f(n) = 2n$ and $i(n) = n$ of the functions f and i, but it does the job: it does indeed associate one and only one member of P with each positive integer n. And the function g so defined does indeed enumerate P: for each member m of P there is a positive integer n for which we have $g(n) = m$.

We have noticed that it is perfectly all right if a list that enumerates a set is redundant, for if we wished, we could always thin it out so as to remove repetitions. It is also perfectly all right if a list has gaps in it, since one could go through and close up the gaps in turn. Thus, a flawless enumeration of the positive integers is given by the following gappy list:

$$1, -, 2, -, 3, -, 4, -, 5, -, 6, -, \ldots$$

The corresponding function (call it 'h') assigns values corresponding to the first, third, fifth, ... entries, but assigns no values corresponding to the gaps (second, fourth, sixth, ... entries). Thus we have $h(1) = 1$, but $h(2)$ is nothing at all, for the function h is undefined for the argument 2; $h(3) = 2$, but $h(4)$ is undefined; $h(5) = 3$, but $h(6)$ is undefined. And so on: h is a *partial function* of positive integers, i.e., it is defined only for positive integer arguments, but not for all such arguments. Explicitly, we might define the partial function h as follows:

$$h(n) = (n+1)/2 \quad \text{if } n \text{ is odd.}$$

Or, to make it clear that we haven't simply forgotten to say what values h assigns to even positive integers, we might put the definition as follows:

$$h(n) = \begin{cases} (n+1)/2 & \text{if } n \text{ is odd,} \\ \text{undefined otherwise.} \end{cases}$$

Now the partial function h is a queer but perfectly acceptable enumeration of the set P of positive integers.

It would be perverse to choose h instead of the simple function i as an enumeration of P; but other sets are most naturally enumerated by partial functions. Thus, the set E of even positive integers is conveniently enumerated by the partial function (say, 'j') that agrees with i for even arguments, and is undefined for odd arguments:

$$j(n) = \begin{cases} n & \text{if } n \text{ is even,} \\ \text{undefined otherwise.} \end{cases}$$

The corresponding gappy list (in decimal notation) is

$$-, 2, -, 4, -, 6, -, 8, \ldots$$

Of course, the function f, defined by

$$f(n) = 2n$$

for all positive integers n, is an equally acceptable enumeration of E, corresponding to the gapless list,

$$2, 4, 6, 8, \ldots$$

Any set S of positive integers is enumerated quite simply by a partial function, s, which is defined as follows:

$$s(n) = \begin{cases} n & \text{if } n \text{ is in the set } S, \\ \text{undefined otherwise.} \end{cases}$$

Thus, the set $\{1, 2, 5\}$ which consists of the numbers 1, 2, and 5 is enumerated by the partial function k which is defined

$$k(n) = \begin{cases} n & \text{if } n = 1 \quad \text{or} \quad n = 2 \quad \text{or} \quad n = 5, \\ \text{undefined otherwise.} \end{cases}$$

We shall see in the next chapter that although every set *of positive integers* is enumerable, there are sets of other sorts which are not enumerable. To say that a set A is enumerable is to say that there is a function all of whose arguments are positive integers and all of whose values are members of A, and that each member of A is a value of this function: for each member a of A there is at least one positive integer n to which the function assigns a as its value. Notice that nothing in this definition requires A to be a set of positive integers. Instead, A might be a set of people (members of the United States Senate, perhaps); it might be a set of strings of symbols (perhaps, the set of all grammatically correct English sentences, where we count the space between adjacent words as a symbol); or the members of A might themselves be sets, as when A is the set $\{P, E,$

$\varnothing\}$. Here, A is a set with three members, each of which is itself a set. One member of A is the infinite set P of all positive integers; another member of A is the infinite set E of all even positive integers; and the third is the empty set $\varnothing$. The set A is certainly enumerable, e.g. by the following finite list:

$$P,\ E,\ \varnothing.$$

Each entry in this list names a member of A, and every member of A is named sooner or later in the list. The list determines a function – call it 'f' – which can be defined by the three statements

$$f(1) = P, \quad f(2) = E, \quad f(3) = \varnothing.$$

To be precise, f is a *partial function* of positive integers, being undefined for arguments greater than 3. Moral: a (partial or total) function of positive integers may have its values in any set whatever. Its arguments must all be positive integers, but its values need not be positive integers.

The set of all finite strings of letters of the alphabet provides an example of an enumerably infinite set which is not a set of positive integers. This set is enumerable because its members can be arranged in a single list, which we shall describe but not exhibit. The first 26 entries in the list are the 26 letters of the alphabet, in their usual order. The next 676 ($= 26^2$) entries are the two-letter sequences, in dictionary order. The following 17576 ($= 26^3$) entries are the three-letter strings, again in dictionary order. And so on, without end. From this description, with the aid of paper, pencil, and patience, you can determine the position in the list of any finite string that interests you. Then in effect we have defined a function of positive integers which enumerates the set in question. The definition does not have the form of a mathematical equation, but it is a perfectly acceptable definition for all that: it does its job, which is to associate a member of the set, as value, with each positive integer, as argument. No matter that the definition was given in English instead of mathematical notation, as long as it is clear and unambiguous.

Since the members of the set of all finite strings of letters are inscriptions, we can plausibly speak of arranging those members themselves in a list. But we might also speak of the entries in the list as *names of themselves* so as to be able to continue to insist that in enumerating a set, it is *names* of members of the set that are arranged in a list.

In conclusion, let us straighten out our terminology.

A *function* is an assignment of *values* to *arguments*. The set of all those arguments to which the function assigns values is called the *domain* of the function. The set of all those values which the function assigns to its arguments is called the *range* of the function. In the case of functions whose arguments are positive integers, we distinguish between *total* functions and *partial* functions. A total function of positive integers is one whose domain is the whole set P of positive integers. A partial function of positive integers is one whose domain is something less than the whole set P. From now on, when we speak simply of a *function of positive integers*, we should be understood as leaving it open, whether the function is total or partial. This is a departure from the usual terminology, in which 'function' (of positive integers) always means *total function*. Also, from now on we shall regard as a genuine partial function of positive integers the weird function e whose domain is the empty set $\varnothing$. *This function is undefined for all arguments!* Now a set is *enumerable* if and only if it is *the range of some function of positive integers*. The empty set is enumerable because it is the range of e.

Example. The set of positive rational numbers is enumerable. A positive rational number is a number that can be expressed as a ratio of positive integers: in the form m/n where m and n are positive integers. As a preliminary to enumerating them, we organize them in a rectangular array as in Table 1-1(*a*).

TABLE 1-1

(*a*)

$\frac{1}{1}$	$\frac{2}{1}$	$\frac{3}{1}$	$\frac{4}{1}$	$\frac{5}{1}$	...
$\frac{1}{2}$	$\frac{2}{2}$	$\frac{3}{2}$	$\frac{4}{2}$	$\frac{5}{2}$	...
$\frac{1}{3}$	$\frac{2}{3}$	$\frac{3}{3}$	$\frac{4}{3}$	$\frac{5}{3}$	...
$\frac{1}{4}$	$\frac{2}{4}$	$\frac{3}{4}$	$\frac{4}{4}$	$\frac{5}{4}$	...
$\frac{1}{5}$	$\frac{2}{5}$	$\frac{3}{5}$	$\frac{4}{5}$	$\frac{5}{5}$	...
⋮	⋮	⋮	⋮	⋮	

(*b*)

1	2	5	10	17	...
4	3	6	11	18	...
9	8	7	12	19	...
16	15	14	13	20	...
25	24	23	22	21	...
⋮	⋮	⋮	⋮	⋮	

Every positive rational number is represented somewhere in this array. Thus, the number m/n is represented in the mth column, nth row. In fact, every such number is represented infinitely often, for m/n, $2m/2n$, $3m/3n$, ... all represent the same number. In particular, the number 1 is represented by every entry in the main diagonal of Table 1-1(*a*): first row, first column; second row, second column; etc.

To show that the positive rationals are enumerable, we must arrange Table 1-1(*a*) in a single list. This can be done in various ways, e.g. in the pattern shown in Table 1-1(*b*), where the number at each position gives the position in the list of the corresponding entry in Table 1-1(*a*). Then the list, in which each positive rational is represented infinitely often, is this:

$$\frac{1}{1}, \quad \frac{2}{1}, \quad \frac{2}{2}, \quad \frac{1}{2}, \quad \frac{3}{1}, \ldots$$

The two arrays in Table 1-1 also give a perfectly satisfactory definition of a function of positive integers (call it 'r') which enumerates the positive rationals. Thus we have

$$r(1) = \tfrac{1}{1}, \quad r(2) = \tfrac{2}{1}, \quad r(3) = \tfrac{2}{2}, \quad r(4) = \tfrac{1}{2}, \quad r(5) = \tfrac{3}{1}, \ldots$$

or, simplifying,

$$r(1) = 1, \quad r(2) = 2, \quad r(3) = 1, \quad r(4) = \tfrac{1}{2}, \quad r(5) = 3, \ldots$$

The function r is quite adequately defined by the two arrays: given any positive integer n, it is perfectly straightforward to find the value $r(n)$ by building the two arrays up to the point where the number n is reached in the second. The value $r(n)$ will then be given by the corresponding entry in the first array. It is amusing, but by no means necessary, to give a more mathematical-looking definition of r. If you want to try it, here is a hint: every positive integer, n, is somewhere in the interval from one perfect square to the next, i.e., for each n there is exactly one m for which we have $(m-1)^2 < n \leqslant m^2$. Let us use the symbol 'sq' for the function which assigns the value m to the argument n as above. Your problem is to replace the question marks in the following definition by mathematical expressions involving n and the function sq:

$$r(n) = \begin{cases} \mathrm{sq}(n)/? & \text{if} \quad ? \\ ?/\mathrm{sq}(n) & \text{otherwise.} \end{cases}$$

But if this sort of thing paralyzes you, don't bother with it. If you do want to try it, you can check your answer against the solution to Exercise 1.1 below.

Exercises. (Solutions follow)

1.1 Define the function r in the form shown above, where $\mathrm{sq}(n)$ is the smallest number whose square is n or greater. *Hint*: begin by writing out the first few entries in the mth rows and mth columns of Table 1-1.

1.2 Show that the integers ..., -2, -1, 0, 1, 2, ... are enumerable, by arranging them in a single infinite list with a beginning but no end; and write out a mathematical-looking definition of the function which corresponds to your list.

1.3 Show that the set of all the rationals (positive, negative, and zero) is enumerable, by describing a single infinite list (with a beginning but no end) which enumerates them.

1.4 Show that the set of all *finite* strings of pluses and minuses is enumerable, by describing a single list that enumerates them.

1.5 Then show that the set of all *finite* sets of positive integers is enumerable, by describing a way in which finite strings of pluses and minuses can be used as names of finite sets of positive integers.

1.6 Let D be a certain set of sets of positive integers, as follows: a set of positive integers belongs to D if and only if it is definable by a finite number of words in English. Thus, the following sets belong to D because they are defined here by finite numbers of words in English.

E: the set of all even positive integers (7 words).
$\varnothing$: the set which has no members at all (8 words).
$\{1, 2\}$: the set whose only members are the numbers one and two (11 words).
$\{1, \ldots, 999\}$: the set of all positive integers less than one thousand (10 words).

And so on. *Show that the set D is enumerably infinite.*

Solutions

1.1 Bearing in mind that $m = \mathrm{sq}(n)$, the mth rows and columns of Table 1-1 are as follows:

(*a*)

$$\begin{array}{ccccc} & & & & \mathrm{sq}(n)/1 \\ & & & & \mathrm{sq}(n)/2 \\ & & & & \mathrm{sq}(n)/3 \\ & & & & \vdots \\ 1/\mathrm{sq}(n) & 2/\mathrm{sq}(n) & 3/\mathrm{sq}(n) & \ldots & \mathrm{sq}(n)/\mathrm{sq}(n) \ldots \\ & & & & \vdots \end{array}$$

(*b*)

$$\begin{array}{ccccc} & & & & (\mathrm{sq}(n)-1)^2+1 \\ & & & & (\mathrm{sq}(n)-1)^2+2 \\ & & & & (\mathrm{sq}(n)-1)^2+3 \\ & & & & \vdots \\ (\mathrm{sq}(n))^2-0 & (\mathrm{sq}(n))^2-1 & (\mathrm{sq}(n))^2-2 & \ldots & (\mathrm{sq}(n)-1)^2+m \ldots \\ & & & & \vdots \end{array}$$

The required definition then emerges as

$$r(n) = \begin{cases} \dfrac{\mathrm{sq}(n)}{n-(\mathrm{sq}(n)-1)^2} & \text{if} \quad n \leqslant (\mathrm{sq}(n)-1)^2+\mathrm{sq}(n), \\[2ex] \dfrac{(\mathrm{sq}(n))^2+1-n}{\mathrm{sq}(n)} & \text{otherwise.} \end{cases}$$

This is cute, but no more illuminating than the definition of r in terms of the arrays in Table 1-1.

1.2 The simplest list is: $0, 1, -1, 2, -2, 3, -3, \ldots$. Then if the corresponding function is called 'f' we have $f(1) = 0$, $f(2) = 1$, $f(3) = -1$, $f(4) = 2$, etc. Here is a mathematical-looking definition of f:

$$f(n) = \begin{cases} -\dfrac{n-1}{2} & \text{if } n \text{ is odd,} \\[2ex] \dfrac{n}{2} & \text{if } n \text{ is even.} \end{cases}$$

1.3 You already know how to arrange the positive rationals in a single infinite list. Write '0' in front of this list and then write the positive rationals, backwards, with minus signs before them, in front of that. You now have this:

$$\ldots, -3, -\tfrac{1}{2}, -1, -2, -1, 0, 1, 2, 1, \tfrac{1}{2}, 3, \ldots$$

Finally, use the method of Exercise 1.2 to turn this into a proper list:

$$0, 1, -1, 2, -2, 1, -1, \tfrac{1}{2}, -\tfrac{1}{2}, 3, -3, \ldots$$

1.4 Here is an enumeration:

$$+, -, ++, +-, -+, --, +++, ++-, +-+,$$
$$+--, -++, \ldots$$

Description of the list: first list the strings of length 1, then the strings of length 2, then the strings of length 3, and so on; within each group, arrange the strings in dictionary order, thinking of '+' as coming before '−' in the alphabet. (Thus, the next six entries are $-+-$, $--+$, $---$, $++++$, $+++-$, $++-+$.)

1.5 Interpret a finite string as a name of the set to which the number n belongs if there is an nth symbol in the string and it is a plus; and to which the number n does not belong if there are fewer than n symbols in the string, or if there is an nth symbol but it is a minus. Every finite set of positive integers now has a name. In fact, it has infinitely many names, e.g. the empty set $\varnothing$ is named by '$-$' and by '$--$' and by '$---$' etc., and the set $\{1, 2, 5\}$ is named by '$++--+$' and by '$++--+-$' and by '$++--+--$' etc.

1.6 Certainly the set D is infinite, for among its members are all of the sets in the following infinite collection:

$\{1\}$: the set whose only member is the number one
$\{2\}$: the set whose only member is the number two
⋮

Since every positive integer can be named in English, this subset of D is infinite, and therefore D itself is infinite. The fact that D is enumerable follows from the fact that every definition in English of a set of positive integers is a finite string of symbols, each of which is either one of the 26 letters of the alphabet (never mind about capitals or punctuation) or a space (to separate adjacent words), which we can treat as a 27th letter of the alphabet. (If you want to use capitals and punctuation, just add the appropriate symbols to this alphabet. You will still have a finite but large alphabet.) Now the finite strings of signs in the enlarged alphabet can be enumerated: first list the strings of length 1, then the strings of length 2, and so on; and within each group, arrange the strings in dictionary order (after arbitrarily deciding on some dictionary order for the enlarged alphabet). In this enumeration, most of the entries will be gibberish, or will make sense but will fail to be definitions of sets of positive integers. Erase all such entries. This leaves a gappy list which enumerates all the English definitions of sets of positive integers.

2
Diagonalization

Not all sets are enumerable: some are too big. An example is the set of *all sets of positive integers*. This set (call it 'P^*') contains, as a member, each finite and each infinite set of positive integers: the empty set $\varnothing$, the set P of all positive integers, and every set between these two extremes.

To show (following Georg Cantor) that P^* is not enumerable, we give a method which can be applied to *any* list L of sets of positive integers in order to discover a set $\bar{D}(L)$ of positive integers which is not named in the list. If you then try to repair the defect by adding $\bar{D}(L)$ to the list as a new first member, the same method, applied to the augmented list L', will yield a different set $\bar{D}(L')$ which is not in the augmented list!

The method is this. Confronted with any infinite list

$$S_1, S_2, S_3, \ldots \qquad (L)$$

of sets of positive integers, we define a set $\bar{D}(L)$ as follows:

For each positive integer n, n is in $\bar{D}(L)$ if and only if n is *not* in S_n. $\qquad$ (*)

As the notation '$\bar{D}(L)$' indicates, the composition of the set $\bar{D}(L)$ depends on the composition of the list L so that different lists L may yield different sets $\bar{D}(L)$. It should be clear that the statement (*) genuinely defines a set $\bar{D}(L)$ for, given any positive integer, n, we can tell whether n is in $\bar{D}(L)$ if we can tell whether n is in the nth set in list L. Thus, if S_3 happens to be the set E of all even positive integers, the number 3 is not in S_3 and therefore it *is* in $\bar{D}(L)$.

To show that the set $\bar{D}(L)$ which this method yields is never in the given list L, we argue by *reductio ad absurdum*: we suppose that $\bar{D}(L)$ does appear somewhere in list L, say as entry number m, and deduce a contradiction, thus showing that the supposition must be false. Here we go. *Supposition*: For some positive integer m,

$$S_m = \bar{D}(L).$$

(Thus, if 127 is such an m, we are supposing that '$\bar{D}(L)$' and 'S_{127}' are two names of one and the same set: we are supposing that a positive integer belongs to $\bar{D}(L)$ if and only if it belongs to the 127th set in list L.) To deduce a contradiction from this supposition we apply definition (*) to the particular positive integer m: with $n = m$, (*) tells us that

$$m \text{ is in } \overline{D}(L) \text{ if and only if } m \text{ is } \textit{not} \text{ in } S_m.$$

Now a contradiction follows from our supposition: if S_m and $\overline{D}(L)$ are one and the same set we have

$$m \text{ is in } \overline{D}(L) \text{ if and only if } m \text{ is } \textit{not} \text{ in } \overline{D}(L).$$

Since this is a flat self-contradiction, our supposition must be false. For no positive integer m do we have $S_m = \overline{D}(L)$. In other words, the set $\overline{D}(L)$ is named nowhere in list L.

So the method works. Applied to any list of sets of positive integers it yields a set of positive integers which was not in the list. Then no list enumerates all sets of positive integers: the set P^* of all such sets is not enumerable.

The method ('diagonalization') is so important that it will be well to look at it again from a slightly different point of view, which allows the entries in list L to be more readily visualized. Accordingly, we think of the sets S_1, S_2, ... as represented by functions $s_1, s_2, \ldots$ of positive integers which take the numbers 0 and 1 as values. The relationship between the set S_n and the corresponding function s_n is simply this: for each positive integer p we have

$$s_n(p) = \begin{cases} 1 & \text{if } p \text{ is in } S_n, \\ 0 & \text{if } p \text{ is not in } S_n. \end{cases}$$

Then the list L can be visualized as an infinite rectangular array of zeros and ones as in Table 2-1.

TABLE 2-1

	1	2	3	4	...
s_1	$s_1(1)$	$s_1(2)$	$s_1(3)$	$s_1(4)$	...
s_2	$s_2(1)$	$s_2(2)$	$s_2(3)$	$s_2(4)$	...
s_3	$s_3(1)$	$s_3(2)$	$s_3(3)$	$s_3(4)$	...
s_4	$s_4(1)$	$s_4(2)$	$s_4(3)$	$s_4(4)$	...
⋮	⋮	⋮	⋮	⋮	

The nth row of this array represents the function s_n and thus represents the set S_n: that row

$$s_n(1)\, s_n(2)\, s_n(3)\, s_n(4) \ldots$$

is a sequence of zeros and ones in which the pth entry, $s_n(p)$, is 1 or 0 accordingly as the number p is or is not in the set S_n. The entries in the diagonal of the array (upper left to lower right) form a sequence of zeros and ones:

$$s_1(1)\ s_2(2)\ s_3(3)\ s_4(4) \ldots$$

This sequence of zeros and ones ('the diagonal sequence') determines a set of positive integers ('the diagonal set') *which may well be among those listed in L.* In other words, there may well be a positive integer d such that the set S_d is none other than our diagonal set. The sequence of zeros and ones in the dth row of Table 2-1 would then agree with the diagonal sequence, entry by entry:

$$s_d(1) = s_1(1), \quad s_d(2) = s_2(2), \quad s_d(3) = s_3(3), \ldots$$

That is as may be: the diagonal set may or may not appear in the list L, depending on the detailed makeup of that list. What we want is a set we can rely upon *not* to appear in L, no matter how L is composed. Such a set lies near to hand: it is *the antidiagonal set* which consists of the positive integers not in the diagonal set. The corresponding *antidiagonal sequence* is obtained by changing zeros to ones and ones to zeros in the diagonal sequence. We may think of this transformation as a matter of subtracting each member of the diagonal sequence from 1: we write the antidiagonal sequence as

$$1-s_1(1), \quad 1-s_2(2), \quad 1-s_3(3), \quad 1-s_4(4), \ldots$$

This sequence can be relied upon not to appear as a row in Table 2-1 for if it did so appear – say, as the mth row – we should have

$$s_m(1) = 1-s_1(1), \quad s_m(2) = 1-s_2(2), \quad \ldots \quad s_m(m) = 1-s_m(m), \ldots$$

But the mth of these equations cannot hold. (*Proof.* $s_m(m)$ must be zero or one. If zero, the mth equation says that $0 = 1$. If one, the mth equation says that $1 = 0$.) Then the antidiagonal sequence differs from every row of our array, and so the antidiagonal set differs from every set in our list L. This is no news, for the antidiagonal set is simply the set $\bar{D}(L)$. We have merely repeated with the aid of a diagram – Table 2-1 – our proof that $\bar{D}(L)$ appears nowhere in the list L.

Of course, it is rather queer to say that the members of an infinite set 'can be arranged' in a single infinite list. By whom? Certainly not by any human being, for nobody has that much time or paper; and similar restrictions apply to machines. In fact, to call a set enumerable is simply to say that it is the range of some function of positive integers. Thus, the set E of even positive integers is enumerable because there are functions of positive integers which have E as their range, e.g. the function f for which we have $f(n) = 2n$ for each positive integer n. Any such function can then be thought of as a program which a superhuman enumerator can follow in order to arrange the members of the set in a single list. More explicitly, the program (the set of instructions) is this:

> Start counting from 1, and never stop. As you reach each number n, write a name of $f(n)$ in your list. (Where $f(n)$ is undefined, leave the nth position blank.)

But there is no need to refer to the list, or to a superhuman enumerator: anything we need to say about enumerability can be said in terms of the functions themselves, e.g. to say that the set P^* is not enumerable is simply to deny the existence of any function of positive integers which has P^* as its range.

Vivid talk of lists and superhuman enumerators may still aid the imagination, but in such terms the theory of enumerability and diagonalization appears as a chapter in mathematical theology. To avoid treading on living toes we might put the whole thing in a classical Greek setting: Cantor proved that there are sets which even Zeus cannot enumerate, no matter how fast he works, or how long (even, infinitely long).

If a set *is* enumerable, Zeus can enumerate it in one second by writing out an infinite list faster and faster. He spends $\frac{1}{2}$ second writing the first entry in the list; $\frac{1}{4}$ second writing the second entry; $\frac{1}{8}$ second writing the third; and in general, he writes each entry in half the time he spent on its predecessor. At no point *during* the one second interval has he written out the whole list, but when one second has passed, the list is complete! On a time scale in which the marked divisions are sixteenths of seconds, the process can be represented as in Table 2-2.

To speak of writing out an infinite list (e.g. of all the positive integers, in decimal notation) is to speak of such an enumerator either working faster and faster as above, or taking all of infinite time to complete the list (making one entry per second, perhaps). Indeed, Zeus could write out an infinite sequence of infinite lists if he chose to, taking only one second to complete the job. He could simply allocate the first half second to the business of writing out the first infinite list ($\frac{1}{4}$ second for the first entry, $\frac{1}{8}$ second for the next, and so on); he could then write out the whole second list in the following quarter second ($\frac{1}{8}$ second for the first entry, $\frac{1}{16}$ second

TABLE 2-2

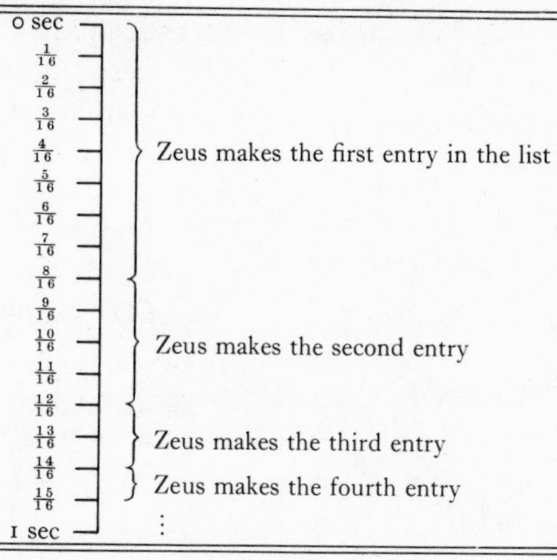

for the next, and so on); and in general, he could write out each subsequent list in just half the time he spent on its predecessor, so that after one second had passed he would have written out every entry in every list, in order. But the result does not count as a single infinite list, in our sense of the term. In our sort of list, each entry must come some *finite* number of places after the first.

As we use the term 'list', Zeus has not produced a list by writing infinitely many infinite lists one after another. But he could perfectly well produce a genuine list which exhausts the entries in all the lists, by using some such device as we used in Chapter 1 to enumerate the positive rational numbers. He need only imagine an infinite rectangular array in which, for each positive integer n, the nth row is the nth of his lists. He can then pick out various entries from various lists in either of the orders indicated in Table 2-3 so as to get a single list in which every entry in each of the original lists appears at one or another *finite* number of places from the beginning.

TABLE 2-3

(*a*)

1	2	5	...
4	3	6	...
9	8	7	...
⋮	⋮	⋮	

(*b*)

1	2	4	7	...
3	5	8	·	
6	9	·		
10	·			
⋮				

But Cantor's *diagonal argument* shows that neither this nor any more ingenious device is available, even to a god, for arranging all the sets of positive integers into a single infinite list. Such a list would be as much an impossibility as a round square: the impossibility of enumerating all the sets of positive integers is as absolute as the impossibility of drawing a round square, even for Zeus.

Exercises

2.1 Given a list L of sets $S_1, S_2, \ldots$, define a set $D(L)$ as follows:

For each positive integer n, n is in $D(L)$ if and only if n is in S_n.

Suppose that L is the enumeration of the finite sets of positive integers that was given in the solution to Exercise 1.5. (*a*) What is $D(L)$? (*b*) Is $D(L)$ in L? (*c*) What is $\bar{D}(L)$?

2.2 Adapt the diagonal argument (see Table 2-1) so as to prove that the set of real numbers in the interval from 0 to 1 is not enumerable. *Hint*: each such number is represented in one or two ways in decimal notation by a decimal point followed by an infinite string of digits, e.g. $0.5 = 0.5000\ldots = 0.4999\ldots$ and $0.123 = 0.123000\ldots = 0.122999\ldots$, but if no zeros or nines appear in it, the decimal representation (as an *infinite* string of digits) is unique.

2.3 Prove by a diagonal argument that the set of all total and partial functions of positive integers that take only positive integers as values is not enumerable.

2.4 Define a *word* as any finite string of letters from a certain enumerably infinite alphabet $a_1, a_2, \ldots$. Prove that each of the following sets is enumerable:

(*a*) the two-letter words,
(*b*) the three-letter words,
(*c*) the n-letter words, for arbitrary but fixed n,
(*d*) all words, irrespective of (*finite*) length.

Solutions

2.1 The successive entries in L are $S_1 = \{1\}$, $S_2 = \varnothing$, $S_3 = \{1, 2\}$, $S_4 = \{1\}$, etc. Only for $n = 1$ does n belong to S_n, for the subscripts n grow much faster than the largest members of S_n (which are never greater than the lengths of the corresponding strings of pluses and minuses). Then

(*a*) $D(L) = \{1\}$.
(*b*) $D(L)$ is in L, e.g. $D(L) = S_1$ and $D(L) = S_4$.
(*c*) $\bar{D}(L) = \{2, 3, 4, \ldots\}$ which is infinite and hence not in L – as we knew anyway, by the diagonal argument.

2.2 Let L be any list in which each entry is a decimal point followed by an unending string of decimal digits. In particular, suppose that the successive digits in the nth entry are $s_n(1), s_n(2), \ldots$, so that the entries in L are the rows of Table 2-1 (where, now, each position in the table can represent any one of the ten digits '0', ..., '9'). Where d is any one of the ten digits, define

$$d' = \begin{cases} 1 & \text{if} \quad d = 2, \\ 2 & \text{otherwise.} \end{cases}$$

(The essential feature of this definition is that $0 \neq d' \neq d$.) Now let the successive digits of $\bar{D}(L)$ be $s_1(1)', s_2(2)', \ldots$. These are not the same as the successive digits of any row of Table 2-1 – say, the nth – for if they were, we should have $s_n(n)' = s_n(n)$, which is impossible by definition of the operation $'$. Then since $\bar{D}(L)$ contains no zeros or nines, the number which it represents is represented by no entry in the list L. Thus, no list L of real numbers in the unit interval can be exhaustive.

2.3 Let $L = f_1, f_2, \ldots$ be an enumeration of such functions. Define the antidiagonal function u ($= \bar{D}(L)$) as follows:

$$u(n) = \begin{cases} 1 & \text{if } f_n(n) \text{ is undefined,} \\ f_n(n) + 1 & \text{otherwise.} \end{cases}$$

Then u is a well defined total function of positive integers which takes only positive integers as values. But u cannot be in L: if we had $u = f_m$ (say) the definition would yield

$$f_m(m) = u(m) = \begin{cases} 1 & \text{if } f_m(m) \text{ is undefined,} \\ f_m(m) + 1 & \text{otherwise.} \end{cases}$$

Then whether or not f_m assigns a value to the argument m, we have a contradiction: if $f_m(m)$ is undefined it *is* defined (and equal to 1), while if $f_m(m)$ is defined and equal to n (say) we have $n = n + 1$.

2.4 The straightforward solution uses the same trick (again and again) that was used in enumerating the positive rational numbers.
(*a*) Form a rectangular array of all two-letter words as in Table 2-4(*a*) and weave your way through them, e.g. in the triangular pattern of Table 2-3(*b*) to get the single list

$$a_1a_1,\ a_1a_2,\ a_2a_1,\ a_1a_3,\ a_2a_2,\ a_3a_1, \ldots$$

TABLE 2-4

(*a*)

	a_1	a_2	a_3	...
a_1	a_1a_1	a_1a_2	a_1a_3	...
a_2	a_2a_1	a_2a_2	a_2a_3	...
a_3	a_3a_1	a_3a_2	a_3a_3	...
⋮	⋮	⋮	⋮	

(*b*)

	a_1a_1	a_1a_2	a_2a_1	a_1a_3	...
a_1	$a_1a_1a_1$	$a_1a_1a_2$	$a_1a_2a_1$	$a_1a_1a_3$	...
a_2	$a_2a_1a_1$	$a_2a_1a_2$	$a_2a_2a_1$	$a_2a_1a_3$	...
a_3	$a_3a_1a_1$	$a_3a_1a_2$	$a_3a_2a_1$	$a_3a_1a_3$	...
⋮	⋮	⋮	⋮	⋮	

(*b*) To get the three-letter words, notice that each of these is obtainable by writing a two-letter word after a one-letter word. Form the array of Table 2-4(*b*) and weave in the triangular pattern of Table 2-3(*b*) to get the list

$$a_1a_1a_1,\ a_1a_1a_2,\ a_2a_1a_1,\ a_1a_2a_1,\ a_2a_1a_2, \ldots$$

(*c*) To list the four-letter words, weave through an array which has the three-letter words on top and the one-letter words down the side. And so on. By iterating the process sufficiently often, you can list the n-letter words for any particular n.

(*d*) To list all words of whatever finite length, weave through an array in which the first row is the list of one-letter words $a_1, a_2, \ldots$ the second row is the list of two-letter words (*a*), the third row is the list of three-letter words (*b*), etc.

But there is a simpler, direct method for obtaining a gappy list of all words of whatever finite length on the alphabet $a_1, a_2, \ldots$. To determine the position in the list of any such word, replace each occurrence of a_1 in it by '12', replace each occurrence of a_2 in it by '122', and in general replace each occurrence of a_n in it by a *one* followed by *n twos*. The result is a string of digits which, read in the decimal notation, gives the position of the word in the list. (The list then starts with eleven gaps. The first entry is the one-letter word a_1, in position twelve. The next entry is the word a_2, in position 122. The third entry is the word a_1a_1, in position 1212. The fourth is a_3, in position 1222.) The list of n-letter words for any fixed n can be obtained similarly (and more simply).

3
Turing machines

No human being can write fast enough, or long enough, or small enough† to list all members of an enumerably infinite set by writing out their names, one after another, in some notation. But humans can do something equally useful, in the case of certain enumerably infinite sets: They can give explicit instructions for determining the *n*th member of the set, for arbitrary finite *n*. Such instructions are to be given quite explicitly, in a form in which they could be followed by a computing machine, or by a human who is capable of carrying out only very elementary operations on symbols. The problem will remain, that for all but a finite number of values of *n* it will be physically impossible for any human or any machine to actually carry out the computation, due to limitations on the time available for computation, and on the speed with which single steps of the computation can be carried out, and on the amount of matter in the universe which is available for forming symbols. But it will make for clarity and simplicity if we ignore these limitations, thus working with a notion of computability which goes beyond what actual men or machines can be sure of doing. The essential requirement is that our notion of computability not be too narrow, for shortly we shall use this notion to prove that certain functions are not computable, and that certain enumerable sets are not effectively (mechanically) enumerable – even if physical limitations on time, speed, and amount of material could somehow be overcome.

The notion of computation can be made precise in many different ways, corresponding to different possible answers to such questions as the following. 'Is the computation to be carried out on a linear tape, or on a rectangular grid, or what? If a linear tape is used, is the tape to have a beginning but no end, or is it to be endless in both directions? Are the squares into which the tape is divided to have addresses (like addresses of houses on a street) or are we to keep track of where we are by writing special symbols in suitable squares (as one might mark a trail in the woods by marking trees)?' And so on. Depending on how such questions are

† There is only a finite amount of paper in the world, so you'd have to write smaller and smaller without limit, to get an infinite number of symbols down on paper. Eventually, you'd be trying to write on molecules, on atoms, on electrons, ...

answered, computations will have one or another appearance, when examined in detail. But our object is not to give a faithful representation of the individual steps in computation, and in fact, *the very same set of functions turns out to be computable, no matter how these questions are answered.* We shall now answer these and other questions in a certain way, in order to get *a* characterization of the class of computable functions. A moderate amount of experience with this notion of computability will make it plausible that the net effect would have been the same, no matter in what plausible way the questions had been answered. Indeed, given any other plausible, precise characterization of computability that has been offered, one can prove by careful, laborious reasoning that our notion is equivalent to it in the sense that any function which is computable according to one characterization is also computable according to the other. But since there is no end to the possible variations in detailed characterizations of the notions of computability and effectiveness, one must finally accept or reject the *thesis* (which does not admit of mathematical proof) that the set of functions computable in our sense is identical with the set of functions that men or machines would be able to compute by whatever effective method, if limitations on time, speed, and material were overcome. Although this thesis ('Church's thesis') is unprovable, it *is* refutable, *if false.* For if it is false, there will be some function which is computable in the intuitive sense, but not in our sense. To show that such a function is computable in the intuitive sense, it would be sufficient to give instructions for computing its values for all (arbitrary) arguments, and to see that these instructions are indeed effective. (Presumably, we are capable of seeing *in particular cases,* that sets of instructions are effective.) To show that such a function is not computable in our sense, one would have to survey all possible Turing machines and verify (by giving a suitable proof) that none of them compute that function. (You will see examples of such proofs shortly, in Chapters 4 and 5; but the functions shown there not to be Turing computable have not been shown to be effectively computable in *any* intuitive sense, so *they* won't do as counterexamples to Church's thesis.) Then if Church's thesis is false, it is refutable by finding a counterexample; and the more experience of computation we have without finding a counterexample, the better confirmed the thesis becomes.

Now for the details. We suppose that the computation takes place on a tape, marked into squares, which is unending in both directions – either because it is actually infinite or because there is a man stationed at each end to add extra blank squares as needed. With at most a finite number of exceptions, all squares are blank, both initially and at each subsequent stage of the calculation. If a square is not blank, it has exactly one of a finite number of symbols $S_1, \ldots, S_n$ written in it. It is convenient to think of a blank square as having a certain symbol – the *blank* – written in it. We designate the blank by 'S_0'.

At each stage of the computation, the human or mechanical computer is *scanning* some one square of the tape. The computer is capable of telling what symbol is written in the scanned square. It is capable of erasing the symbol in the scanned square and writing a symbol there. And it is capable of movement: one square to the right, or one square to the left. If you like, think of the machine quite crudely, as a box on wheels which, at any stage of the computation, is over some square of the tape. The tape is like a railroad track; the ties mark the boundaries of the squares; and the machine is like a very short car, capable of moving along the track in either direction, as in Figure 3-1. At the bottom of the car there is a device that can read what's written between the ties; erase such stuff; and write symbols there.

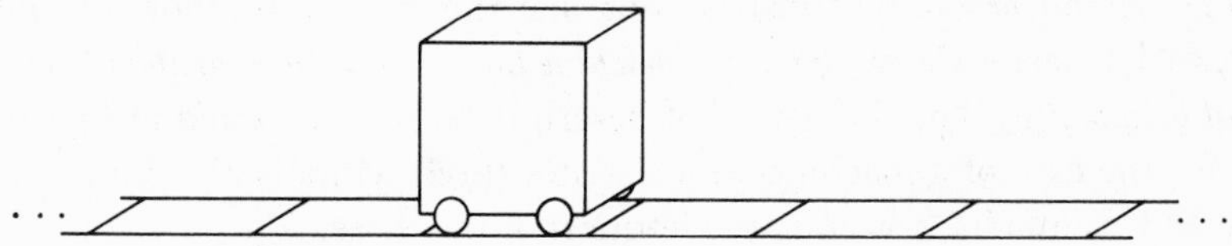

Figure 3-1

The machine is designed in such a way that at each stage of the computation it is in one of a finite number of internal *states*, $q_1, \ldots, q_m$. Being in one state or another might be a matter of having one or another cog of a certain gear uppermost, or of having the voltage at a certain terminal inside the machine at one or other of m different levels, or what have you: We are not concerned with the mechanics or the electronics of the matter. Perhaps the simplest way to picture the thing is quite crudely: Inside the box there is a man, who does all the reading and writing and erasing and moving. (The box has no bottom: the poor mug just walks along between the ties, pulling the box with him.) The man has a list of m instructions written down on a piece of paper. *He is in state q_i when he is carrying out instruction number i.*

Each of the instructions has conditional form: It tells the man what to do, depending on whether the symbol being scanned (the symbol in the scanned square) is S_0 or S_1 or ... or S_n. Namely, there are $n+4$ things he can do:

(1) Halt the computation.
(2) Move one square to the right.
(3) Move one square to the left.
(4) Write S_0 in place of whatever is in the scanned square.
(5) Write S_1 in place of whatever is in the scanned square.
⋮
($n+4$) Write S_n in place of whatever is in the scanned square.

So, depending on what instruction he is carrying out (= what state he is in) and on what symbol he is scanning, the man will perform one or another of these $n+4$ overt acts. Unless he has halted (overt act number 1), he will also perform a covert act, in the privacy of his box, *viz.*, he will determine what the next instruction (*next* state) is to be. Thus, the *present state* and the presently *scanned symbol* determine what overt *act* is to be performed, and what the *next state* is to be.

The overall *program* of instructions that the man is to follow can be specified in various ways, e.g. by a *machine table* or by a *flow graph* or by a *set of quadruples*. The three sorts of descriptions are illustrated in Figure 3-2 for the case of a machine which writes three symbols S_1 on a blank tape and then halts, scanning the leftmost of the three.

Example 3.1. Write $S_1 S_1 S_1$

The machine will write an S_1 in the square it's initially scanning, move left one square, write an S_1 there, move left one more square, write a third S_1 there, and halt. (It halts when it has no further instructions.) There will be three states – one for each of the symbols S_1 that are to be written. In Figure 3-2, the entries in the top row of the machine table (under the horizontal line) tell the man that when he's following instruction q_1 he is to (1) write S_1 and repeat instruction q_1, if the scanned symbol is S_0, but (2) move left and follow instruction q_2 next, if the scanned symbol is S_1. The same information is given in the flow graph by the two arrows that emerge from the node marked 'q_1'; and the same information is also given by the first two quadruples. The significance of a table entry, of an arrow in a flow graph, and of a quadruple, is shown in Figure 3-3.

Unless otherwise stated, it is to be understood that a machine starts in its

Present state	Scanned symbol S_0	S_1
q_1	$S_1 q_1$	$L q_2$
q_2	$S_1 q_2$	$L q_3$
q_3	$S_1 q_3$	

(*a*) *Machine table*

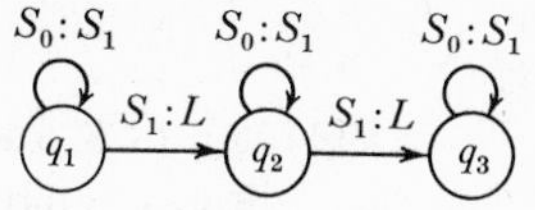

(*b*) *Flow graph*

$q_1 S_0 S_1 q_1$, $q_1 S_1 L q_2$,
$q_2 S_0 S_1 q_2$, $q_2 S_1 L q_3$,
$q_3 S_0 S_1 q_3$

(*c*) *Set of quadruples*

Figure 3-2

Act Next state

(*a*) *Table entry*

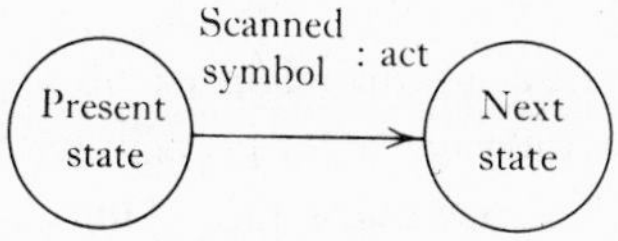

(*b*) *Arrow in a flow graph*

Same sequence as for arrows

(*c*) *Quadruple*

Figure 3-3

lowest-numbered state. The machine we have been considering *halts* when it is in state q_3 scanning S_1, for there is no table entry or arrow or quadruple telling it what to do in such a case.

A virtue of the flow graph as a way of representing the machine program is that if the starting state is indicated somehow (e.g. if it is understood that the leftmost node represents the starting state unless there is an indication to the contrary) then we can dispense with the names of the states: It doesn't matter what you call them. Then the flow graph could be drawn as in Figure 3-4. We can indicate how such a machine operates by writing down its sequence of *configurations*. Each configuration shows what's on the tape at some stage of the computation, shows what state the machine is in at that stage, and shows which square is being scanned. We'll do this by writing out what's on the tape and writing the name of the present state under the symbol in the scanned square:

```
0000000  0000100  0000100  0001100  0001100  0011100
    1        1      2        2       3         3
```

Here we have written the symbols S_0 and S_1 simply as 0 and 1, and similarly we have written the states q_1, q_2, q_3 simply as 1, 2, 3 to save needless

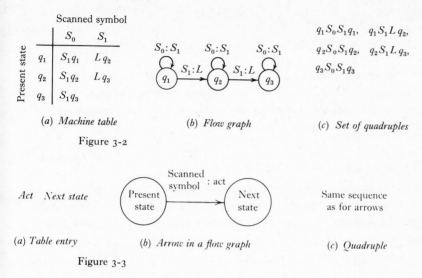

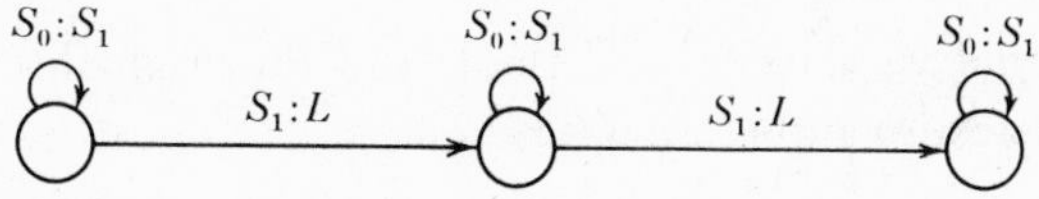

Figure 3-4. Write three S_1s.

fuss. Thus, the last configuration is short for

$$S_0S_0\underset{q_3}{S_1}S_1S_1S_0S_0$$

Of course, the strings of S_0s at the beginning and end of the tape can be shortened or lengthened *ad lib.* without changing the significance: the tape is understood to have as many blanks as you please at each end. Now here is a more complex example.

Example 3.2. Double the number of 1s

This machine starts off scanning the leftmost of a string of 1s on an otherwise blank tape, and winds up scanning the leftmost of a string of twice that many 1s on an otherwise blank tape. Here is the flow graph:

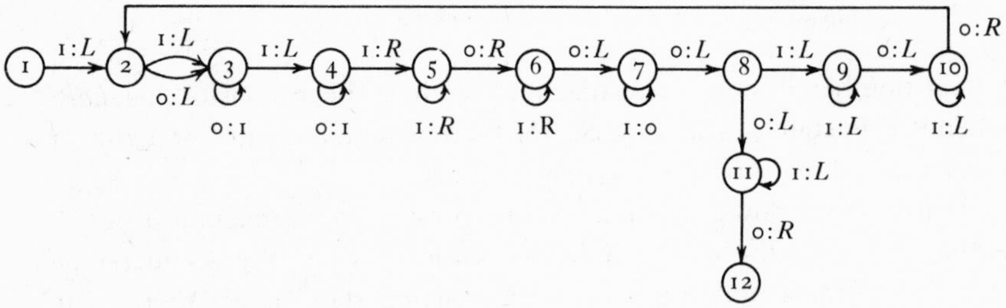

How does it work? In general, by writing double 1s at the left, and erasing single 1s from the right. In particular, suppose that the initial configuration is $\underset{1}{1}11$, so that we start in state 1, scanning the leftmost of a string of three 1s on an otherwise blank tape. The next few configurations are

$$\underset{2}{0}111,\quad \underset{3}{0}0111,\quad \underset{3}{1}0111,\quad \underset{4}{0}10111,\quad \underset{4}{1}10111.$$

So we have written our first double 1 at the left – separated from the original string, 111, by a 0. Next we go right, past the 0 to the right-hand end of the original string, and erase the rightmost 1. Here is how that works, in two stages. Stage 1:

$$1\underset{5}{1}0111,\quad 11\underset{5}{0}111,\quad 110\underset{6}{1}11,\quad 1101\underset{6}{1}1,\quad 11011\underset{6}{1},\quad 110111\underset{6}{0}.$$

Now we know that we have passed the last of the original string of 1s, so (stage 2) we back up and erase one:

$$11011\underset{7}{1},\quad 11011\underset{7}{0}.$$

Now we hop back to our original position, scanning the leftmost 1 in what remains of the original string. When we have gone round the loop once more we shall be in this configuration: $111101\underset{7}{0}$. We'll then go through these:

$11110\underset{8}{1}$, $1111\underset{9}{0}1$, $111\underset{10}{1}01$, $11\underset{10}{1}101$, $1\underset{10}{1}1101$, $\underset{10}{1}11101$, $\underset{10}{0}111101$,

$\underset{2}{1}11101$, $\underset{3}{0}111101$, $\underset{3}{1}111101$, $\underset{4}{0}1111101$, $\underset{4}{1}1111101$.

We now have our six 1s on the tape; but we want to erase the last of the original three and then halt, scanning the leftmost of the six that we wrote down. Here's how:

$\underset{5}{1}1111101$, $1\underset{5}{1}111101$, ..., $11111\underset{5}{1}01$, $111111\underset{5}{0}1$, $1111110\underset{6}{1}$,

$11111101\underset{6}{0}$, $1111110\underset{7}{1}$, $1111110\underset{7}{0}$, $111111\underset{8}{0}$, $11111\underset{11}{1}$,

$1111\underset{11}{1}1$, ..., $\underset{11}{1}11111$, $\underset{11}{0}111111$, $\underset{12}{1}11111$

Now we are in state 12, scanning a 1. Since there is no arrow from that node telling us what to do in such a case, we halt. The machine performs as advertised. (*Note*: the fact that the machine doubles the number of 1s when the original number is three is not a proof that the machine performs as advertised. But our examination of the special case in which there are three 1s initially made no essential use of the fact that the initial number was three: it is readily converted into a proof that the machine doubles the number of 1s no matter how long the original string may be.)

Example 3.3. Write $2n$ 1s on a blank tape and halt, scanning the leftmost 1

There is a straightforward way, using $2n$ states. We made use of the method in Example 3.1. In general, the flow graph would be obtained by stringing together $2n$ replicas of this: 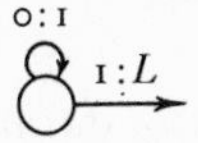

and lopping off the last arrow. But there is another way, using $n+11$ states: write n 1s on the tape in the straightforward way, and then use the method of Example 3.2 to double *that*. Schematically, we put the two graphs together by identifying node 1 in Example 3.2 with node n in the graph of the machine that writes n 1s on a blank tape. If n is large, the straightforward way of writing $2n$ 1s will require many more states than the indirect way which we have just

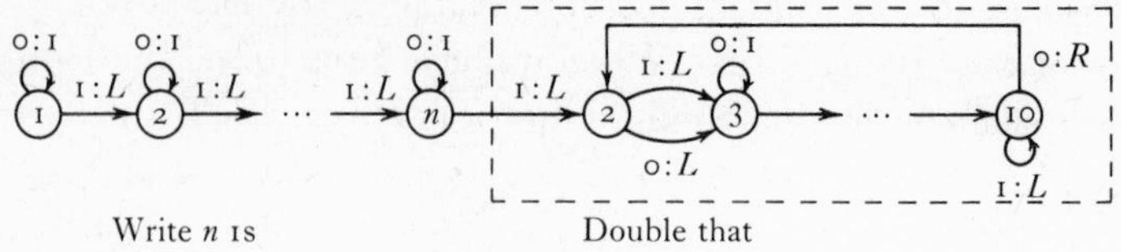

Graph of Example 3.2

described. (If n is greater than 11, $2n$ is greater than $n+11$.) Of course, we could save even more states by repeating the doubling trick, e.g. to write 100 1s we could use 25 states to write 25 of them, use another 11 to double that, and use yet another 11 to get the full 100 by doubling *that*. This does it with 47 states instead of the 100 we would need by the direct method, or the 61 we would need if we used the doubling trick only once.

Exercises. (Solutions follow)

3.1 Design a Turing machine which, started scanning the leftmost of an unbroken string of 1s on an otherwise blank tape, eventually halts, scanning a square on an otherwise blank tape – where the scanned square contains a blank or a 1 depending on whether there were an even or an odd number of 1s in the original string.

3.2 Design a Turing machine which, started scanning the leftmost of an unbroken string of 1s and 2s on an otherwise blank tape (in any order, and perhaps with all 1s or all 2s) eventually halts – at which point the contents of the tape indicate (following some convention which you are free to stipulate) whether or not there were exactly as many 1s as 2s in the original string.

3.3 *Add 1 in decimal notation.* Initially the machine scans the rightmost of an unbroken string of decimal digits on an otherwise blank tape. When it halts, the tape holds the decimal representation of 1 + the original number. Design a simple machine which does this. (Give the machine table – a graph would be messy.)

3.4 *Multiply in monadic notation.* Initially, the tape is blank except for two solid blocks of 1s, separated by a single blank. Say there are p 1s in the first block and q in the second. Design a machine which, started scanning the leftmost 1 on the tape, eventually halts, scanning the leftmost of an unbroken block of pq 1s on an otherwise blank tape.

3.5 *Add in in monadic notation.* (Identical with Exercise 3.4 but with '$p+q$' in place of 'pq'.)

3.6 *Not all functions from positive integers to positive integers are Turing computable.* You know via an Exercise in Chapter 2 that there are more such functions than there are Turing machines, so there must be some that no Turing machines compute. Explain a bit more fully.

3.7 *Productivity.* Consider Turing machines which use only the symbol 1 in addition to the blank. Initially the tape is blank. If the machine eventually halts, scanning the leftmost of an unbroken string of 1s on an otherwise blank tape, its *productivity* is said to be the length of that string. But if the machine never halts, or halts in some other configuration, its productivity is said to be 0. Now define $p(n) =$ *the productivity of the most productive n-state Turing machines*. Prove that

(*a*) $p(1) = 1$;
(*b*) $p(47) \geqslant 100$;
(*c*) $p(n+1) > p(n)$;
(*d*) $p(n+11) \geqslant 2n$.

(For solutions, see Chapter 4.)

3.8 *The busy beaver problem* (after Tibor Rado, *Bell System Technical Journal*, May 1962). According to Exercise 3.6, uncomputable functions exist. But what would such a thing look like? Actually, you have just met one: the function p defined in Exercise 3.7. The proof that it is uncomputable turns on the fact that the 'busy beaver problem' is unsolvable. This is the problem of designing a Turing machine which uses only the symbol 1 in addition to the blank and which, started in state 1 scanning the leftmost of an unbroken string of n 1s on an otherwise blank tape, eventually halts scanning the leftmost of an unbroken string of $p(n)$ 1s on an otherwise blank tape. In fact, no such machine can exist. To prove that, assume that there *is* such a machine (with k states, say) and deduce a contradiction: That takes some ingenuity. *Hint*: if there were such a machine, then by stringing two replicas of it onto an n-state machine which writes n 1s on a blank tape (right to left) you can get an $n+2k$ state machine which has productivity $p(p(n))$; and from this, together with the last two things you proved in Exercise 3.7, you can deduce a contradiction.

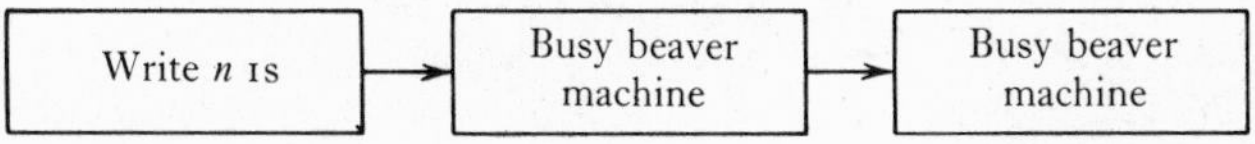

(For solution, see Chapter 4.)

3.9 *Uncomputability via diagonalization.* Every Turing machine which uses only the symbol 1 in addition to the blank computes some function from positive integers to positive integers according to the following convention. To compute the value which the function assigns to the argument n, start the machine in its lowest-numbered state, scanning the leftmost of a string of n 1s on an otherwise blank tape. The function is defined for the argument n if and only if the machine eventually halts, scanning the leftmost of an unbroken string of 1s on an otherwise blank tape, in which case the value of the function is the length of the string. Since the Turing machines are enumerable (cf. Exercise 3.6), so are the functions they compute. Define an 'antidiagonal' function and prove that no Turing machine computes it. (For solution, see Chapter 5.)

3.10 *The halting problem.* Let $M_1, M_2, \ldots$ be an enumeration of all Turing machines which use only the symbol 1 in addition to the blank. The *self-halting problem* is that of specifying a uniform effective procedure in some format or other for computing a total or partial function s for which we have $s(n) = 1$ if and only if M_n never halts, after being started in its initial state, scanning the leftmost of an unbroken string of n 1s on an otherwise blank tape. The full *halting problem* is that of specifying a uniform effective procedure for computing some function h for which we have $h(m, n) = 1$ if and only if M_m never halts, after being started in its initial state, scanning the leftmost of an unbroken string of n 1s on an otherwise blank tape.

(*a*) Prove that the self-halting problem is unsolvable.

(*b*) Prove the corollary: the halting problem is unsolvable.

Hint: Church's thesis must be used, sooner or later. Sooner *and* later: prove that one of the functions shown in Exercise 3.6 to be computable by no Turing machine is computable in an unusual format (using two Turing machines at once) if the self-halting problem is solvable.

Solutions

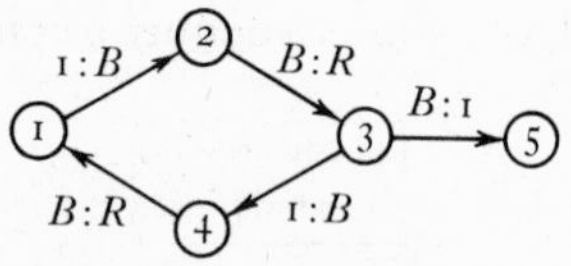

(We have written 'B' for S_0, the blank.)

3.1 If there were 0 or 2 or 4 or ... 1s to begin with, this machine halts in state 1, scanning a blank on a blank tape; if there were 1 or 3 or 5 or ..., it halts in state 5, scanning a 1 on an otherwise blank tape.

3.2 When this machine halts, the tape is blank if and only if there were equal numbers of 1s and 2s in the original string.

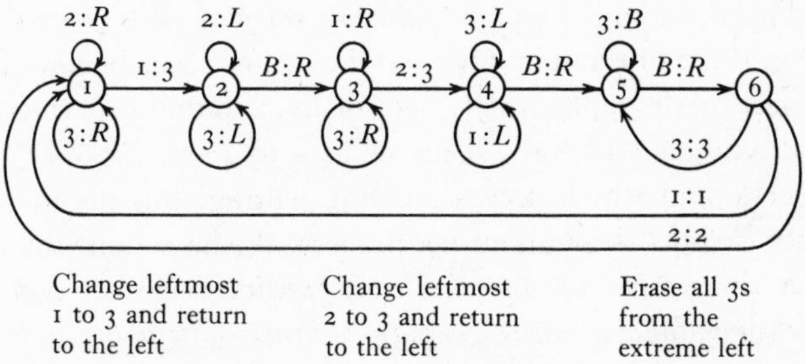

There are other ways of doing this, e.g. ways in which the machine uses no symbols other than 1 and 2. The simplest way of doing it which is compatible with the letter, but not the spirit of the problem statement, is to adopt the following convention regarding the way in which the contents of the tape when the machine halts indicate whether or not there were equal numbers of 1s and 2s initially: there were equal numbers of 1s and 2s initially if and only if there are equal numbers of 1s and 2s on the tape when the machine halts. The machine can then halt immediately, so that the final tape contents are identical with the initial contents. The graph of such a machine would look like this: ①

3.3 There are 11 possible scanned symbols, among which *B* is the blank and 0 is not the blank but the cipher zero; 1, ..., 9 are the usual ciphers. In the body of the table, each pair represents the *act to be per-*

Machine table for adding 1 in decimal notation

Scanned symbol	Present state 1	2
B	1 2	
0	1 2	*L* 1
1	2 2	
2	3 2	
3	4 2	
4	5 2	
5	6 2	
6	7 2	
7	8 2	
8	9 2	
9	0 2	

formed and the *next state*, in that order. The many blanks in column 2 represent situations in which the machine has no instructions, and therefore halts.

3.4 There are lots of equally efficient ways of multiplying in monadic notation. Many of them involve use of extra symbols beyond the B and 1 that appear on the tape initially and finally, and that is perfectly all right: they are symbols which the machine uses to keep track of where it is in the process, and which it erases before halting. But the method shown uses only B and 1 throughout the process. We have done that in order to illustrate this point: *any function from positive integers to positive integers which is computable at all is computable in monadic notation by a machine which uses only the symbols B and 1.*

Here is how this machine works. The first block, of p 1s, is used as a *counter*, to keep track of how many times the machine has added q 1s to the group at the right. To start, the machine erases the leftmost of the p 1s and sees if there are any 1s left in the counter group. If not, $pq = q$, and all the machine has to do is position itself over the leftmost 1 on the tape, and halt. But if there are any 1s left in the counter the machine goes into its *leapfrog routine*: in effect, it moves the block of q 1s (the 'leapfrog group') q places to the right along the tape, e.g. with $p = 2$ and $q = 3$ the tape looks like this initially

11B111

and looks like this after going through the leapfrog routine:

B1$BBBB$111

The machine will then note that there is only one 1 left in the counter, and will finish up by erasing that 1, moving right two squares, and changing all Bs to 1s until it comes to a 1, at which point it continues on to the leftmost 1 and halts. This is how the leapfrog routine works:

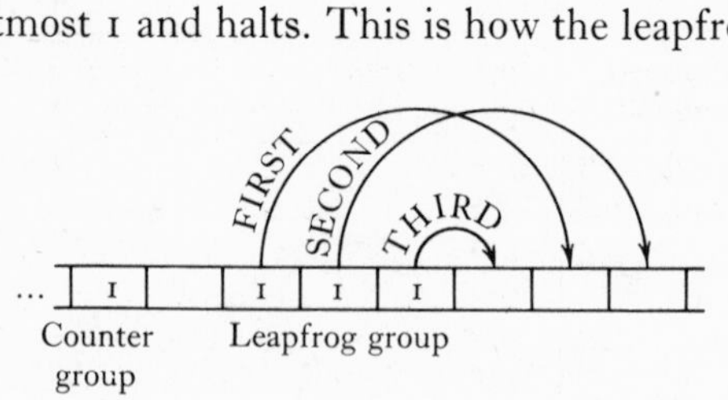

In general, the leapfrog group consists of a block of 0 or 1 or ... or q 1s, followed by a blank, followed by the remainder of the q 1s. The blank is there to tell the machine when the leapfrog game is over: without it the

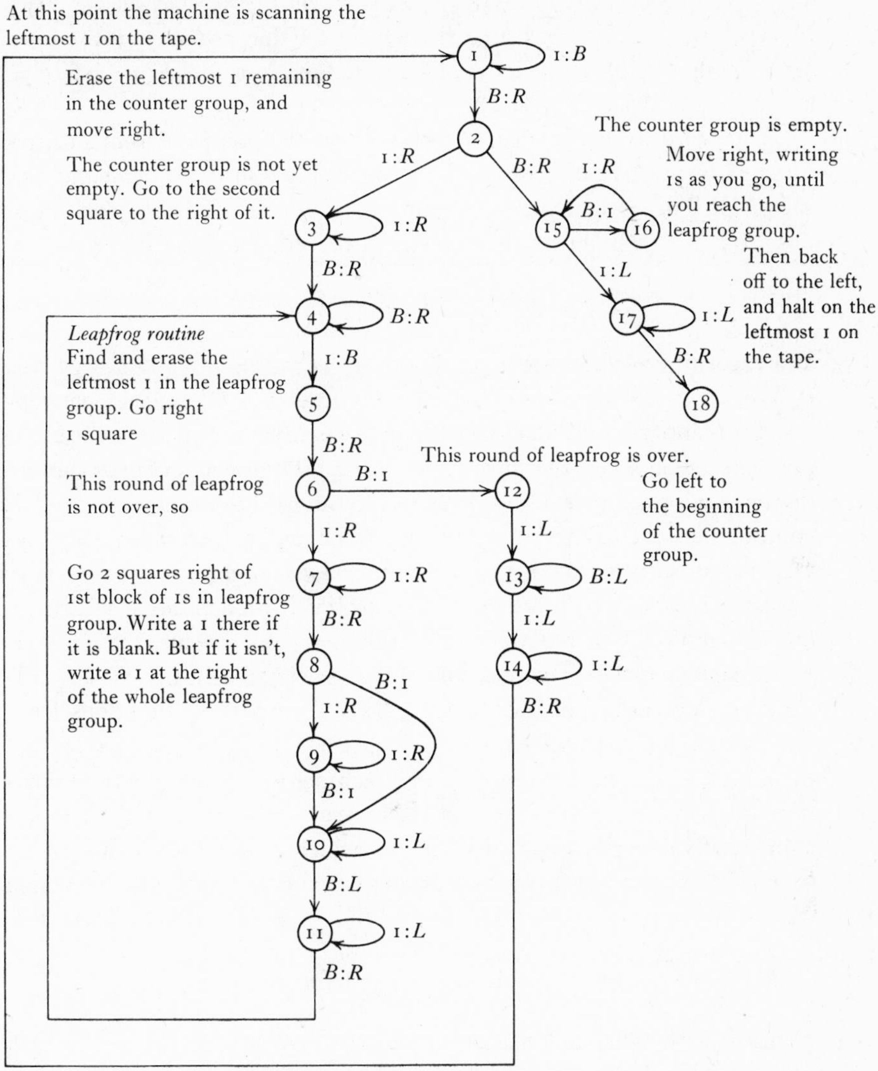
At this point the machine is scanning the leftmost 1 on the tape.
Erase the leftmost 1 remaining in the counter group, and move right.
The counter group is not yet empty. Go to the second square to the right of it.
The counter group is empty. Move right, writing 1s as you go, until you reach the leapfrog group. Then back off to the left, and halt on the leftmost 1 on the tape.
Leapfrog routine
Find and erase the leftmost 1 in the leapfrog group. Go right 1 square
This round of leapfrog is not over, so
This round of leapfrog is over. Go left to the beginning of the counter group.
Go 2 squares right of 1st block of 1s in leapfrog group. Write a 1 there if it is blank. But if it isn't, write a 1 at the right of the whole leapfrog group.
1:B
B:R
1:R
B:R
1:R
B:1
1:R
B:R
1:L
1:L
B:R
B:R
1:B
B:R
B:1
1:R
1:L
1:R
B:L
B:R
1:L
B:1
1:L
1:R
B:R
1:R
B:1
1:L
B:L
1:L
B:R

Flow graph for multiplying in monadic notation

group of q 1s would keep moving right along the tape forever. (In playing leapfrog, the portion of the q 1s to the left of the blank in the leapfrog group functions as a counter: it controls the process of adding 1s to the portion of the leapfrog group to the right of the blank. That is why there are two big loops in the flow graph: one for each counter-controlled subroutine.)

3.5 The object is to erase the leftmost 1, fill the gap between the two blocks of 1s, and halt scanning the leftmost 1 that remains on the tape. Here is one way of doing it, in quadruple notation: $q_1S_1S_0q_1$; $q_1S_0Rq_2$; $q_2S_1Rq_2$; $q_2S_0S_1q_3$; $q_3S_1Lq_3$; $q_3S_0Rq_4$.

3.6 According to Exercise 2.3 the set of all functions from positive integers to positive integers is not enumerable. On the other hand, the set of all Turing machines *is* enumerable, for any Turing machine is describable by a finite string of letters of the infinite alphabet

$$;,\ R,\ L,\ S_0,\ q_1,\ S_1,\ q_2,\ S_2,\ q_3,\ \ldots$$

and according to Exercise 2.4(*d*) (with $a_1 = ;$, $a_2 = R$, $a_3 = L$, $a_4 = S_0$, $a_5 = q_1, \ldots$) the set of all such strings is enumerable. (We need a convention for identifying the starting state, e.g. we might require that the starting state be assigned the lowest q-number.) Then whatever convention one may adopt, for deciding what (if anything) the function is which is computed by each Turing machine in our list, our list of Turing machines yields a (possibly gappy) list of *all* the functions from positive integers to positive integers which are computed by Turing machines according to our convention. One possible convention is the one specified in Exercise 3.4, according to which we may take it that Turing machines whose quadruples contain any of the symbols S_2, S_3, ... compute *no* functions from positive integers to positive integers. Note that if $f_1, f_2, f_3, \ldots$ is a gapless list of all functions computed by Turing machines according to our convention, the antidiagonal function u of Exercise 2.3 is an example of a function from positive integers to positive integers which is computed in our format by no Turing machine. Another example is the partial function t which we define:

$$t(n) = \begin{cases} 1 & \text{if } f_n(n) \text{ is undefined,} \\ \text{undefined} & \text{if } f_n(n) \text{ is defined.} \end{cases}$$

(For further details, see Chapter 5.)

3.10(*a*) If the self-halting problem were solvable, then by Church's thesis there would be some Turing machine S which computes s in some format, e.g. in the format of Exercise 3.9. The function t of Exercise 3.6 would then be computable in the following format. To determine $t(n)$, start machines S and M_n in their initial states, scanning the leftmost of unbroken strings of n 1s on their otherwise blank tapes.

Case 1: *M_n never halts.* Then S will eventually halt, scanning a 1 on its otherwise blank tape. At that time we shall know that M_n will never halt, and thus we shall know that the corresponding function f_n is undefined for the argument n, so that $t(n) = 1$.

Case 2: M_n eventually halts. When it does, we can determine by examining the tape whether or not $f_n(n)$ is defined, and will thus know whether $t(n)$ is undefined or equal to 1. Then if the self-halting problem is solvable, t is computable in an unusual format, and hence, by Church's thesis, t is computed by some Turing machine, in contradiction to what we have proved in Exercise 3.6.

3.10(*b*) If the halting problem were solvable, i.e., if h were computable, then the self-halting problem would be solvable, i.e., s would be computable, for we have $s(n) = h(n, n)$ for every positive integer n.

4
Uncomputability via the busy beaver problem

In this chapter we exhibit an uncomputable function by working out Exercises 3.7 and 3.8; we show that the busy beaver problem is unsolvable. In the next chapter we obtain an uncomputable function t by applying the method of diagonalization to a list of the Turing computable functions of one argument. The uncomputable function p of the present chapter will be obtained more directly.

We confine our attention to Turing machines which read and write only two symbols: B and 1. Any such machine M can be thought of as computing some total or partial function f from positive integers to positive integers as follows.

> To discover the value (if any) which f assigns to the argument n, start M in its lowest-numbered state, scanning the leftmost of a block of n 1s on an otherwise blank tape. If M eventually halts in a *standard configuration*, i.e. scanning the leftmost of a block of 1s on an otherwise blank tape, $f(n)$ is the number of 1s in that block. But if M never halts, or halts in some nonstandard configuration, $f(n)$ is undefined.

It is easy to extend this definition so as to allow the arguments and values of f to be any *natural numbers* (= non-negative integers = 0 together with the positive integers): simply allow the number of 1s in the initial and final blocks to be 0. Thus, if M eventually halts, scanning a square on a completely blank tape, we have $f(n) = 0$ where n is the number of 1s in the initial block. On the other hand, we can compute $f(0)$ by starting M in its lowest-numbered state, scanning a square of a completely blank tape.

Where a machine M computes a *total* function f, we define the *productivity* of M as $f(0)$, *viz.*, the number of 1s on the tape when M halts (as it surely will, and in a standard configuration), having been started in its lowest-numbered state, scanning a square on a blank tape, More generally, where M computes a function f which may be total or partial, we define

$$M\text{'s } \textit{productivity} = \begin{cases} f(0) & \text{if } f(0) \text{ is defined,} \\ 0 & \text{otherwise.} \end{cases}$$

This agrees with the definition in Exercise 3.7:

Initially, the tape is blank and M is in its lowest-numbered state. If M eventually halts, scanning the leftmost of an unbroken string of 1s on an otherwise blank tape, its *productivity* is said to be the length of that string. But if M never halts, or halts in some other configuration, its productivity is said to be 0.

Now we define the function p as in Exercise 3.7:

$p(n)$ = *the productivity of the most productive n-state machines.*

The busy beaver problem (set and proved unsolvable by Tibor Rado, *Bell System Technical Journal*, May 1962, pp. 877–84) is the problem of designing a Turing machine which computes the function p – a machine which reads and writes only the symbols B and 1. The proof that this problem is unsolvable uses the facts about p which are mentioned in Exercise 3.7 (*c*) and (*d*). Parts (*a*) and (*b*) of that exercise, to which we now turn, are exercises designed to familiarize the reader with the notion of productivity.

Proof that $p(1) = 1$ (Exercise 3.7(*a*))

Note that there are infinitely many different 1-state Turing machines which read and write no symbols other than B and 1, but these fall into just 25 distinct classes, within each of which all machines can be relied upon to behave identically: the machines in any one class differ only with regard to the *name* (q_1 or q_2 or …) of the single state. Then the 25 classes correspond to the 25 different flow graphs in which there is just one node (unlabelled) and 0 or 1 or 2 arrows (from that node back to itself).

In particular, there is one such flow graph with no arrows at all. The corresponding machines halt immediately with the tape still blank, and thus have productivity 0.

The 1-state machines whose flow graphs have two arrows, labelled 'B:–' and '1:–' (where each – may be R or L or B or 1) never halt, so their productivities are all 0. They never halt because no matter what symbol they are scanning, there is always an instruction for them to follow, even if it is an instruction like 'print a blank on the (blank) square you are scanning and return to the state you are in'. There are $4 \times 4 = 16$ such flow graphs, corresponding to the 4 ways in which each of the two labels can be completed.

There are eight 1-state flow graphs with just one arrow, labelled 'B:–' or '1:–'. If the label has form '1:–', the machine halts immediately since the tape is blank, and thus has productivity 0; of the four graphs in which the arrow has a label of form 'B:–', three represent machines which never halt, and thus have productivity 0. These are the graphs whose solitary arrow is labelled 'B: R' or 'B: L' (*touring* machines) or 'B: B'. The only remaining graph is the one shown in Figure 4-1. Machines

Figure 4-1. Flow graph of the most productive 1-state machines.

with that graph print a single 1 on a blank tape and then halt, scanning that 1. Then these machines have productivity 1. Since all others have productivity 0, the productivity of the most productive 1-state machines which read and print only the symbols B and 1 is seen to be 1, and we have $p(1) = 1$.

Proof that $p(47) \geqslant 100$ (Exercise 3.7(b))

At the end of our discussion of Example 3.3 we described a 47-state machine which has productivity 100. For all we know there may be 47-state machines whose productivities are greater than 100, but we have established that the productivity of the most productive 47-state machines is *at least* 100: $p(47) \geqslant 100$. To get quite clear about what our machine looks like, let us abbreviate the graph of the 25-state machine which writes 1s on a blank tape and then halts in a standard configuration as in Figure 4-2. And let us abbreviate the graph of the 12-state machine in Example 3.2 which doubles the number of 1s on the tape as in Figure 4-3. (The

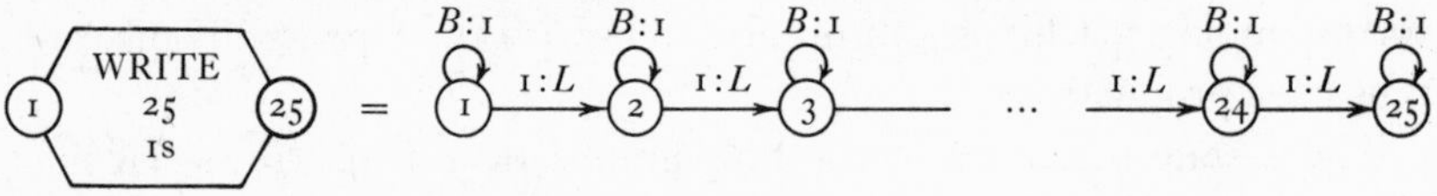

Figure 4-2. $f(0) = 25$.

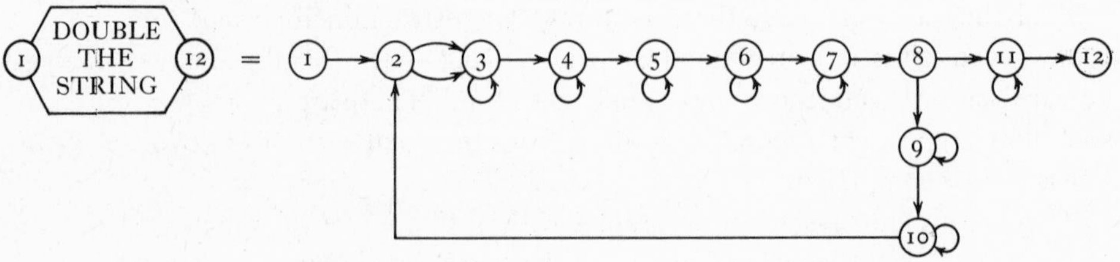

Figure 4-3. $f(n) = 2n$.

positions of some of the nodes have been altered, and the labels have been left off the arrows.) Now the 47-state machine with productivity 100 is obtained by plugging together two replicas of the doubling machine, and plugging the input of that combination into the output of the 'Write 25 1s machine', where *plugging together* is a matter of superimposing the starting node of one machine on the halting node of the other: *identifying* the two nodes. Plugging the three machines of Figure 4-4 together in order, we have the machine of Figure 4-5: a 47-state machine with

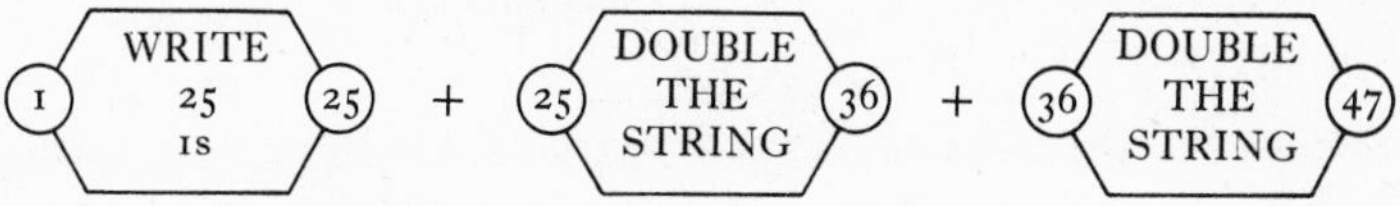

Figure 4-4. Plugging three machines together.

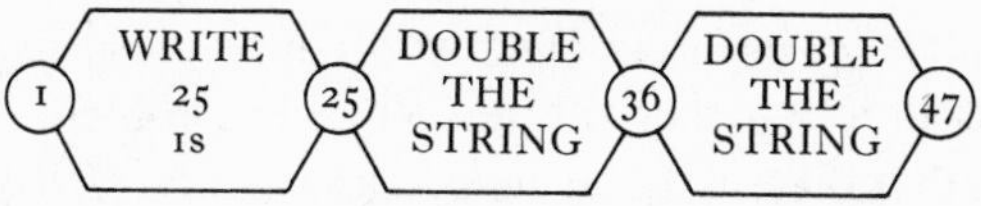

Figure 4-5. $f(0) = 100$.

productivity 100. If you are puzzled by the numbering of the states, think of the 12 states of (say) the middle machine in Figure 4-4 as having the numbers $25+0$, $25+1, \ldots, 25+11$, and imagine that the successive nodes in Figure 4-3 are renumbered $0, 1, \ldots, 11$ instead of $1, 2, \ldots, 12$.

Proof that $p(n+1) > p(n)$ (Exercise 3.7(c))

Choose any of the most productive n-state machines and add one more state as in Figure 4-6. The result is an $(n+1)$-state machine of producti-

Figure 4-6. $f(0) = p(n)+1$.

vity $p(n)+1$. There may be $(n+1)$-state machines with even greater productivity than this, but we have established that the productivity of the most productive $(n+1)$-state machines is greater (by *at least* 1) than the productivity of the most productive n-state machines.

Proof that $p(n+11) \geqslant 2n$ (Exercise 3.7(*d*))

In 3.3 we designed an $(n+11)$-state machine of productivity $2n$: the scheme is shown here as Figure 4-7. Since there may well be $(n+11)$-state machines with even greater productivity we are not entitled to conclude that $p(n+11) = 2n$; but we *are* entitled to conclude that $p(n+11)$ is $2n$ *or more*.

Figure 4-7. $f(0) = 2n$.

Proof that the busy beaver problem is unsolvable (Exercise 3.8)

The proof is by *reductio ad absurdum*: we deduce an absurdity (*viz.*, $0 \geqslant 1$) from the assumption that the busy beaver problem *is* solvable. The assumption is that there is a Turing machine (call it '*BB*') with, say, k states, which computes the function p. This machine is to read and write only the symbols B and 1. Started in state 1 scanning the leftmost of an unbroken string of n 1s on an otherwise blank tape, it eventually halts in state k, scanning the leftmost of an unbroken string of $p(n)$ 1s on an otherwise blank tape.

To begin, note that if there is such a machine as *BB*, then there is a machine with $n+2k$ states which has productivity $p(p(n))$: see Figure 4-8. Thus, if *BB* exists, the productivity of the *most* productive $(n+2k)$-state machines must be *at least* $p(p(n))$:

$$p(n+2k) \geqslant p(p(n)) \quad \textit{if BB exists.} \tag{1}$$

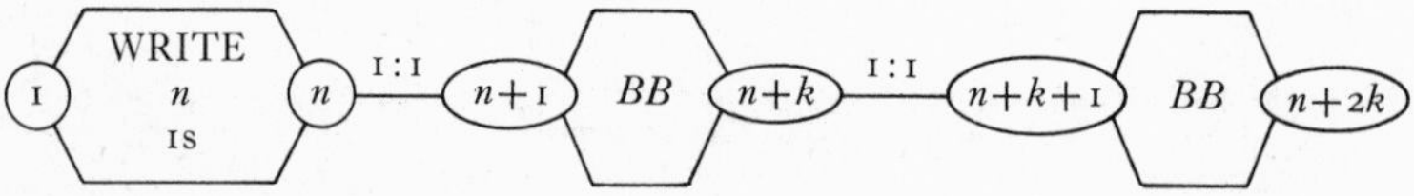

Figure 4-8. $f(0) = p(p(n))$.

Whether or not *BB* exists, we know that $p(n+1) > p(n)$ for every positive integer n, and from this it follows that for any positive integers i and j,

$$p(i) > p(j) \quad \textit{if} \quad i > j. \tag{2}$$

Note that the denials of the statements '$p(i) > p(j)$' and '$i > j$' can be written '$p(j) \geqslant p(i)$' and '$j \geqslant i$', respectively: if the first number is *not* greater than the second, then the second *is* greater than or equal to the first. Then from (2) by contraposition we have

$$j \geqslant i \quad if \quad p(j) \geqslant p(i) \tag{3}$$

for all positive integers i and j. Now from (3) and (1) – setting $j = n+2k$ and $i = p(n)$ – we have

$$n+2k \geqslant p(n) \quad \textit{if BB exists} \tag{4}$$

for every positive integer n.

Being true for each particular n, (4) remains true when n is increased by 11:

$$n+11+2k \geqslant p(n+11) \quad \textit{if BB exists.} \tag{5}$$

But we have just proved (Exercise 3.7(*d*)) that

$$p(n+11) \geqslant 2n \tag{6}$$

whether or not *BB* exists. Combining inequalities (5) and (6) we have for all positive integers n

$$n+11+2k \geqslant 2n \quad \textit{if BB exists.} \tag{7}$$

Subtracting n from both sides of (7) we have

$$11+2k \geqslant n \quad \textit{if BB exists} \tag{8}$$

for every positive integer n. In particular, with $n = 12+2k$ we have $11+2k \geqslant 12+2k$ or, subtracting $11+2k$ from both sides,

$$0 \geqslant 1 \quad \textit{if BB exists} \tag{9}$$

Thus the assumption that *BB* exists yields the promised absurdity.

Then the function p is computed by no Turing machine in the standard form specified at the beginning of this chapter. To get from this to the conclusion that p is simply uncomputable we must convince ourselves that if a function is computable at all, it is computable by some Turing machine in our standard form. The thesis (a form of Church's thesis) is this:

If a total function f is computable in the intuitive sense then there is a Turing machine which reads and writes no symbols but B and 1 and which, started in its lowest-numbered state scanning the leftmost of an unbroken string of n 1s on an otherwise blank tape will eventually halt, scanning the leftmost of an unbroken string of $f(n)$ 1s on an otherwise blank tape.

(It is immaterial whether we assert this thesis for all natural numbers *n*, or restrict it to positive integers.) Lacking a precise general characterization of computability, we cannot *prove* this thesis; but we can convince ourselves of its plausibility by such considerations as those to be given in Chapters 6–8.

In conclusion, let us have another look at the function *p* in the light of Church's thesis. The question is, 'Why isn't *p* computable in some intuitive sense?' After all, there are only finitely many different graphs of *n*-state machines if we don't trouble to number any of the nodes except for node 1, the starting node. Then for each *n* we can (in imagination, anyway) set all the *n*-state machines going, starting in state 1 on a blank tape, and await developments. As time passes, one or another of the machines may halt, at which point we can see whether it is scanning the leftmost of an unbroken string of 1s on an otherwise blank tape. If it does halt in that standard position, we find its productivity by counting the number of 1s in the string, but if it halts in a non-standard position we know that its productivity is 0. But there is a catch: some of the *n*-state machines may never halt, so that no matter how long we wait, it may be that the productivity of one or more of the *n*-state machines cannot be determined in the way we have just sketched. *Those* machines will have productivity 0 because they never halt.

Zeus, working faster and faster in the manner indicated in Table 2-2, could simulate every step of an unending computation in 1 second; or in $\frac{1}{2}$ second, or in $\frac{1}{4}$ second, ..., as he pleases. Then *he* can simulate infinitely many computations, be they individually finite or infinite, in 1 second: he spends $\frac{1}{2}$ second on the first, $\frac{1}{4}$ second on the second, and so on. Zeus would not be baffled by the non-halters: *he* could compute *p*.

But we, and our machines, can only carry out finite numbers of operations in finite periods of time, be they ever so long. To be usable by us, a systematic procedure for identifying non-halters must always terminate after one or another finite number of steps, no matter to which machine we apply it. If we had such a procedure we could apply it to the graphs of all the *n*-state machines while those machines are working (having been started in their lowest-numbered states, on blank tapes). After some finite period of time, each machine will either have halted or have been identified as a non-halter, so that for each *n*, there would be some period of time after which we would know the productivity of every *n*-state machine. For each *n*, we would then be able to compute *p*(*n*): the function *p* would be computable in an intuitive sense if there were a systematic procedure for identifying non-halters – a procedure which always terminates after one or another finite number of steps, no matter what non-halter it may be applied to.

The *halting problem* is the problem of designing a systematic method – an effective procedure – for identifying Turing machines which never halt, once started in their lowest-numbered states on blank tapes. We have seen that if the halting problem is solvable, the function p is computable in some intuitive sense. We have also seen that the function p is computable by no Turing machine in standard form. Conclusion:

If Church's thesis is true, the halting problem is unsolvable.

Later (in Chapter 10) we shall make use of this fact to show that if Church's thesis is true, the decision problem for first-order logic is unsolvable.

Question. In Figure 4-8, why not simply plug the three machines together, in the manner shown in Figures 4-4 and 4-5? (Answer follows.)

Answer. First, see why the plugging trick worked, in Figures 4-4 and 4-5. The halting state of the 'Write 25 1s' machine could be superimposed on the starting state of the doubling machine because those two states look like this, graphically:

They can be superimposed without giving the machine conflicting instructions about what to do in state 25:

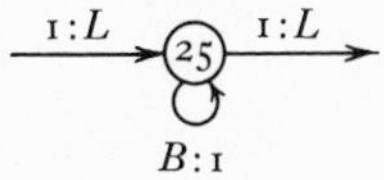

Similarly for the halting state of the first doubler and the starting state of the second:

But we have no assurance that in its starting state, BB has no instruction about what to do in case it is scanning a blank. (Note that the starting state may well have other functions: the machine may return to it at various later stages of its operation, if it occurs in closed loops, as does the starting state of the multiplier in Exercise 3.4.) Thus, for all we know, the halting state of the 'Write n 1s' machine and the starting state of BB might clash, e.g. as follows:

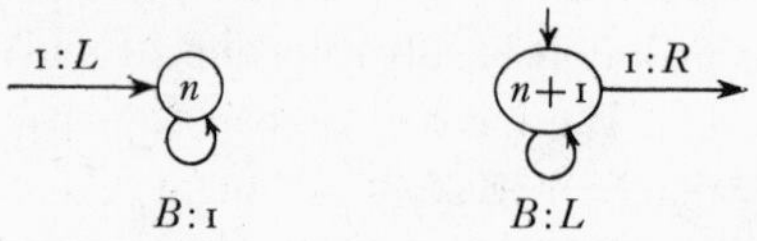

Superimposing these states, the machine would have conflicting instructions about what to do in state n ($= n+1$) when scanning a blank.

5
Uncomputability via diagonalization

In this chapter we exhibit various uncomputable functions as in Exercises 3.8, 3.9 and 3.10. In calling these functions simply 'uncomputable' instead of 'uncomputable by Turing machines' we are presupposing Church's thesis, evidence for which will be given in Chapters 6–8. Throughout this chapter, 'function' will mean *function from positive integers to positive integers*.

The bare existence of uncomputable functions is quickly proved, in a rough and ready way, for we know (Exercise 2.4(*d*)) that the set of all Turing machines (represented by finite strings of quadruples) can be enumerated, whereas (Exercise 2.3) the set of all functions cannot. If the set of Turing computable functions were identical with the set of all functions, that set would be both enumerable and not. Therefore, the two sets are distinct: the former is a proper subset of the latter.

The foregoing proof is rough and ready in that we have not yet precisely defined the notion of Turing computability of functions from positive integers to positive integers. Turing machines operate on symbols, not numbers. Before we can speak of Turing machines as computing numerical functions, we must specify the notation in which the numerical arguments and values are to be represented on the machine's tape, e.g. as in the adder and multiplier of Exercises 3.4 and 3.5. The choice of monadic notation makes for simplicity, but it is not essential: The decimal notation, or various others, would serve as well. What *is* essential is that such particulars as notation *be* specified, in one way or another. Our specifications are as follows.

(*a*) The arguments of the function will be represented in monadic notation by blocks of 1s, separated by single blanks, on an otherwise blank tape. Then at the beginning of the computation of (say) $3+2$, the tape will look like this: $111B11$.

(*b*) Initially, the machine will be scanning the leftmost 1 on the tape. Then if the machine's initial state is 1, the initial configuration in computing $3+2$ will be this: $\underset{1}{1}11B11$.

(*c*) If the function which is to be computed assigns a value to the arguments which are represented initially on the tape, the machine will eventually *halt in a standard final configuration*, i.e., scanning the leftmost of a block of 1s on an otherwise blank tape. The number of 1s in that block will be the value which the function assigns to the given arguments. Thus, the final configuration in the computation of 3 + 2 by the adder in Exercise 3.4 will be $\underset{4}{1}1111$.

(*d*) If, on the other hand, the function assigns no value to the given arguments, the machine will not eventually halt in a standard final configuration: It will either go on running forever, or will eventually halt in some such nonstandard final configuration as $\underset{1}{B}111$ or $\underset{1}{1}B1$ or $1\underset{1}{1}1$, etc.

The initial configurations described in (*a*) and (*b*) above will be called 'standard' initial configurations, just as the final configurations described in (*c*) are called 'standard' final configurations.

With these specifications, any Turing machine can be seen to compute a function of one argument, a function of two arguments, and in general, a function of *n* arguments, for each positive integer *n*. Thus, consider the machine which is specified by the single quadruple '$q_1 1 1 q_2$' with initial state q_1. Started in a standard initial configuration, it halts immediately, leaving the tape unaltered. If there was only a single block of 1s on the tape initially, its final configuration will be standard, and thus this machine computes the identity function, *i*, of one argument: $i(n) = n$ for each positive integer *n*. Then the machine computes a certain total function of one argument, but if there are two or more blocks of 1s on the tape initially, the final configuration will not be standard. Accordingly, the machine computes the extreme partial function of two arguments which is undefined for all pairs of arguments. And in general, for *n* arguments, this machine computes the extreme partial function which assigns values to no *n*-tuples of arguments.

On the other hand, the machine of Figure 5-1 computes, for each *n*, the total function which assigns the same value, *viz.*, 1, to each *n*-tuple of argument. Started in state 1 in a standard initial configuration, this machine erases the first block of 1s (cycling between states 1 and 2) and goes to state 3, scanning the second square to the right of the first block. If it sees a blank there, it knows it has erased the whole tape, and so prints a single 1 and halts in state 4, in a standard final configuration. But if it sees a 1 there, it reenters the cycle between states 1 and 2, erasing the second block of 1s and inquiring again, in state 3, whether the whole tape is blank, or whether there are one or more blocks still to be dealt with.

In this chapter, we shall be concerned with functions of a single argument. We show that the set of all such functions cannot be enumer-

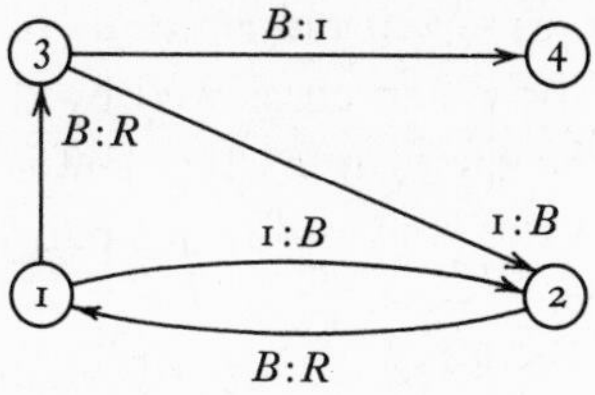

Figure 5-1. $f(x_1, ..., x_n) = 1$.

ated, while the set of functions of one argument which are computed by Turing machines can be enumerated. And by attending to the details of that enumeration, we define a particular function, t, of one argument, which is not computed by any Turing machine. According to our specifications, a machine M computes the function f of one argument if and only if for each $n = 1, 2, 3, ...$ for which $f(n)$ is defined, and for no other ns, M eventually halts in a standard configuration with $f(n)$ 1s on its tape when it is started in a standard configuration with n 1s on its tape. The function f is said to be *Turing computable* if and only if it is computed by some machine M, according to these specifications.

We enumerate the Turing computable functions of one argument by enumerating the Turing machines. To this end, we specify each machine by writing its quadruples in a single string, with no spaces separating adjacent quadruples, beginning with a quadruple whose first symbol is the machine's initial state. Example: the machine of figure 5-1, which has initial state q_1, is represented by the string,

$$q_1 S_0 R q_3 q_1 S_1 S_0 q_2 q_2 S_0 R q_1 q_3 S_0 S_1 q_4 q_3 S_1 S_0 q_2.$$

Clearly, each Turing machine is specified in this way by some finite string of letters of the enumerably infinite alphabet

$$R, L, S_0, q_1, S_1, q_2, S_2, q_3, S_3, ... \tag{1}$$

Not every word on this alphabet specifies a Turing machine. A word specifies a Turing machine if and only if it meets these conditions:

(a) Its length is exactly divisible by 4.
(b) Only the symbols q_1, q_2, q_3, etc. occur in positions 1, 4, 5, 8, 9, 12, ..., $4n$, $4n+1$, ...
(c) Only the symbols S_0, S_1, S_2, etc. occur in positions 2, 6, 10, ..., $4n+2$, ...
(d) Only the symbols R, L, S_0, S_1, ... occur in positions 3, 7, 11, ..., $4n+3$, ...
(e) No configuration of form q_iS_j occurs more than once in the word. (2)

Condition (*e*) is a bit arbitrary. It serves to rule out contradictory instructions, but that purpose would have been served as well by this variant, which allows repetitions of the same quadruple: If a configuration of form $q_i\, S_j$ occurs more than once in the word, the next pair of symbols must be the same at each occurrence.

Now by Exercise 2.4(*d*) we know that the set of all words on the enumerably infinite alphabet (1) is enumerable: to speak vividly, it is possible to arrange them in a single (possibly, gappy) list, $w_1, w_2, w_3, \ldots$. If from this list we delete the words which fail to specify Turing machines, we are left with a gappy list in which each Turing machine is named at least once, and nothing else is named. Now (still speaking vividly) let us close the gaps to obtain a gapless list $M_1, M_2, M_3, \ldots$ of all Turing machines. Since every Turing machine determines a definite function from positive integers to positive integers, this list of machines determines a list

$$f_1, f_2, f_3, \ldots \tag{3}$$

of all Turing computable functions of one argument, where for each n, f_n is the total or partial function of one argument which machine M_n computes.

Then as soon as we specify a particular enumeration $w_1, w_2, w_3, \ldots$ of the words on the alphabet (1), we shall have specified a particular enumeration (3) of the Turing computable functions of a single argument. Let us enumerate the *w*s by the method which was described at the end of Chapter 2. This gives us a gappy list in which the first entry is w_{12}, *viz.* R. In general, the position of a word in this list is given, in decimal notation, by the string obtained from that word by replacing each occurrence of R by 12, of L by 122, of S_0 by 1222, and so on: The *n*th letter in alphabet (1) is replaced by a 1 followed by n 2s.

Having determined the sequence of *w*s, it is now trivial (but laborious) to determine the sequence of *M*s. The first of the *w*s which determines a Turing machine (i.e., which meets conditions (2)) is $q_1S_0Rq_1$. This is w_n for $n = 1\,222\,212\,221\,212\,222$: A quadrillion and then some. Then M_1 is as shown in Figure 5-2(*a*). In any standard initial configuration it is scanning a 1, and therefore halts immediately, in its initial position. Then

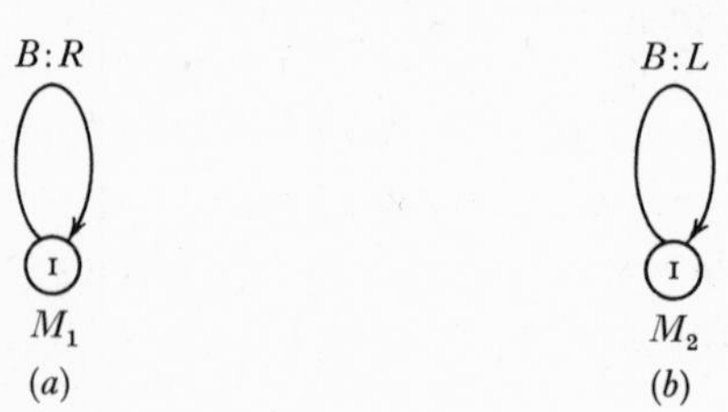

Figure 5-2

f_1 is i, the identity function: we have $f_1(n) = n$ for each n. Furthermore, f_2 is the same function as f_1, for the second of the ws which represent Turing machines is '$q_1 S_0 L q_1$', which is w_n for $n = 12\,222\,122\,212\,212\,222$. Then M_2 is as in Figure 5-2(*b*): It too, halts immediately, in a standard final configuration, when it is started in a standard configuration with a single block of 1s on the tape.

We have now enumerated the Turing computable functions of one argument by enumerating the machines which compute them. The fact that such an enumeration is possible shows that there must exist uncomputable functions of a single argument. The point of actually specifying one such enumeration is to be able to exhibit a particular such function, *viz.*, the function u which is defined as follows. (Cf. Exercises 2.3 and 3.6.)

$$u(n) = \begin{cases} 1 & \text{if } f_n(n) \text{ is undefined,} \\ f_n(n)+1 & \text{otherwise.} \end{cases} \tag{4}$$

Now u is a perfectly genuine total function of one argument, but it is not Turing computable, i.e., u is neither f_1 nor f_2 nor f_3, etc. *Proof.* Suppose that u *is* one of the Turing computable functions – the mth, let us say. Then for each positive integer n, either $u(n)$ and $f_m(n)$ are both defined and equal, or neither of them is defined. But consider the case $n = m$:

$$u(m) = f_m(m) = \begin{cases} 1 & \text{if } f_m(m) \text{ is undefined,} \\ f_m(m)+1 & \text{otherwise.} \end{cases} \tag{5}$$

Then whether f_m is or is not defined for the argument m, we have a contradiction: Either $f_m(m)$ is undefined, in which case (5) tells us that it has the value 1 (and is thus both defined and undefined), or $f_m(m)$ is defined and equal to its own successor (from which we deduce that $0 = 1$). Since we have derived a contradiction from the assumption that u appears somewhere in the list $f_1, f_2, \ldots, f_m, \ldots$, we may conclude that the supposition is false: We conclude that the function u is not Turing computable.

Although no Turing machine computes the function u, we can readily determine its first few values. Thus, since $f_1 = f_2 = i =$ the identity function, we have $f_1(1) = 1$ and $f_2(2) = 2$, so that by (4) we have $u(1) = 2$ and $u(2) = 3$. Note that there is no contradiction here, or in the definition (4) of u. Equation (5) was contradictory because it incorporated the assumption that u is Turing computable: Computed by machine M_m for some positive integer m. The contradiction in (5) refutes that assumption, but does not impeach definition (4).

Now it may seem that we can actually compute $u(n)$ for any positive integer n – if we don't run out of time. Certainly, we have discovered the values that u assigns to the arguments 1 and 2, and it may seem that it must be perfectly routine to do this for as many subsequent arguments as time permits. Here, the real question is, whether we ever really need luck or ingenuity to discover the values which u assigns to its arguments. It may seem that we do not, for it is straightforward to discover which quadruples determine M_n for $n = 1, 2, 3, \ldots$. (This is straightforward, but, eventually, humanly impossible because the duration of the trivial calculations, for large n, exceeds the lifetime of a human being and, in all probability, the lifetime of the human race. But in our notion of computability, we ignore the fact that human life is limited.) The essential question concerns the business of determining whether machine M_n, started scanning the leftmost of an unbroken string of n 1s on an otherwise blank tape, does or does not eventually halt scanning the leftmost of an unbroken string of 1s on an otherwise blank tape. Is *that* perfectly routine? If so, the function u is computable after all – although not by any Turing machine.

But *is* it perfectly routine, to discover whether M_n eventually halts in standard position as above? Certainly, it is perfectly routine to follow the operations of M_n, once the initial configuration has been specified; and if M_n does eventually halt, we shall eventually get that information by following its operations. But what if M_n is destined never to halt, given the initial configuration? Must there be some point in the routine process of following its operations at which it becomes clear that it will never halt? In simple cases this is certainly so, e.g. it is obvious that the quadruple $q_1S_1S_1q_1$ represents a machine which never halts if it starts scanning a square with a 1 in it, and it is obvious that the machine described by the string $q_1S_1S_0q_1q_1S_0Lq_1$ will never halt. But when we consider more complicated cases it is by no means evident that there is a uniform, mechanical procedure for discovering whether or not a given machine, started in a given configuration, will ever halt. Thus, consider the mutliplier of Exercise 3.4. It is described by various strings of 124 symbols on the alphabet (1). Imagine encountering one such string in the list w_1, $w_2, \ldots$. It is routine to verify that the string describes a Turing machine, and one can easily enough derive from that string a flow graph like the one shown in Chapter 3 but without the annotation, and without the accompanying text which explains the point and the workings of the leapfrog routine and the rest. Is there a uniform method which, in this and in much more complicated cases, allows one to determine whether the machine eventually halts in a standard final configuration, once the initial tape configuration has been specified? If so, the function u is computable, although not by any Turing machine. But if not, we have failed in our attempt to see u as a function which is computable in some way, but not in Turing's way.

The crucial problem is to specify a uniform routine which, given any machine's quadruples and a description of any initial tape configuration, determines after some finite number of steps whether the machine would eventually halt in a standard configuration, having been started in the given initial configuration. If there is such a routine, then Turing's notion of computability is too narrow: there are then computable functions which are not Turing computable, and in particular, u is such a function. But if Turing's notion of computability is broad enough to encompass all functions which are computable in any intuitive sense, this problem is unsolvable.

A closely related problem, likewise unsolvable if Turing's notion of computability is universal, is the halting problem of Exercise 3.10: specify a uniform effective procedure for discovering whether or not Turing machines eventually halt (at all – whether in standard final configuration or in some other configuration) when started in their initial states scanning the leftmost 1 in an unbroken string of them on an otherwise blank tape. It is noteworthy that whether or not Church's thesis is to be accepted, the following function h, which is computable (in the intuitive sense) if and only if the halting problem is solvable, can be proved not to be Turing computable: $h(m, n) = 2$ or 1 accordingly as machine number m does or does not eventually halt after being started in its initial state scanning the leftmost of an unbroken string of n 1s on an otherwise blank tape. Therefore *if Church's thesis is correct, the halting problem is unsolvable.*

The proof that h is not Turing computable which we now give is a more direct alternative to the proof given in Exercise 3.10: we do not now deduce the uncomputability of h from that of u, say, but establish it directly, by a diagonal argument. Toward that end we have specified h completely, and as a total function, whereas in Exercise 3.10 we specified only the conditions under which h assumes the value 1. The argument is by *reductio ad absurdum*: we show that if h is computed by some Turing machine H, then there must exist another Turing machine M_m such that for all positive integers n, M_m halts if and only if machine M_n does not, after being started in its initial state, scanning the leftmost of an unbroken string of n 1s. But of course there can be no such machine M_m, for if there were, it would have to halt if and only if it never halts, after being started in its initial state scanning the leftmost of an unbroken string of m 1s on an otherwise blank tape.

The truth of the claim that if h is computed by some Turing machine H then there exists a Turing machine M_m of the sort described can be seen by considering Figure 5-3. The 'H' box in Figure 5-3 represents the

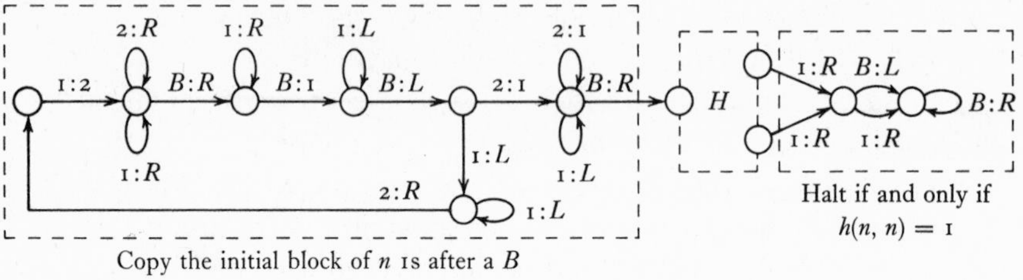

Figure 5-3. Graph of M_m.

graph of H. The node on the left represents its initial state. Since h is total there will be one or more nodes in the graph of H from which emerge no arrows with 1s before their colons. All the states in which H might halt are represented by such nodes. (So, perhaps, are some states in which H does not halt because in computing values of h, H never scans a 1 while in those states.) We have supposed, for definiteness, that there are just two such states, which are indicated by the nodes at the right of the box. The other states and arrows of H are hidden inside the box.

Figure 5-3 thus repesents the graph of a machine M_m which contains the machine H as a part and which, when started in its initial state (= leftmost node) scanning the leftmost of a block of n 1s on an otherwise blank tape, first recopies that block to the right, leaving a blank between the two blocks, and then sets H to work on the pair of blocks. When H is through, the whole machine halts if and only if H has left exactly one 1 on the tape. (H will always end by scanning the left of either 1 or 11 on an otherwise blank tape.)

Thus M_m halts if and only if M_n does not (if and only if $h(n, n) = 1$) when started in its initial state scanning the leftmost of an unbroken string of n 1s on an otherwise blank tape. Setting $m = n$, we see that M_m cannot exist, and thus H cannot exist: the function h is not Turing computable, whether or not Church's thesis is true. In other words, the halting problem is solvable by no Turing machine. If Church's thesis is true, it is absolutely unsolvable.

Exercises

5.1 What are the functions of one and of three arguments which are computed by the adder of Exercise 3.5?

5.2 Identify M_3 and M_4, and determine the values $u(3)$ and $u(4)$.

5.3 Define the function t as follows:

$$t(n) = \begin{cases} 1 & \text{if } f_n(n) \text{ is undefined,} \\ \text{undefined} & \text{if } f_n(n) \text{ is defined.} \end{cases}$$

Prove (without using Church's thesis) that t is not Turing computable.

5.4 Define $f(m, n) = f_m(n)$ for each $m, n = 1, 2, \ldots$. Is f computable?

Solutions

5.1 The adder of Exercise 3.5 computes the identity function of one argument, and computes the extreme partial function of three arguments which assigns values to no triples of positive integers.

5.2 M_3 is $q_1 S_0 R q_2$ and M_4 is $q_1 S_0 S_0 q_1$. Both compute the identity function of one argument, so $u(3) = 4$ and $u(4) = 5$.

5.3 If t were Turing computable – say, by machine M_k in our enumeration – we should have

$$t(k) = f_k(k) = \begin{cases} 1 & \text{if } f_k(k) \text{ is undefined,} \\ \text{undefined} & \text{if } f_k(k) \text{ is defined.} \end{cases}$$

Then the hypothesis that t is Turing computable leads to a contradiction (without use of Church's thesis) and is therefore false.

5.4 Yes. (Use Church's thesis.)

6
Abacus computable functions are Turing computable

Put very broadly, Church's thesis says that any mechanical routine for symbol manipulation can be carried out in effect by some Turing machine or other. Thus, multiplication of positive integers, which we tend to think of as a mechanical routine in two dimensions involving the ciphers 0, 1, ..., 9, can be performed one-dimensionally by a Turing machine which reads and prints the ten ciphers, the blank, and perhaps a few extra symbols which it uses in the course of the computation for book-keeping purposes but which do not appear on the tape initially or when the machine halts. But in Exercise 3.4 we saw how multiplication can also be carried out by a Turing machine which has these special characteristics:

(1) The two numbers to be multiplied are represented by blocks of 1s, separated by a single blank, on an otherwise blank tape, and the product is represented by a single block of 1s on an otherwise blank tape, when the machine halts.

(2) The machine starts and halts in standard position, i.e. scanning the leftmost 1 on the tape.

(3) Throughout the computation, the machine never goes more than two squares to the left of the leftmost 1 that was on the tape initially. Then the tape need only be infinite in one direction: to the right.

(4) The machine reads and writes no symbols other than B and 1, i.e. the machine's quadruples are words on the alphabet

$$R, L, B, 1, q_1, q_2, q_3, \ldots$$

Let us now reserve the term 'Turing computable' for functions which are computable by Turing machines in accordance with these four restrictions and with the further restriction (5) that the leftmost 1 in the block of 1s representing the value of the function appear in the very same square of the tape in which, initially, the leftmost 1 appeared. The evidence for Church's thesis which we shall adduce in this and the following two chapters is a survey of the Turing computable functions: a survey which shows them to comprise a remarkably broad class, which it is plausible to suppose inclusive of all functions from positive integers to positive integers that are effectively computable at all.

The restriction to monadic notation (instead of, say, decimal notation) and to the standard position (scanning the leftmost 1) for starting and halting is inessential; but *some* assumptions had to be made about initial and final positions of the machine, and these assumptions seem especially simple. It is even possible to describe a process whereby any decimal computation can be transformed into one that satisfies conditions (1)–(5) above: a process whereby the flow graph of a decimal machine which computes a certain function can be converted step by step into the flow graph of a machine which computes the same function in a format which meets conditions (1)–(5).

In broadest outline, the process is this. Code the ciphers 0, 1, ..., 9 as strings of 1s: 1, 11, ..., 1111111111. Then e.g. decimal 213 would look like this: 111*B*11*B*1111. It is not hard to design a monadic-to-coded decimal converter, i.e. a machine which, started in standard position scanning a string of *n* 1s, eventually halts in standard position scanning the coded decimal form of the number *n*. (Thus, for $n = 10$, the initial tape contents would be 11111111111 and the final contents would be 11*B*1.) Similarly, one can design a coded-decimal-to-monadic converter. And one can modify the flow graph of the given decimal machine so as to obtain the flow graph of a machine which behaves in essentially the same way, except that it works with the *coded* forms of the ten ciphers, and meets condition (3). (That might be arranged by systematically modifying the graph so that every movement of the machine to the left on the tape is preceded by a routine which moves all 1s on the tape a single square to the right. And *that* is a matter of replacing each arrow labelled $B{:}L$ or $1{:}L$ by a graph which executes the required general movement to the right and then returns to the main routine at the appropriate position.) Now the required standardized routine is obtained by stringing together, in order, a monadic-to-coded-decimal converter, the transformed graph of the given decimal machine, and a coded-decimal-to-monadic converter.

It would be laborious to specify this process in full detail; nor, after all that work, would we have succeeded in showing that monadic notation can serve as well as *any* other, for purposes of Turing computation. No end of notations might be invented, and there is no hope of proving that everything computable in any of them is computable in monadic notation. It is for this reason that we adopt the monadic notation at the outset: *define* Turing computability as computability in a format which satisfies conditions (1)–(5); and interpret

Church's thesis: all computable functions are Turing computable

in the light of that definition.

As an important part of the evidence for Church's thesis we show how the operations of the ordinary sort of high-speed digital computer can be simulated by a Turing machine which meets requirements (1)–(5). By a digital computer of the 'ordinary' sort we mean a device which has an unlimited amount of *random-access storage*. In contrast to a Turing machine, which stores information symbol by symbol on squares of a one-dimensional tape along which it can move a single step at a time, a machine of the seemingly more powerful 'ordinary' sort has access to an unlimited number of *registers* R_0, R_1, R_2, ..., in each of which can be written numbers of arbitrary size. Thus, the usual sort of machine is a bit like a generalized Turing machine which can read and write an infinite variety of symbols S_0, S_1, S_2, ..., any one of which can appear in any square of its tape. (The symbols S_n might represent the number n.) But the usual sort of machine has still another feature – random access – which allows it to go directly to register R_n without inching its way, square by square, along the tape. That is, each register (R_n, say) has its own *address* (the number n) which allows the machine to carry out such instructions as

> compute the sum of the numbers in registers R_m and R_n, and store it in register R_p,

which we abbreviate:

$$[m]+[n] \rightarrow p$$

(In general, $[n]$ is the number in register R_n, and the number at the right of an arrow identifies the register in which the result of the operation at the left of the arrow is to be stored.)

It should be noted that our 'usual' sort of computing machine is really quite unusual in one respect: although real digital computing machines often have random-access storage, there is always a finite upper limit on the size of the numbers which can be stored, e.g. a real machine might have the ability to store any of the numbers 0, 1, ..., 10 000 000 in each of its registers, but no number greater than ten million. Thus, it is entirely possible that a function which is computable by one of our idealized 'usual' machines is not computable by any real machine, simply because, for certain arguments, the computation would require more capacious registers than any real machine possesses. (Indeed, addition is a case in point: there is no finite bound on the sizes of the numbers one might think of adding, and hence no finite bound on the sizes of the registers needed for the arguments and the sum.) But this is in line with our objective of abstracting from technological limitations so as to arrive at a notion of computability which is *not too narrow*. We seek to show that certain functions are uncomputable in an absolute sense: uncomputable even by our idealized machines, and, therefore, uncomputable by any past, present, or future real machine.

One last point: instead of working with functions whose arguments and values are positive integers 1, 2, 3, ... we shall now work with functions whose arguments and values are natural numbers (non-negative integers) 0, 1, 2, ... We do this simply in order to conform with current fashion and thus ease comparison of what we say here with what is said in most books on the subject. The change is simply a matter of interpreting a solid block of one or more 1s on a Turing machine tape as representing the natural number $n-1$ instead of the positive natural number n. Thus, in a computation of $0+2$ in our standard format the tape would have $1B111$ on it initially and would have 111 on it finally.

In order to avoid discussion of electronic or mechanical details we may imagine the 'usual' sort of machine in crude, stone age terms. Each register may be thought of as a roomy, numbered box capable of holding any number of stones: none or one or two or ..., so that $[n]$ will be the number of stones in box number n. The 'machine' is operated by a man who is capable of carrying out two sorts of operations:

add a stone to the pile in box number n

and

remove a stone from the pile in box number n if there are any there.

A program can be specified by a flow chart which indicates the sequence in which these two sorts of operations are to be performed. Typically, the program will have one or more *branch points* at which the next operation will be one thing or another depending on whether the box from which a stone is to be removed (if possible) is already empty. Thus, branch points are associated with elementary operations of the second sort, above. Following Joachim Lambek, we shall refer to such an 'ordinary' machine as an (infinite) *abacus*.

The elementary operations will be symbolized as in Figure 6-1.

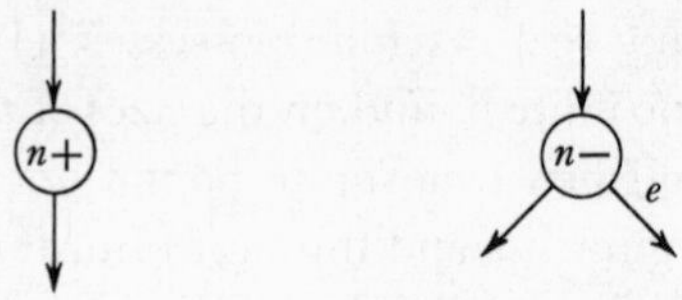

Add a stone to the pile in box n

If box n is empty, come out on the arrow marked 'e', but if not, remove a stone and come out on the other arrow

Figure 6-1. Elementary operations in abaci.

Flow charts can then be built up as in the following examples.

Example 6.1. Empty box n

Flow chart *Block diagram*

Example 6.2. Empty box m into box n (where $m \neq n$)

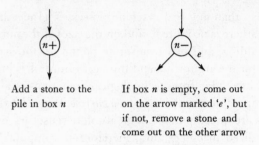

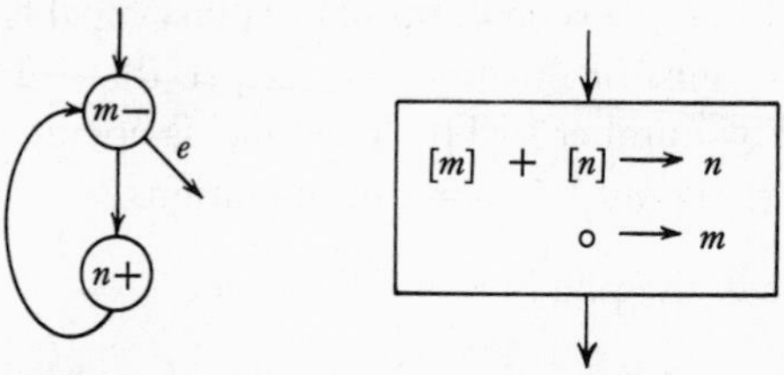

Flow chart *Block diagram*

(If $m = n$, the program halts – exits on the 'e' arrow – either at once or never, accordingly as the box is empty or not, initially.) When (as intended) $m \neq n$, the *effect* of the program is the same as that of carrying stones from box m to box n until box m is empty, but there is no way of instructing our cave man to do precisely that. What he *can* do is ($m-$) take stones out of box m, one at a time, and throw them away on the ground, and ($n+$) pick stones up off the ground, one at a time, and put them in box n. There is no assurance that the stones he puts in box n are the very ones he removed from box m, but we need no such assurance in order to be assured of the desired *effect* as described in the block diagram, *viz.*,

$[m]+[n] \rightarrow n$: the number of stones in box n after *this* move equals the sum of the numbers in m and in n before the move

and then

$0 \rightarrow m$: the number of stones in box m after this move is 0.

In general, the number on the right of an arrow identifies a box (a storage register, a pile) and the number on the left of that arrow is the number of stones that will be in that box (storage register, pile) after the move in question has been carried out.

In future, we shall assume (unless the contrary possibility is explicitly allowed) that when boxes are referred to by letters 'm', 'n', 'p', etc., distinct letters represent distinct boxes.

Example 6.3. Add box *m* into box *n* without loss from *m*

If the program is to have this effect, box p must be empty to begin with. In that case, p will be empty at the end, as well.

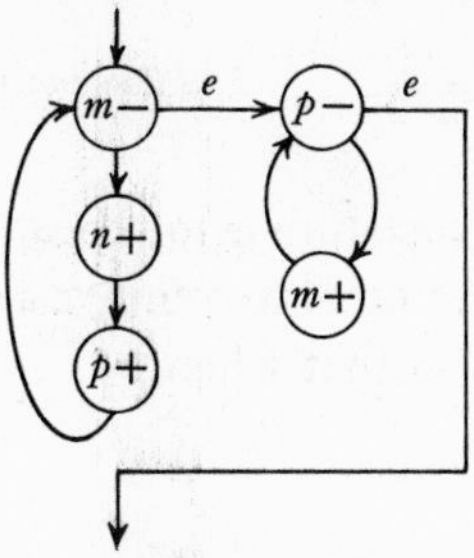

Flow chart

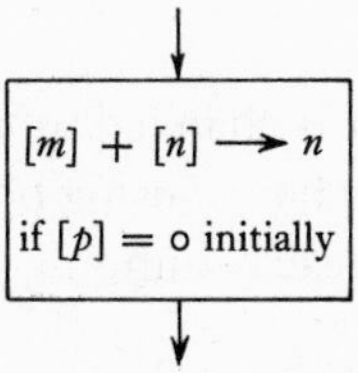

Block diagram

A more general description of the operation of this program, in which no special assumption is made about $[p]$, is this:

$$[m]+[n] \rightarrow n,$$
$$[m]+[p] \rightarrow m,$$
$$0 \rightarrow p.$$

Here, as always, the vertical order represents a *sequence* of moves, from top to bottom, Thus, box p is emptied *after* the other two moves are made. (The order of the first two moves is arbitrary: the effect would be the same, if their order were reversed.)

Example 6.4. Multiplication

The numbers to be multiplied are in distinct boxes m_1 and m_2; two other boxes, n and p, are empty to begin with. The product appears in box n. Instead of constructing a flow chart *de novo*, we use the block diagram of Example 6.3 as shorthand for the flow chart of that example. It is then straightforward to draw the full flow chart, as at the right, where the 'm'

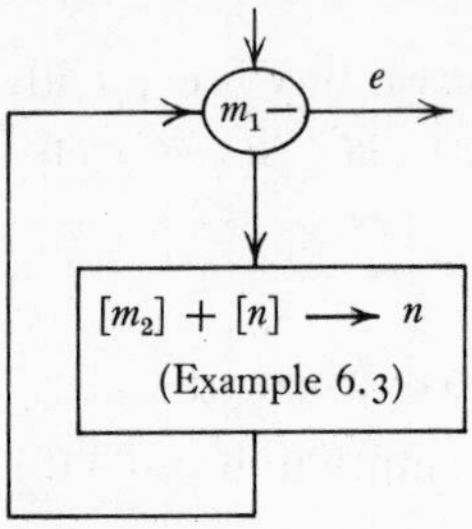

Abbreviated flow chart

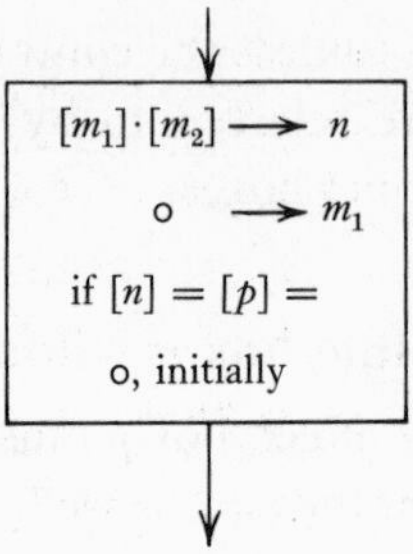

Block diagram

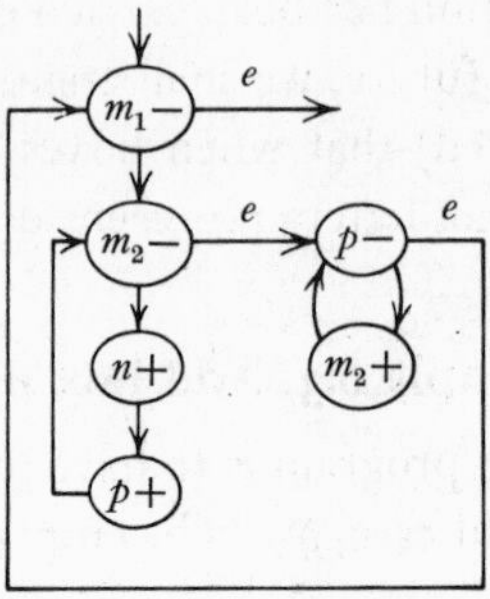

Full flow chart

of Example 6.3 is changed to 'm_2'. The procedure is to repeatedly dump $[m_2]$ stones into box n, using box m_1 as a counter: we remove a stone from box m_1 before each dumping operation, so that when box m_1 is empty we have

$$\underbrace{[m_2]+[m_2]+\ldots+[m_2]}_{[m_1]}$$

stones in box n.

Example 6.5. Exponentiation

Just as multiplication is repeated addition, so exponentiation is repeated multiplication: to raise x to the power y, multiply together y xs,

$$x^y = \underbrace{x \cdot x \cdot \ \ldots \ \cdot x}_{y}.$$

The program is perfectly straightforward, once we arrange to modify the multiplication program of Example 6.4 so as to have $[m_2]\cdot[n] \rightarrow n$: see the abbreviated flow graph for exponentiation. The required modification of the routine of Example 6.4 is obtained (following Peter Tovey) by appending the routine of Example 6.2: see the diagram for cumulative

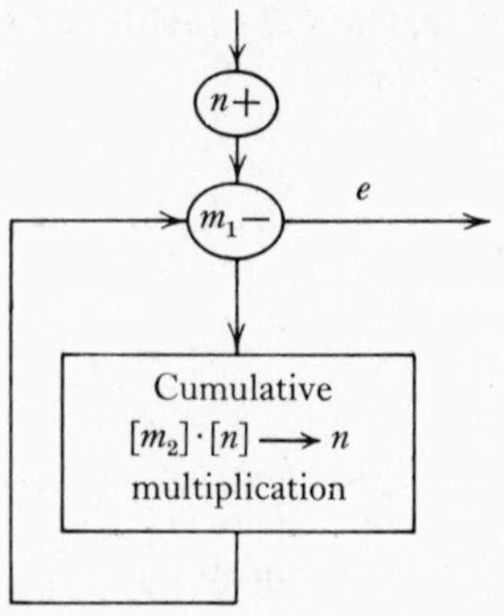

Abbreviated flow graph for exponentiation

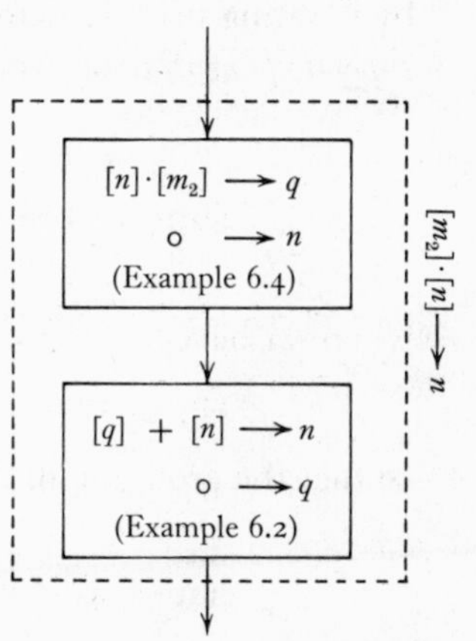

Cumulative multiplication

multiplication. Provided boxes n, p and q are empty initially, the program for exponentiation has the effect,

$$[m_2]^{[m_1]} \to n,$$
$$0 \to m_1$$

in strict analogy to the program for multiplication in Example 6.4: see the block diagram there. Structurally, the abbreviated flow graphs for multiplication and exponentiation differ only in that for exponentiation, we need to put a single stone in box n at the beginning. If $[m_1] = 0$ we have $n = 1$ when the program terminates (as it will at once, without going through the multiplication routine). This corresponds to the convention that

$$x^0 = 1.$$

for any natural number x. But if $[m_1]$ is positive, $[n]$ will finally be a product of $[m_1]$ factors $[m_2]$, corresponding to repeated application of the rule

$$x^{y+1} = x \cdot x^y$$

which is implemented by means of cumulative multiplication, using box m_1 as a counter.

It should now be clear that the initial restriction to two elementary sorts of acts, $n+$ and $n-$, does not prevent us from computing fairly complex functions, e.g., all functions in the series that begins *sum*, *product*, *power*, ..., and where the $n+1$st member of the series is obtained by iterating the nth member. Apparently, the next member of the series, *super-exponentiation*, would be defined

$$\sup(x, y) = \underbrace{x^{x^{x^{\cdot^{\cdot^{\cdot^{x}}}}}}}_{y}$$

where we have

$$\sup(x, 0) = 1, \ \sup(x, y+1) = x^{\sup(x, y)}$$

so that the grouping in the general scheme is (e.g. for $y = 4$)

$$\sup(x, 4) = x^{x^{x^{x}}} = x^{\left(x^{\left(x^{x}\right)}\right)}$$

(The grouping does matter: $3^{(3^3)}$ is 3^{27}, which is the product of 27 threes, but $(3^3)^3$ is 27^3, which is the product of three twenty-sevens, or of only 9 threes.) Apparently the flow graph for a superexponentiator can be obtained from that of an exponentiator just as the flow graph of an exponentiator was obtained from that of a multiplier in Example 6.5.

Having proved that some very complex functions are abacus computable, we now present some evidence for Church's thesis: we prove that every abacus computable function is Turing computable. We do this by giving a method which, applied to the flow graph of any abacus program, yields the flow graph of a Turing machine which computes the same function that the abacus does. The format of the Turing machine computation will be the standard one described at the beginning of this chapter. But before describing our method for transforming abacus flow graphs into equivalent Turing machine flow graphs we must standardize certain features of abacus computations: we must know where to look, initially, for the arguments of the function which the abacus computes, and where to look, finally, for the value. The following conventions will do as well as any:

Format for abacus computation of $f(x_1, \ldots, x_r)$. Initially, the arguments are the numbers of stones in the first r boxes, and all other boxes are empty: $x_1 = [1], \ldots, x_r = [r], 0 = [r+1] = [r+2] = \ldots$. Finally, the value of the function is the number of stones in some previously specified box n (which may but need not be one of the first r). Thus, $f(x_1, \ldots, x_r) = [n]$ when the computation halts, i.e., when we come to an arrow in the flow graph which terminates in no node. If the computation never halts, $f_1(x_1, \ldots, x_r)$ is undefined.

Note that our computation routines for addition, multiplication, and exponentiation in Examples 6.3–6.5 are essentially in this form, with $r = 2$ in each case. They were formulated in a general way, so as to leave open the question of just which boxes are to contain the arguments and value, e.g. in the adder of Example 6.3 we only specified that the arguments are to be stored in distinct boxes numbered m and n, that the value will be found in box n, and that a third box, numbered p, and initially empty, will be used in the course of the computation. But now we must specify m, n and p, subject to the restriction that m and n shall be 1 and 2 (in either order) and p shall be some number greater than 2. Then we might settle on $n = 1$, $m = 2$, $p = 3$ to get a particular program for addition in the standard format (Figure 6-2).

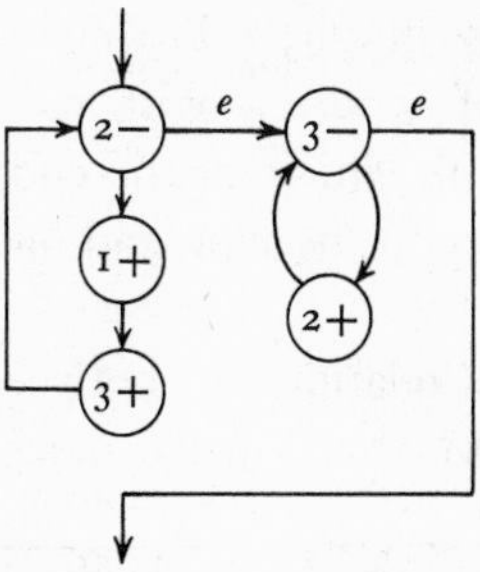

Figure 6-2. Addition (Example 6.3) in standard format.

The standard format associates a definite function from natural numbers to natural numbers with each abacus, once we specify the number r of arguments and the number n of the box in which the values will appear. Similarly the standard format for Turing machine computations associates a definite function from positive integers to positive integers with each Turing machine, once we specify the number r of arguments. Observe that once we have specified the graph of an abacus A in standard form, then for each register n that we might specify as holding the result of the computation there are infinitely many functions A_n^r which we have specified as computed by the abacus: one function for each possible number of arguments. Thus if A is determined by the simplest graph for addition, in Example 6.2, with $n = 1$ and $m = 2$, we have

$$A_1^2(x, y) = x + y$$

for all natural numbers x and y, but we also have the identity function $A_1^1(x) = x$ of one argument, and for three or more arguments we have

$A_1^r(x_1, x_2, ..., x_r) = x_1 + x_2$. Indeed for $r = 0$ arguments we may think of A_1 as computing a 'function' of a sort, *viz.*, the number $A_1^0 = 0$ of 1s in box 1 when the computation halts, having been started with all boxes ('except for the first r') empty. Of course, the case is entirely parallel for Turing machines, each of which computes a function of r arguments in standard format for each $r = 0, 1, 2, ...$.

Having settled on standard formats for the two kinds of computation, we can turn to the problem of designing a method for converting the flow graph of an abacus A_n with n designated as the box in which the values will appear into the graph of a Turing machine which computes the same functions: for each r, the Turing machine will compute the same function A_n^r of r arguments that the abacus computes. Our method will specify a Turing machine flow graph which is to replace each node of type $n+$ with its exiting arrow, in the abacus flow graph; a Turing machine flow graph which is to replace each node of type $n-$ with its two exiting arrows; and a Turing machine flow graph which, at the end, makes the machine erase all but the nth block of 1s on the tape and halt, scanning the leftmost of the remaining 1s.

It is important to be clear about the relationship between boxes of the abacus and corresponding parts of the Turing machine's tape.

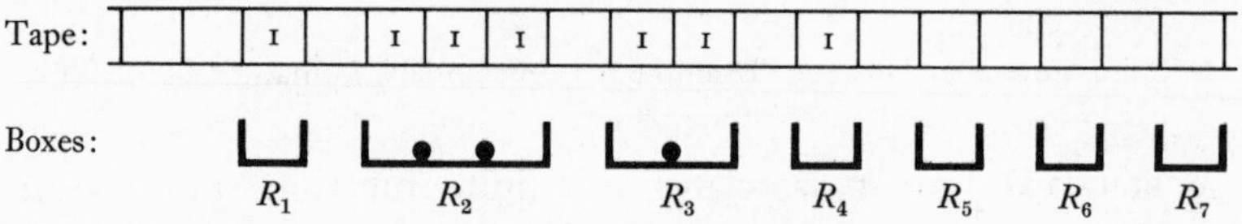

Figure 6-3. Correspondence between boxes and portions of the tape (● = one stone).

Example: in computing $A_n^4(0, 2, 1, 0)$, the initial tape and box configurations would be as shown in the accompanying figure. Boxes containing one or two or ... stones are represented by blocks of two or three or ... 1s on the tape (see Figure 6-3). Single blanks separate portions of the tape corresponding to successive boxes. Empty boxes are always represented by single squares, which may be blank (see R_5, R_6, R_7, ...) or contain a 1 (see R_1 and R_4). The 1 is mandatory if there are any 1s further to the right on the tape, and is mandatory initially for empty argument boxes (R_1 through R_r) and the blank is mandatory initially for R_{r+1}, R_{r+2}, ... Then at any stage of the computation we can be sure that when in moving to the right/left we encounter two successive blanks, there are no further 1s to be found anywhere to the right/left on the tape. The exact portion of the tape which represents a box will wax and wane with the contents of that box as the program progresses, and will shift to the right or left on the tape as stones are added to or removed from lower-numbered boxes.

The first step in our method for converting abacus flow graphs into equivalent Turing machine flow graphs can now be specified: replace each $s+$ node

by a copy of the $s+$ *flow graph* shown in Figure 6-4. The first $2s$ nodes of the $s+$ graph take the Turing machine from its standard position, scanning the leftmost 1 on the tape, to the blank immediately to the right of the sth block of 1s. (In seeking the sth block, the machine substitutes the 1-representation for the B-representation of any empty boxes.) On leaving node sb, the machine writes a 1, moves 1 square right, and does one thing or another (node x) depending on whether it is then scanning a blank or a 1. If it is scanning a blank there can be no more 1s to the right, and it therefore returns to standard position. But if it is scanning a 1 at that point it has more work to do before returning to standard position, for there are more blocks of 1s to be dealt with, to the right on the tape. These must be shifted one square right, by erasing the first 1 in each block and filling the blank to the block's right with a 1 – continuing this

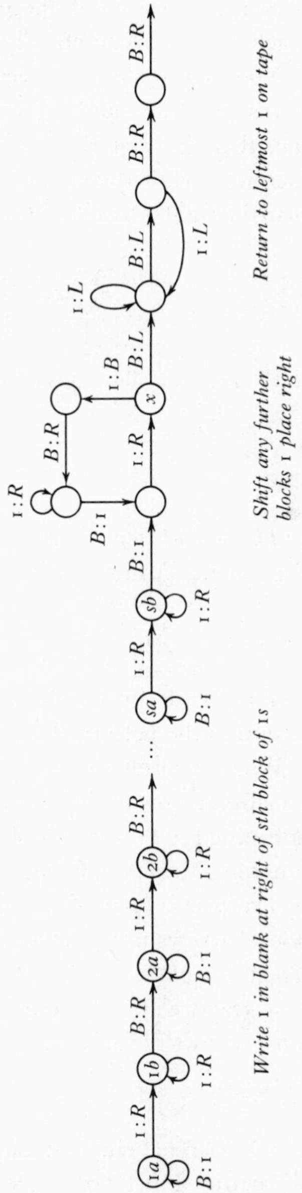

Figure 6-4. The $s+$ flow graph.

routine until it finds a blank to the right of the last blank it has replaced by a 1. At that point there can be no further 1s to the right, and the machine returns to standard position.

Note that node 1*a* is needed in case the number r of arguments is 0: in case the 'function' A_n^0 which the abacus computes is a number. Note, too, that the first $s-1$ pairs of nodes (with their efferent arrows) are identical, while the last pair is different only in that the arrow from node *sb* to the right is labelled 'B:1' instead of 'B:R'.

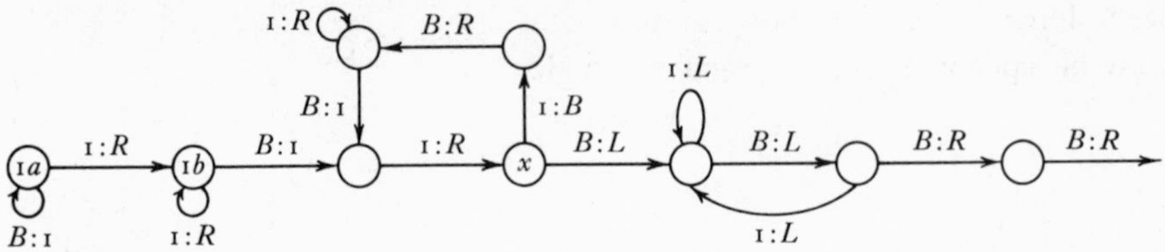

Figure 6-5. The $s+$ flow graph in case $s = 1$.

The second step in our method of converting abacus flow graphs into equivalent Turing machine flow graphs can now be specified: replace

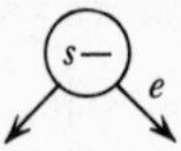

each $s-$ node by a copy of the $s-$ *flow graph* which you are invited to design in the pattern shown by Figure 6-6.

When the first and second steps of the method have been carried out, the abacus flow graph will have been converted into something which is not quite the flow graph of a Turing machine which computes the same functions that the abacus does. The graph will (probably) fall short in two respects. *First*, if the abacus ever halts, there must be one or more 'loose' arrows in the abacus graph: arrows which terminate in no node. Then as a glance at the $s+$ flow graph and the block diagram of the $s-$ flow graph will show, our Turing machine graph will have one or more loose arrows of form

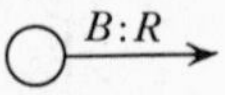

which terminate in no node. But in a proper Turing machine flow graph, all arrows must terminate in nodes. We could repair this defect by simply drawing isolated nodes at the ends of loose arrows

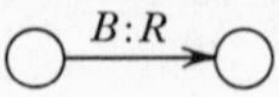

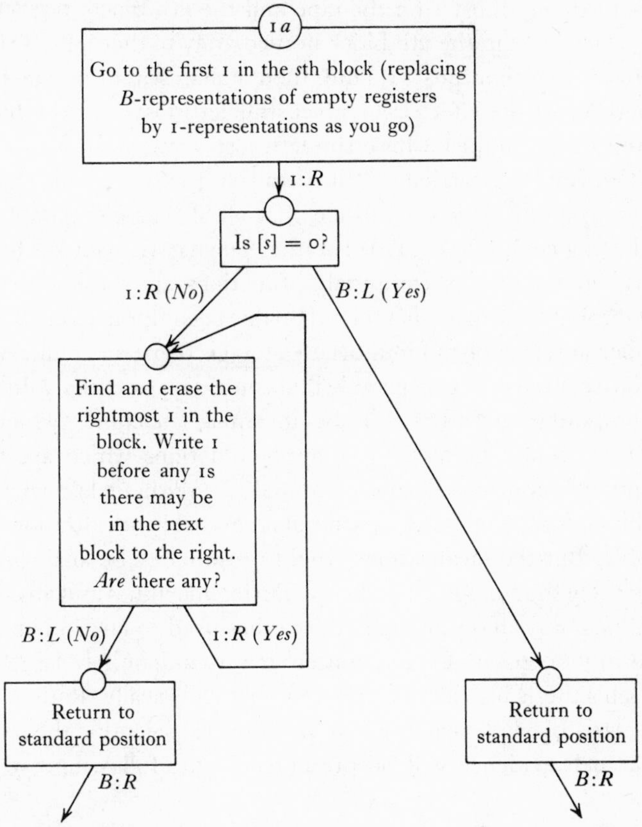

Figure 6-6. Block diagram for *s*-flow graph.

but a *second* shortcoming would remain: in computing $A^r_n(x_1, \ldots, x_r)$ the Turing machine would halt scanning the leftmost 1 on the tape, *but the value of the function would be represented by the nth block of* 1s *on the tape*. Even if $n = 1$, we cannot depend on there being no 1s on the tape after the first block, so our method requires one more step.

The third step: after completing the first two steps, redraw all loose arrows so that they terminate in the input node of a *mop-up graph* which you are invited to design. This graph makes the machine (which will be scanning the leftmost 1 on the tape at the beginning of this routine) erase all but the first block of 1s, if $n = 1$, and halt scanning the leftmost of the remaining 1s. But if $n \neq 1$, it erases everything on the tape except for both the leftmost 1 on the tape and the nth block, repositions all 1s but the leftmost in the nth block immediately to the right of the leftmost 1, erases the rightmost 1, and then halts, scanning the leftmost 1. In both cases the effect is to place the leftmost 1 in the block representing the value just where the leftmost 1 was initially.

When you have carried out the details of the second and third of these moves, you will have established that all abacus computable functions are Turing computable. This is rather persuasive evidence for Church's thesis, for the abacus computable functions are the ones that modern digital computing machines would be able to compute if restrictions on the number and size of random-access storage registers could be removed. Of course it remains conceivable that some day, essentially different sorts of computing machines will be invented: machines which, suitably idealized, would be able to compute functions which are not abacus computable and not Turing computable. (Without knowing how such machines would operate we cannot know what idealizations would be suitable. But the idealizations would presumably be analogous to these concerning the storage capacities of Turing machines and abaci: a Turing machine always has more tape than it has used so far, and an abacus has room in its registers for indefinitely many, indefinitely large numbers.) Church's thesis has not been proved, mathematically. Rather, it has been supported by evidence, much as an empirical scientific theory might be. More such evidence will be presented in the following two chapters.

Exercises

6.1 Draw the $s-$ flow graph (second step of the method for converting abacus flow graphs to equivalent Turing machine flow graphs).

6.2 Draw the mop-up graph (third step of the method).

6.3 *The abacus halting problem.* Let $h(x, y) = 0$ or 1 accordingly as abacus number x in some list of all abaci does or does not eventually halt after being started with y stones in R_1 and all other registers empty. Prove (without using the analogue of Church's thesis for abaci) that h is not abacus computable. *Hint*: ape the argument at the end of Chapter 5.

6.4 *The busy abacus problem.* If an abacus never halts after being started with all registers empty, its productivity is said to be 0, and otherwise its productivity is said to be the number of stones in R_1 when it halts. Let $p(n) =$ the productivity of the most productive n node abaci. Prove that p is not abacus computable.

Solutions

6.1 The top block of the block diagram for the $s-$ flow graph contains a graph identical with the material from node $1a$ to node sa (inclusive) of the $s+$ flow graph. The arrow labelled '$1:R$' from the bottom of this block corresponds to the one that goes right from node sa in the $s+$ flow graph.

The '*Is* [s] $= 0$?' box contains nothing but the shafts of the two emergent arrows: they originate in the node shown at the top of that block.

The 'Return to standard position' blocks contain replicas of the material to the right of node x in the $s+$ graph: the '$B:L$' arrows entering those boxes correspond to the '$B:L$' arrow from node x.

The only novelty is in the remaining block: 'Find and erase the...' That block contains the graph shown below.

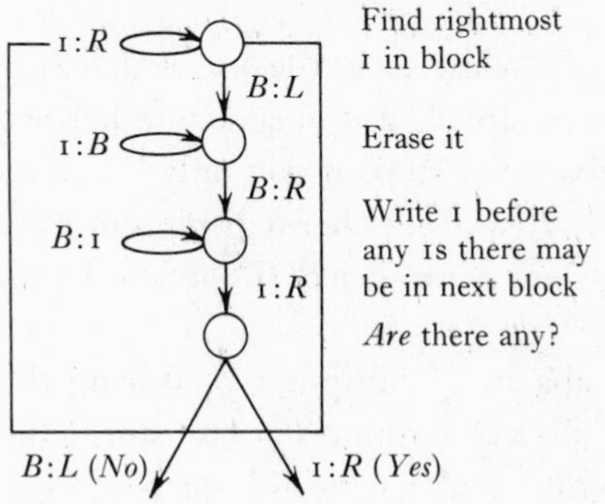
1:R
B:L
1:B
B:R
B:1
1:R
Find rightmost 1 in block
Erase it
Write 1 before any 1s there may be in next block
Are there any?
B:L (No)
1:R (Yes)

6.2 *Mop-up graph, for $n \neq 1$.*

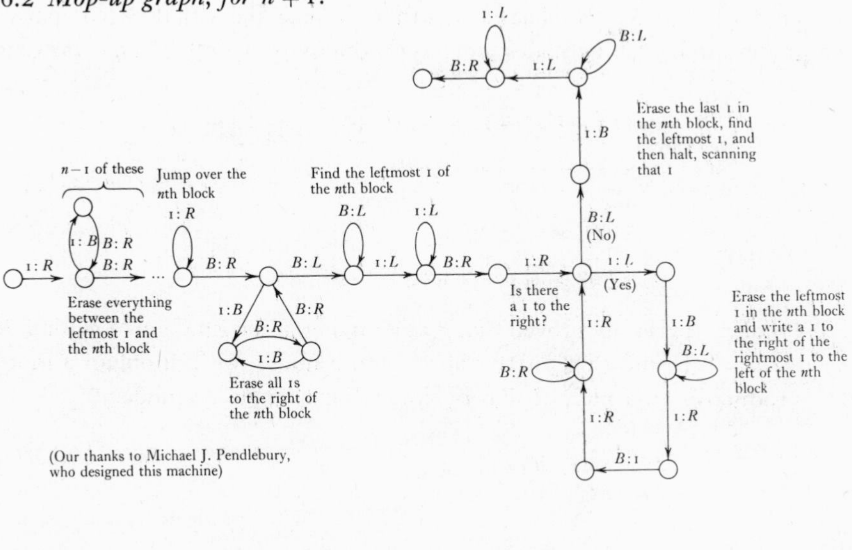
n − 1 of these
Jump over the nth block
Find the leftmost 1 of the nth block
Erase everything between the leftmost 1 and the nth block
Erase all 1s to the right of the nth block
Is there a 1 to the right?
Erase the last 1 in the nth block, find the leftmost 1, and then halt, scanning that 1
Erase the leftmost 1 in the nth block and write a 1 to the right of the rightmost 1 to the left of the nth block
(No)
(Yes)

(Our thanks to Michael J. Pendlebury, who designed this machine)

6.3 Suppose that $h = H_n^2$ for some abacus H, i.e., suppose that H computes h and stores the value of the function in R_n. Then for any x, the abacus shown in the accompanying graph eventually halts after being started with x stones in R_1 if and only if abacus number x never halts after being started with x stones in R_1 – where in each case, all other registers are empty initially. (Note that with the last node deleted, the abacus in question computes the function $h(x, x)$.) If there is such an

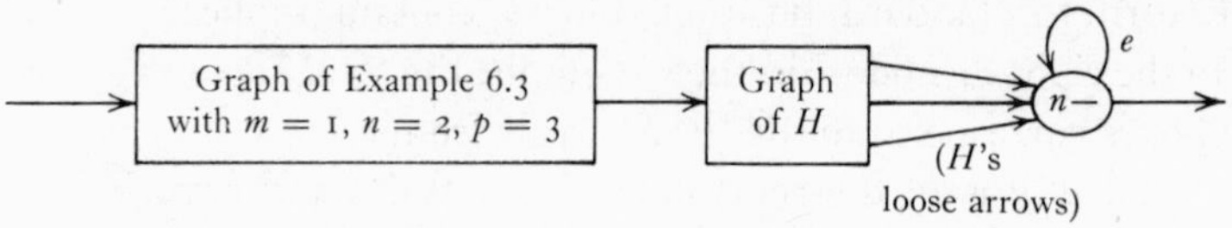

Graph of abacus number m

abacus as H then there is such an abacus as the one we have graphed, and it must appear in our list of all abaci: it is abacus number m, say. But if we set $x = m$ we see that abacus m stops if and only if it does not, after being started with m stones in R_1 and all other registers empty. Then there is no such abacus: and no abacus computes h: the abacus halting problem is unsolvable.

6.4 Suppose that some abacus P computes p, storing the values in register 1. (If some other abacus computes p but stores the values in register n, we can convert into an abacus of the sort we want by adding three more modes: one to empty R_1 and then two as in Example 6.2 to empty R_n into R_1.) If P has k nodes then the machine with the accompanying graph has $n+2k$ nodes and has productivity $p(p(n))$. Then if P exists

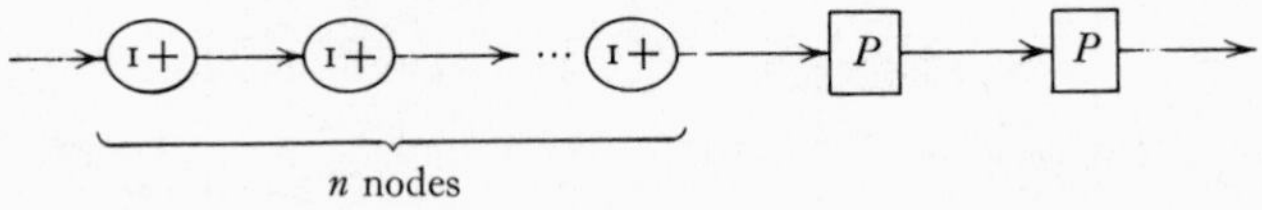

we have

$$p(n+2k) \geqslant p(p(n))$$

for all n. It is easily proved that $p(n)$ is an increasing function of n, for if A is one of the most productive abaci with n nodes, we can obtain a more productive $n+1$ node abacus by appending a single 1+ node.

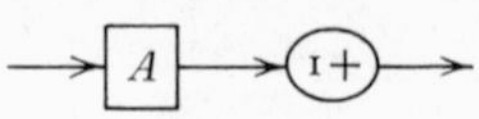

Then we have

$$n+2k \geqslant p(n) \qquad (*)$$

for all n, if P exists. With n nodes, we can put n stones in R_1, and with an additional 5 nodes we can copy those n stones in R_2 by the routine of Example 6.3. Now by appending another replica of the routine of Example 6.3, we obtain an $n+10$ node abacus which has productivity $2n$. Thus, for all n we have

$$p(n+10) \geqslant 2n$$

whether or not P exists. Putting '$n+10$' for 'n' in (*) and combining with the last inequality we have

$$n+10+2k \geqslant 2n$$

for all n if P exists. Then if P exists we have

$$10+2k \geqslant n$$

for *all n*. Putting '$11+2k$' for 'n' we then have

$$0 \geqslant 1$$

if P exists. Then there is no such abacus as P: the busy abacus problem is unsolvable.

Note. The treatment of abaci in this and the following chapters is drawn from Joachim Lambek 'How to program an infinite abacus', *Canadian Mathematical Bulletin* **4** (1961), 295–302, with a correction noted in **5** (1962), 297. Lambek's abaci are simplifications of machines studied by Z. A. Melzak 'An informal arithmetical approach to computability and computation', *Canadian Mathematical Bulletin* **4** (1961), 279–293.

7
Recursive functions are abacus computable

The functions in the sequence sum, product, power, ... noted in Chapter 6 belong to the class of *primitive recursive functions*. We now give a precise definition of that class and of the more extensive class R of *recursive* functions. We shall then prove that all functions in R belong to the class A of abacus computable functions, so that we have

$$R \underset{7}{\subseteq} A \underset{6}{\subseteq} T,$$

where the subscripts identify the chapters in which the inclusions are proved. This will provide further support for Church's thesis, for as we shall see, the class of recursive functions is very broad indeed – so broad as to make it plausible that all functions computable in any intuitive sense are recursive.

To fix ideas, consider the third function in the foregoing sequence – the function exp which, applied to an argument pair (x, y), yields the yth power of x,

$$\exp(x, y) = x^y.$$

Here we have

$$\begin{aligned} x^0 &= 1, \\ x^1 &= x, \\ x^2 &= x \cdot x, \\ &\vdots \\ x^y &= \underbrace{x \cdot x \cdot \ldots \cdot x}_{y \text{ times}}, \\ x^{y+1} &= \underbrace{x \cdot x \cdot \ldots \cdot x \cdot x}_{y+1 \text{ times}} = x \cdot x^y. \end{aligned}$$

The first and last of these equations,

$$x^0 = 1, \quad x^{y+1} = x \cdot x^y,$$

suffice for computation of all values of the function.

Example. Computation of 5^3

$$5^3 = 5^{2+1} = 5 \cdot 5^2 \text{ (setting } x = 5 \text{ and } y = 2 \text{ in the last equation)},$$
$$5^2 = 5^{1+1} = 5 \cdot 5^1 \text{ (setting } x = 5 \text{ and } y = 1 \text{ in the last equation)},$$
$$5^1 = 5^{0+1} = 5 \cdot 5^0 \text{ (setting } x = 5 \text{ and } y = 0 \text{ in the last equation)},$$
$$5^0 = 1 \qquad \text{(setting } x = 5 \text{ in the first equation)}.$$

Combining, we have

$$5^3 = 5 \cdot 5^2 = 5 \cdot 5 \cdot 5^1 = 5 \cdot 5 \cdot 5 \cdot 5^0 = 5 \cdot 5 \cdot 5 \cdot 1.$$

We have thus reduced the problem of computing a power to that of computing a product. *That* can be reduced in a similar way to the problem of computing sums, and that in turn can be reduced to a matter of repeatedly adding 1. The overall result of these computations will be the information that

$$5^3 = \underbrace{1+1+\ldots+1}_{125\text{ `1's}}.$$

Here we have 5^3 expressed in a form of monadic notation, where the numbers zero and one are denoted by the signs '0' and '1', and where the sign for the number $n+1$ thereafter is obtained by appending the sign '+1' to the sign for the number n.

To make it clear that we are not smuggling the operation of summation into our very notation for numbers, we shall use the symbol 's' for the successor function, and denote the successive natural numbers as follows:

$$0, \quad s(0), \quad s(s(0)), \ldots$$

Thus, our official monadic notation explicitly incorporates the sign 's' for the one-place function which, applied to any natural number argument, assumes the next larger natural number as value. It is then clear that e.g. the pair of equations

$$x+0 = x, \quad x+s(y) = s(x+y),$$

conveys information which was not already implicit in the notation for numbers.

Example

Computation of $2+3$, i.e., $s(s(0))+s(s(s(0)))$.

$$\underbrace{s(s(0))}_{x}+\underbrace{s(s(s(0)))}_{y} = s(\underbrace{s(s(0))}_{x}+\underbrace{s(s(0))}_{y}) \quad \text{(second equation)},$$

$$\underbrace{s(s(0))}_{x}+\underbrace{s(s(0))}_{y} = s(\underbrace{s(s(0))}_{x}+\underbrace{s(0)}_{y}) \quad \text{(second equation)},$$

$$\underbrace{s(s(0))}_{x}+\underbrace{s(0)}_{y} = s(\underbrace{s(s(0))}_{x}+\underset{y}{0}) \quad \text{(second equation)},$$

$$\underbrace{s(s(0))}_{x}+0 = \underbrace{s(s(0))}_{x} \quad \text{(first equation)}.$$

Combining, we have

$$\begin{aligned} s(s(0))+s(s(s(0))) &= s(s(s(0))+s(s(0))) \\ &= s(s(s(s(0))+s(0))) \\ &= s(s(s(s(s(0))+0))) \\ &= s(s(s(s(s(0))))). \end{aligned}$$

Having seen how the thing goes in uncompromisingly mechanical terms, we can summarize by writing

$$s0, \quad ss0, \quad sss0, \ldots$$

as shorthand for

$$s(0), \quad s(s(0)), \quad s(s(s(0))), \ldots$$

and so aiding the human eye by displaying only the main pairs of parentheses. With these abbreviations, the equations can be written

$$x+0 = x, \quad x+sy = s(x+y),$$

and the computation proceeds as follows:

$$\begin{aligned} ss0+sss0 &= s(ss0+ss0), \\ ss0+ss0 &= s(ss0+s0), \\ ss0+s0 &= s(ss0+0), \\ ss0+0 &= ss0. \end{aligned}$$

Combining we have

$$
\begin{aligned}
sso + ssso &= s(sso + sso)\\
&= s(s(sso + so))\\
&= s(s(s(sso + o)))\\
&= s(s(s(sso))) = ssssso,
\end{aligned}
$$

where the last step is a matter of mentally rewriting '*sso*' as '*s*(*s*(o))' and and then abbreviating '*s*(*s*(*s*(*s*(*s*(o)))))' as '*ssssso*'.

Another common notation for the successor function is an accent, written *after* the argument. Thus, suppressing parentheses, the successive natural numbers would be denoted

$$o, \quad o', \quad o'', \ldots$$

in this notation, and the equations for summation would be

$$x + o = x, \quad x + y' = (x + y)'.$$

The point of using the fussy notation in which we write '*s*(*s*(*s*(o)))' instead of '3' etc. is that we are applying to a mode of computation with which we are very familiar the same sort of minute analysis which we applied to Turing and abacus computations. Clarity about such minutiae requires us to be wary of familiar shortcuts until such time as we have established their validity within the present framework. To achieve such clarity, we must be quite explicit about the structure of that framework. There are various equivalent ways of doing that. Therefore, any one of those ways will have certain features which seem unnatural because arbitrary. Choice of the '*s*' notation is a trivial first step toward explicitness. Full explication requires us to make several further choices, which we now do in what seems the simplest way and is certainly the commonest way.

We shall define the class of primitive recursive functions by specifying an initial stock of functions as belonging to the class, and then specifying two sorts of operations which, applied to members of the class, yield functions which are to be understood as belonging to the class as well. To complete the definition we stipulate that nothing belongs to the class unless it is either in the initial stock or is obtainable from functions in that stock by a finite sequence of applications of operations of the two sorts.

The initial stock of functions consists of the *zero function*, the *successor function*, and the various *identity functions*.

The zero function, z, is a function of one argument. To each natural number as argument, it assigns the same value, *viz.*, the natural number zero:

$$z(x) = 0.$$

Thus, $z(0) = z(1) = z(2) = \ldots = 0$

or, in our fussy notation,

$$z(0) = z(s(0)) = z(s(s(0))) = \ldots = 0.$$

We have already encountered the successor function, s, of one argument, which assigns to each natural number as argument its successor as value. Thus, $s(0) = 1$, $s(1) = 2$, $s(2) = 3, \ldots$. Translated into our fussy notation, these equations become vacuous ('$s(0) = s(0)$' etc.) since the properties of s are packed into that notation.

The identity function of one argument, id, assigns to each natural number as argument that same number as value:

$$\mathrm{id}(x) = x.$$

There are two identity functions of two arguments: id_1^2 and id_2^2. For any pair of natural numbers as arguments, these pick out the first and second, respectively, as values:

$$\mathrm{id}_1^2(x, y) = x, \quad \mathrm{id}_2^2(x, y) = y.$$

In general, for each positive integer n, there are n identity functions of n arguments, which pick out as values the first, the second, ..., the nth of the arguments:

$$\mathrm{id}_i^n(x_1, \ldots, x_i, \ldots, x_n) = x_i.$$

An obvious alternative notation for id, the identity function of one argument, is 'id_1^1'. Identity functions are often called *projection functions*. (In terms of analytic geometry, $\mathrm{id}_1^2(x, y)$ and $\mathrm{id}_2^2(x, y)$ are the projections x and y of the point (x, y) on the X-axis and the Y-axis respectively.)

From this initial stock of *basic functions* – z, s, and the identity functions in the infinite list id, id_1^2, id_2^2, $\mathrm{id}_1^3, \ldots$ – we may form new primitive recursive functions by operations of two sorts: *composition* and *primitive recursion*. The *primitive recursive functions* are simply the basic functions together with those functions obtainable from them by finite numbers of applications of operations of the two sorts, in any order.

The first sort of operation, composition (= *substitution*), is familiar and straightforward. If f is a function of m arguments and each of $g_1, \ldots, g_m$ is a function of n arguments, then the function h where

$$\boxed{h(x_1, \ldots, x_n) = f(g_1(x_1, \ldots, x_n), \ldots, g_m(x_1, \ldots, x_n))} \quad \text{(Cn)}$$

is the function obtained from $f, g_1, \ldots, g_m$ by *composition*. One might indicate this in shorthand:

$$h = \text{Cn}[f, g_1, \cdots, g_m].$$

Example

The function h of three arguments which always takes the successor of its third argument as value is obtained from the pair s, id_3^3 by composition: $h = \text{Cn}[s, \text{id}_3^3]$. In terms of the format (Cn) we have $n = 3$ and $m = 1$, i.e., the special case of the format with which we are dealing is

$$h(x_1, x_2, x_3) = f(g_1(x_1, x_2, x_3)),$$

where $f = s$ and $g_1 = \text{id}_3^3$.

Another example. Cn [*s*, *s*]

This is a function of one argument, x, of which the value is always $x + 2$, i.e., $s(s(x))$. In terms of the boxed format (Cn) we have $m = n = 1$, so that we are dealing with the special case

$$h(x_1) = f(g_1(x_1)),$$

where $f = g_1 = s$.

Of course, not all of the functions used in composition need be basic.

Example. $h(x) = x+3$

Here, h is obtainable by composition of the functions s, $\text{Cn}[s, s]$ where the second is not basic, but *is* obtainable by composition of basic functions. Thus, h is primitive recursive: $h = \text{Cn}[s, \text{Cn}[s, s]]$.

Observe that such notations as '$\text{Cn}[s, \text{id}_3^3]$' are genuine function symbols: they belong to the same grammatical category as 'h'. Then just as we can write '$h(x_1, x_2, x_3) = s(x_3)$', so can we write

$$\text{Cn}[s, \text{id}_3^3](x_1, x_2, x_3) = s(x_3).$$

The foregoing equation is clearly true: the function $\text{Cn}[s, \text{id}_3^3]$ does apply to triples of natural numbers, and its value is always the successor of its third argument.

The second sort of operation for generating new primitive recursive functions is called 'primitive recursion'. The format is this:

$$\boxed{h(x, \mathrm{o}) = f(x), \quad h(x, s(y)) = g(x, y, h(x, y))} \quad \text{(Pr)}$$

Where the boxed equations hold, h is said to be definable by primitive recursion from the functions f, g, In shorthand:

$$h = \mathrm{Pr}\,[f, g].$$

The definitions of sum, product, and power above are approximately in this format. By fussing over them, we can put them exactly into the format (Pr).

Example. $x+\mathrm{o} = x, x+s(y) = s(x+y)$

To begin, let us replace the operation symbol '+' by the function symbol 'sum':

$$\mathrm{sum}\,(x, \mathrm{o}) = x, \quad \mathrm{sum}\,(x, s(y)) = s(\mathrm{sum}\,(x, y)).$$

To put these into the boxed format (Pr) we must find functions f and g for which we have

$$f(x) = x, \quad g(x, y, -) = s(-)$$

for all natural numbers x, y, and –. Such functions lie ready to hand: $f = \mathrm{id}$, and $g = \mathrm{Cn}\,[s, \mathrm{id}^3_3]$. Further, since both of these are primitive recursive (being basic or obtained from basic functions by composition), *sum* is seen to be primitive recursive as well. In the boxed format we have

$$\mathrm{sum}\,(x, \mathrm{o}) = \mathrm{id}\,(x), \quad \mathrm{sum}\,(x, s(y)) = \mathrm{Cn}\,[s, \mathrm{id}^3_3]\,(x, y, \mathrm{sum}\,(x, y))$$

and in shorthand we have

$$\mathrm{sum} = \mathrm{Pr}\,[\mathrm{id}, \mathrm{Cn}\,[s, \mathrm{id}^3_3]].$$

Another example. prod = Pr [z, Cn [sum, id^3_1, id^3_3]]

To verify this claim we relate it to the boxed formats (Cn) and (Pr). In terms of (Pr) the claim is that the equations

$$\mathrm{prod}\,(x, \mathrm{o}) = z(x), \quad \mathrm{prod}\,(x, s(y)) = g(x, y, \mathrm{prod}\,(x, y))$$

hold for all natural numbers x and y, where (setting $h = g$, $f =$ sum, $g_1 = \mathrm{id}_1^3$, and $g_2 = \mathrm{id}_3^3$ in the boxed (Cn) format) we have

$$\begin{aligned} g(x_1, x_2, x_3) &= \mathrm{Cn}\,[\mathrm{sum}, \mathrm{id}_1^3, \mathrm{id}_3^3]\,(x_1, x_2, x_3) \\ &= \mathrm{sum}\,(\mathrm{id}_1^3\,(x_1, x_2, x_3), \mathrm{id}_3^3\,(x_1, x_2, x_3)) \\ &= x_1 + x_3 \end{aligned}$$

for all natural numbers x_1, x_2, x_3. Overall, then, the claim is that the equations

$$\mathrm{prod}\,(x, 0) = 0, \quad \mathrm{prod}\,(x, s(y)) = x + \mathrm{prod}\,(x, y)$$

hold for all natural numbers x and y, which is true:

$$x \cdot 0 = 0, \quad x \cdot (y+1) = x + x \cdot y.$$

Exercise 7.1

Verify the truth of the description

$$\mathrm{sup} = \mathrm{Pr}\,[\mathrm{Cn}\,[s, z], \ \mathrm{Cn}\,[\exp, \mathrm{id}_1^3, \mathrm{id}_3^3]]$$

of the superexponentiation function which was defined in Chapter 6.

Exercise 7.2

Write out a true description exp = ... of the exponentiation function which shows that exponentiation (and therefore sup as well) is primitive recursive.

Our rigid format for primitive recursive serves for functions of two variables such as sum and prod, but we shall sometimes wish to use such a scheme to define functions of a single variable, and functions of more than two variables. Where there are three or more variables $x_1, \ldots, x_n, y$ instead of the two variables x, y which appear in (Pr), the modification is achieved by viewing each of the five occurrences of x in the boxed format as shorthand for $x_1, \ldots, x_n$. Thus, with $n = 2$, the format is

$$h(x_1, x_2, 0) = f(x_1, x_2)$$

$$h(x_1, x_2, s(y)) = g(x_1, x_2, y, h(x_1, x_2, y)).$$

If there is only one variable, y, we take the format to be

$$h(0) = 0 \quad \text{or} \quad s(0) \quad \text{or} \quad s(s(0)) \text{ or } \ldots$$

$$h(s(y)) = g(y, h(y)).$$

Here the *x*s have disappeared altogether, and the 'function' *f* of 0 arguments is a constant: 0 or 1 or 2 or ...

Exercise 7.3

The *factorial* of *y* is the product of all the positive integers up to and including *y* (but is taken to be 1 when *y* is 0). Define *fac* by primitive recursion.

We now verify that all primitive recursive functions are abacus computable. To compute a function of *n* arguments on an abacus, we specify *n* registers or boxes in which the arguments are to be stored initially (represented by piles of rocks) and we specify a box in which the value of the function is to appear (as a pile of rocks) at the end of the computation. To facilitate comparison with computations by Turing machines in standard form, let us assume that the arguments appear in boxes 1, ..., *n*, and that the value is to appear in box $n+1$; and let us suppose that initially, all boxes except for the first *n* are empty. Now the programs of Figure 7-1 compute the basic functions.

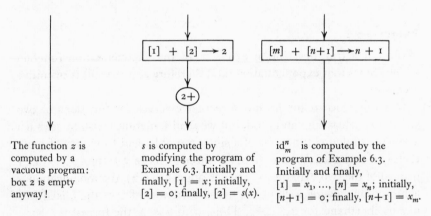

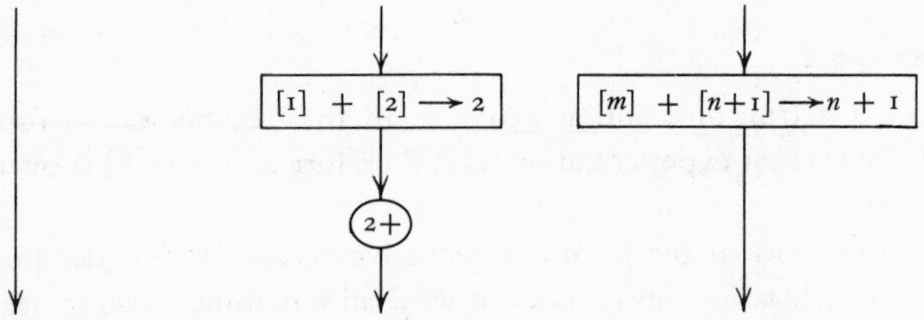

The function *z* is computed by a vacuous program: box 2 is empty anyway!

s is computed by modifying the program of Example 6.3. Initially and finally, $[1] = x$; initially, $[2] = 0$; finally, $[2] = s(x)$.

id^n_m is computed by the program of Example 6.3. Initially and finally, $[1] = x_1, \ldots, [n] = x_n$; initially, $[n+1] = 0$; finally, $[n+1] = x_m$.

Figure 7-1. Abacus computation of the basic functions.

It only remains to show how, once we are given programs which compute functions $f, g_1, \ldots, g_m$, we can arrange them into a program which computes their composition, $h = \mathrm{Cn}[f, g_1, \ldots, g_m]$, and how, given programs which compute functions *f* and *g*, we can arrange them into a program which computes the function *h* obtained from them by primitive recursion, $h = \mathrm{Pr}[f, g]$.

Composition. Suppose that we are given $m+1$ programs

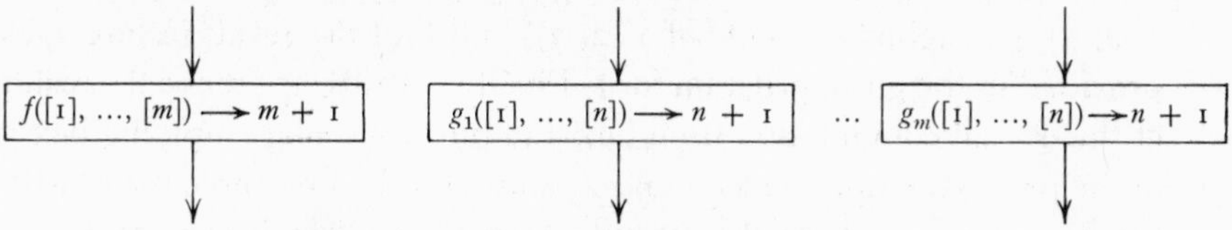

from which we seek to construct a single program

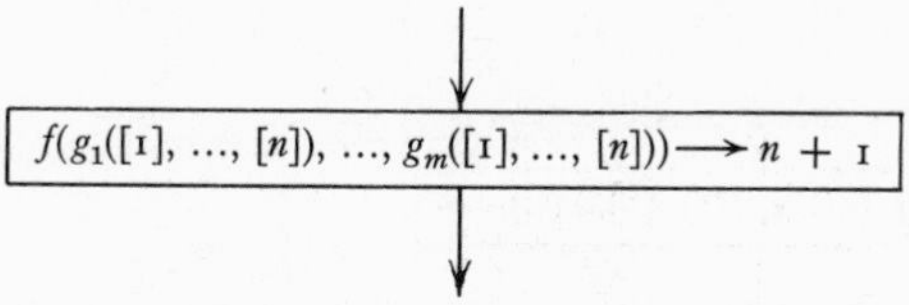

The thing is perfectly straightforward: it is a matter of shuttling the results of the subcomputations around so as to be in the right boxes at the right times. Here is how it is done for the case $m = 2$, $n = 3$; it will be obvious from that, how to treat any other case. First, we identify $m+n = 5$ registers, none of which are used in any of the $m+1$ given programs. Let us call these registers p_1, p_2, q_1, q_2, and q_3. They will be used for temporary storage. Recall that in the single program which we want to construct, the $n = 3$ arguments are stored initially in boxes 1, 2, and 3; all other boxes are empty, initially; and at the end, we want the n arguments back in boxes 1, 2, 3, and want the value, $f(g_1([1],[2],[3]), g_2([1],[2],[3]))$, in box number $n+1 = 4$. To arrange that, all we need are the three given programs and the program of Example 6.2:

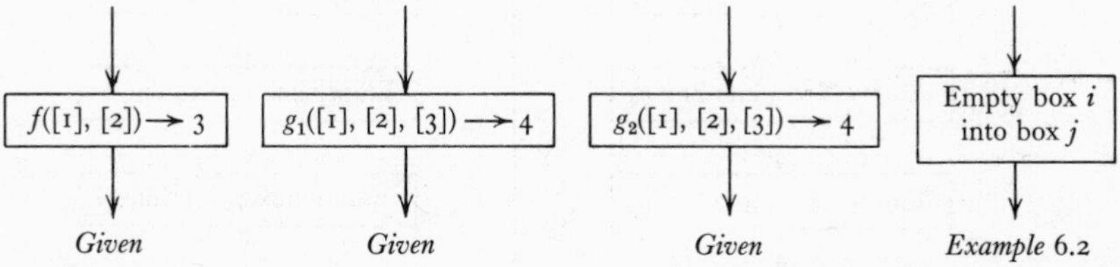

We shall simply compute $g_1([1],[2],[3])$ and store the result in box p_1 (which figures in none of the given programs, remember): then compute g_2 of the arguments in boxes 1, 2, and 3, and store the result in box p_2; then we shall store the arguments in boxes 1, 2, and 3 in boxes q_1, q_2, and q_3, and empty boxes 1 through 4; then get the results of the first two computations out of boxes p_1 and p_2 (which we empty), and put them in boxes 1 and 2, so that we can compute $f([1], [2]) = f(g_1(\text{original contents of } 1, 2, 3), g_2(\text{original contents of } 1, 2, 3))$ and find the result in box 3, as provided in the given program for f. Finally, we tidy up; move the result of the overall computation from box 3 to box $n+1 = 4$, emptying box 3 in the process; empty boxes 1 and 2; and refill the first three boxes with the $n = 3$ arguments of the overall computation, which were stored in boxes q_1, q_2, and q_3. Now everything is as it should be, and we have a flow chart of which the block diagram is this:

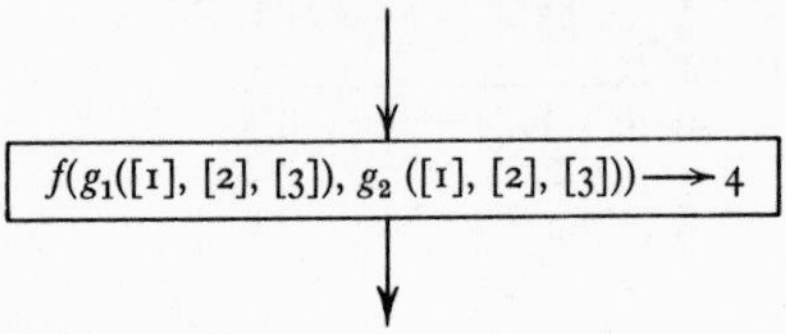

The structure of the flow chart is shown in Figure 7-2.

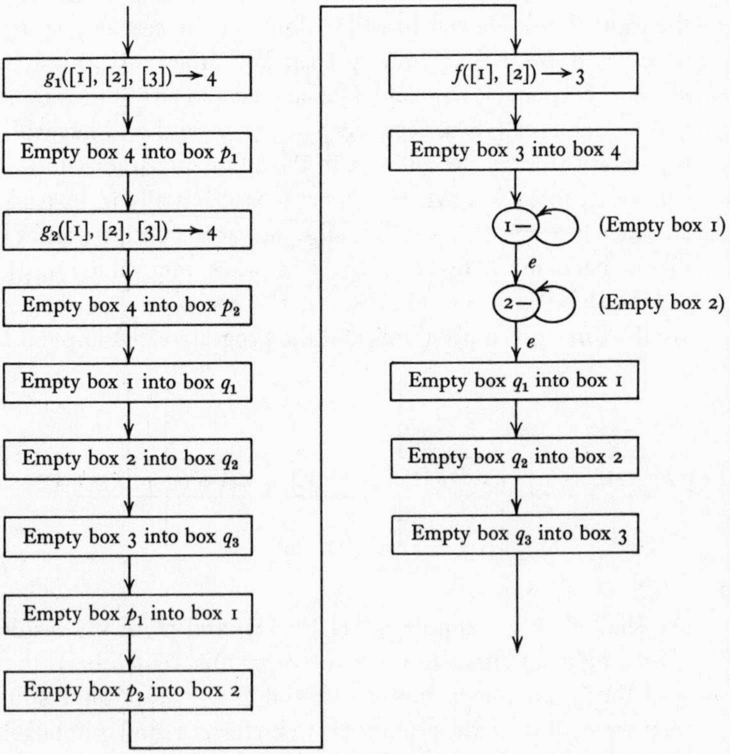

Figure 7-2. Composition of programs in case $m = 2$, $n = 3$.

It only remains to prove that if we have programs to compute the functions f and g, we can always design a program to compute the function h which is obtained from them by primitive recursion. We consider the case where h is a function of two arguments, as in the boxed format. The cases in which h has only one argument, or has more than two arguments, are treated similarly. Then the given flow graphs have these block diagrams:

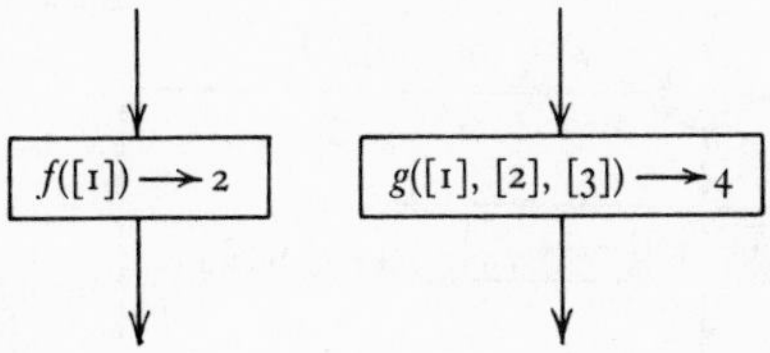

Using these, we want to design a flow graph with this block diagram:

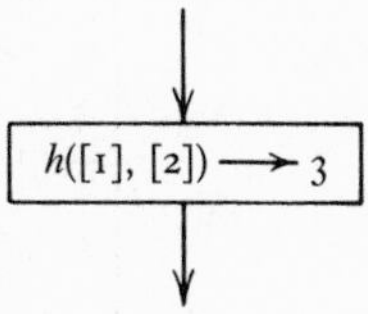

The thing is easily done, as in Figure 7-3. Initially $[1] = x$, $[2] = y$, and $[3] = [4] = \ldots = 0$. We use register number p (which is not used in the f or g programs) as a counter: we put y into it at the beginning, and after each stage of the computation we see whether $[p] = 0$. If so, the computation is essentially finished; if not, we subtract 1 from $[p]$ and go through another stage. In the first three steps (Figure 7-3) we calculate $f(x)$ and then see whether y was 0. If so, the first of the boxed equations is operative: $h(x,y) = h(x,0) = f(x)$, and the computation is finished, with the result in box 3, as required, If not, we successively compute $h(x,1)$, $h(x,2)\ldots$ (see the cycle in Figure 7-3) until the counter (box p) is empty. At that point the computation is finished, with $h(x,y)$ in box 3, as required. Given programs to compute functions f and g of 1 and 3 arguments respectively, this program computes the function $h = \Pr[f,g]$ of 2 arguments.

We have now proved that all primitive recursive functions are abacus computable. But there are still more functions that abaci can compute: not only the primitive recursive functions, but all (total and partial) *recursive* functions. These consist of the basic functions ($z, s, \text{id}, \text{id}^2_1, \ldots$) together with all functions obtainable from them by finite numbers of applica-

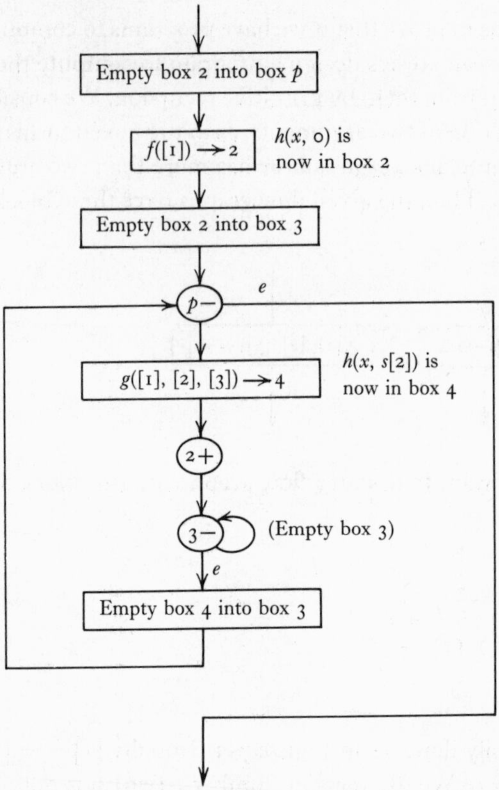

Figure 7-3. Primitive recursion.

tions of *three* sorts of operations: composition, primitive recursion, and *minimization*. Applied to a *total* function f of $n+1$ arguments, the operation of minimization yields a function h of n arguments as follows:

$$h(x_1, \ldots, x_n) = \begin{cases} \text{the smallest } y \text{ for which } f(x_1, \ldots, x_n, y) = 0, \text{ if any,} \\ \text{undefined if } f(x_1, \ldots, x_n, y) = 0 \text{ for no } y. \end{cases}$$

We shall write '$h = \text{Mn}[f]$' to indicate that h is obtainable from f by minimization.

Example

Mn [sum] is the partial function of one argument which takes the value 0 when its argument is 0 and is otherwise undefined, for $x+y = 0$ if and only if $x = y = 0$. (Remember that we are dealing with *natural* numbers: the xs and y cannot assume negative values, here.)

Another example

Mn [prod] = z, for $x \cdot y = 0$ whenever $y = 0$, regardless of the value of x. Then for every x, Mn [prod] $(x) = 0$.

In general – where the function f need not be total – we define Mn [f] as follows.

$$\mathrm{Mn}[f](x_1, \ldots, x_n) = \begin{cases} y, \text{ in case } f(x_1, \ldots, x_n, y) = 0 \text{ while } f(x_1, \ldots, x_n, t) \text{ is} \\ \quad \text{defined and positive for all } t \text{ less than } y, \\ \text{undefined, in case there is no such } y. \end{cases}$$

Where the function f is total, this definition agrees with that given earlier; but if the earlier definition were applied to partial functions it would sometimes give different results from the present one.

Example

$f(0)$ is undefined and $f(1) = 0$. Then Mn[f] is undefined (according to the definition which we have adopted), but would be defined (and equal to 1) if the earlier definition had been adopted without the restriction to total functions.

Observe that all primitive recursive functions are total: the basic functions are defined for all natural number arguments, and the operations of composition and primitive recursion always yield total functions when applied to total functions. But as we have seen in the case of Mn [sum], the operation of minimization can yield partial functions even when applied to total functions.

Exception. Mn [f], where f is *regular*. A function f of $n+1$ arguments is said to be regular when, for every n-tuple of natural numbers $x_1, \ldots, x_n$, there is a natural number y for which we have $f(x_1, \ldots, x_n, y) = 0$.

To establish that all recursive functions are abacus computable, it only remains to show how, given an abacus program which computes a function f, we can construct an abacus program which computes the function Mn [f], as in Figure 7-4, where we set $n = 1$ for definiteness, so that the function f has two arguments, and Mn [f] has one. The program of Figure 7-4 computes Mn [f] (x), where x is the number of stones in box 1 initially. Initially box 2 is empty, so that if $f(x, 0) = 0$, the program will halt with the correct answer, Mn [f]$(x) = 0$, in box 2. (Box 3 will be empty.) Otherwise, box 3 will be emptied and a single rock placed in box 2, preparatory to computation of $f(x, 1)$. If this value is 0, the program halts, with the correct value, Mn [f]$(x) = 1$, in box 2. Otherwise,

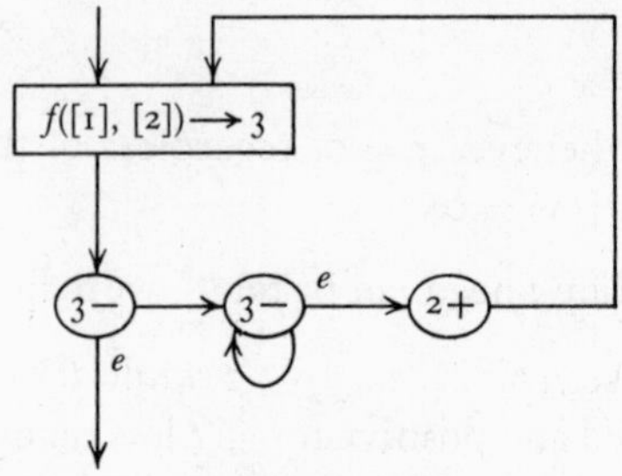

Figure 7-4. Minimization.

another rock is placed in box 2, and the procedure continues until such time (if any) as we have enough rocks (y) in box 2 to make $f(x, y) = 0$.

We have now proved that all recursive functions are abacus computable and hence computable by Turing machines in standard form. To draw the moral for Church's thesis it is necessary to grasp how wide the class of recursive functions is. To get an inkling of that, let us get some sense of how wide the class of primitive recursive functions is.

Examples of primitive recursive functions

7.1 $\text{sum} = \text{Pr}[\text{id}, \text{Cn}[s, \text{id}_3^3]]$ $(\text{sum}(x, y) = x+y)$.

7.2 $\text{prod} = \text{Pr}[z, \text{Cn}[\text{sum}, \text{id}_1^3, \text{id}_3^3]]$ $(\text{prod}(x, y) = x \cdot y)$.

7.3 $\text{exp} = \text{Pr}[\text{Cn}[s, z], \text{Cn}[\text{prod}, \text{id}_1^3, \text{id}_3^3]]$ $(\text{exp}(x, y) = x^y)$.

7.4 $\text{fac} = \text{Pr}[s(0), \text{Cn}[\text{prod}, \text{Cn}[s, \text{id}_1^2], \text{id}_2^2]]$ $(\text{fac}(y) = y!)$.

7.5 $\text{pred} = \text{Pr}[0, \text{id}_1^2]$ (*predecessor*, except that $\text{pred}(0) = 0$).

7.6 $\text{dif} = \text{Pr}[\text{id}, \text{Cn}[\text{pred}, \text{id}_3^3]]$ $(\text{dif}(x, y) = x \dot{-} y = x - y$ if $x \geqslant y$; otherwise, $x \dot{-} y = 0)$.

More informally, the foregoing definitions may be written:

7.1 $x+0 = x, \quad x+y' = (x+y)'$.

7.2 $x \cdot 0 = 0, \quad x \cdot y' = x + x \cdot y$.

7.3 $x^0 = 1, \quad x^{y'} = x \cdot x^y$.

7.4 $0! = 1, \quad y'! = y' \cdot y!$.

7.5 $\text{pred}(0) = 0, \quad \text{pred}(y') = y$.

7.6 $x \dot{-} 0 = x, \quad x \dot{-} y' = \text{pred}(x \dot{-} y)$.

We continue in this informal style:

7.7 $|x-y| = (x \dot{-} y) + (y \dot{-} x)$.

7.8 $\mathrm{sg}(y) = 1 \dot{-} (1 \dot{-} y)$ (*signum* (0) = 0, otherwise, $\mathrm{sg}(y) = 1$).

7.9 $\overline{\mathrm{sg}}(y) = 1 \dot{-} y$ ($\overline{\mathrm{sg}}(0) = 1$; otherwise, $\overline{\mathrm{sg}}(y) = 0$).

7.10 *Definition by cases.* Suppose that the function f is defined in the form

$$f(x,y) = \begin{cases} g_1(x,y) & \text{if} \quad C_1, \\ \vdots & \\ g_n(x,y) & \text{if} \quad C_n, \end{cases}$$

where $C_1, \ldots, C_n$ are mutually exclusive, collectively exhaustive conditions on x, y, and the functions $g_1, \ldots, g_n$ are primitive recursive. The *characteristic function* of a condition C_i on x, y is a function c_i which takes the value 1 for argument pairs (x, y) which satisfy the condition, and takes the value 0 for all other argument pairs. Now if the characteristic functions $c_1, \ldots, c_n$ of the conditions $C_1, \ldots, C_n$ in the definition are primitive recursive, so is the function f, for it can be defined by composition out of the gs and cs *via*

$$\boxed{f(x, y) = g_1(x, y)c_1(x, y) + \ldots + g_n(x, y)c_n(x, y)}$$

This works because for each pair of natural number arguments x, y, all but one of the cs must assume the value 0, while the nonzero characteristic function, say, c_i, will assume the value 1 and will correspond to the condition C_i which the numbers x, y actually satisfy.

7.11 *An example of definition by cases*: $\max(x,y)$ = the larger of the numbers x, y. This can be defined by cases as follows:

$$\max(x,y) = \begin{cases} x & \text{if} \quad x \geqslant y, \\ y & \text{if} \quad x < y \end{cases}$$

in the format boxed in 7.10, $g_1 = \mathrm{id}_1^2$ and $g_2 = \mathrm{id}_2^2$. To find the characteristic function c_1 of the condition $x \geqslant y$, observe that $y \dot{-} x = 0$ if and only if $x \geqslant y$. Then $\overline{\mathrm{sg}}(y \dot{-} x) = 1$ if $x \geqslant y$ and $\overline{\mathrm{sg}}(y \dot{-} x) = 0$ otherwise, *viz.*, if $x < y$. Then $\overline{\mathrm{sg}}(y \dot{-} x)$ is the characteristic function of the condition $x \geqslant y$; and apparently $\mathrm{sg}(y \dot{-} x)$ is the characteristic function of the condition $x < y$. Then the definition is

$$\max(x,y) = x \cdot \overline{\mathrm{sg}}(y \dot{-} x) + y \cdot \mathrm{sg}(y \dot{-} x)$$

whereby the function max is defined by composition of primitive recursive functions.

7.12 $\min(x,y) =$ the smaller of $x, y = y \cdot \overline{\mathrm{sg}}(y \dot{-} x) + x \cdot \mathrm{sg}(y \dot{-} x)$.

7.13 *General summation*:

$$g(x_1, \ldots, x_n, y) = f(x_1, \ldots, x_n, 0) + \ldots + f(x_1, \ldots, x_n, y) = \sum_{i=0}^{y} f(x_1, \ldots, x_n, i).$$

This function is primitive recursive if f is, with these recursion equations:

$$g(x_1, \ldots, x_n, 0) = f(x_1, \ldots, x_n, 0),$$

$$g(x_1, \ldots, x_n, y') = f(x_1, \ldots, x_n, y') + g(x_1, \ldots, x_n, y).$$

Example: $n = 0$, $f =$ the square. Then $g(y) = 0^2 + \ldots + y^2$, and the recursion equations are: $g(0) = 0$, $g(y') = y'^2 + g(y)$.

7.14 *General product*:

$$g(x_1, \ldots, x_n, y) = f(x_1, \ldots, x_n, 0) \cdot \ldots \cdot f(x_1, \ldots, x_n, y) = \prod_{i=0}^{y} f(x_1, \ldots, x_n, i).$$

Here the recursion equations are as in 7.13, except that '$\cdot$' appears in place of '$+$' in the second. *Example*: the product of the first y positive integers. If we set $g(0) = 1$, the second recursion equation will be $g(y') = y' \cdot g(y)$. Thus, $f(0) = 1$ but $f(y') = y'$.

7.15 *Logical composition of conditions*. If C is a condition of which c is the characteristic function, then $\mathrm{Cn}[\overline{\mathrm{sg}}, c]$ is the characteristic function of the condition $-C$ (the *denial* of C) which holds when C fails and fails when C holds. Thus, if C is a condition on x, y, the characteristic function of $-C$ will be the function $\bar{c}$ for which we have $\bar{c}(x, y) = \overline{\mathrm{sg}}(c(x, y))$. Similarly, suppose that $C_1, \ldots, C_n$ are conditions on x, y of which the corresponding characteristic functions are $c_1, \ldots, c_n$. Then the characteristic function of the condition $C_1 \& \ldots \& C_n$ (the *conjunction* of the Cs) will simply be the product of the cs: $c_1(x, y) \cdot \ldots \cdot c_n(x, y)$ will be 1 if all of the conditions $C_1, \ldots, C_n$ are met, and will be 0 if even one of the conditions fails. Since all truth-functional modes of composition of conditions can be built out of denial and conjunction, the characteristic function of any such compound condition will be primitive recursive as long as the characteristic functions of the constituent conditions are primitive recursive. *Example*: *disjunction*. Since $C_1 \vee C_2$ is the denial of the conjunction of the denials of the Cs, $-(-C_1 \& -C_2)$, we have

$$d(x, y) = \overline{\mathrm{sg}}(\overline{\mathrm{sg}}(c_1(x, y)) \cdot \overline{\mathrm{sg}}(c_2(x, y)))$$

if d is the characteristic function of the disjunction $C_1 \vee C_2$.

7.16 *Bounded quantification*. Given a condition $C(x, y)$ on x, y, we can form other conditions by *bounded universal quantification*, $\forall i (i \leqslant y \rightarrow C(x, i))$ ('$C(x, i)$ holds for every i from 0 to y') and *bounded existential quantification*, $\exists i(i \leqslant y \& C(x, i))$('$C(x, i)$ holds for some i from 0 to y'). If the characteristic function $c(x,y)$ of condition $C(x,y)$ is primitive recursive, so are the characteristic functions u and e of the conditions obtained by bounded quantification, for we have

$$u(x,y) = \prod_{i=0}^{y} c(x,i), \quad e(x,y) = \mathrm{sg}\left(\sum_{i=0}^{y} c(x,i)\right).$$

(See Examples 7.13 and 7.14.)

7.17 *Minimization with primitive recursive bound.* Suppose that $f(x_1, \ldots, x_n, y)$ is primitive recursive and *regular*. Then $\mathrm{Mn}[f]$ is a total recursive function which need not be primitive recursive. (It *will* be primitive recursive if there is another, equivalent definition in which the Mn operator is not used.) Suppose, however, that we have a primitive recursive function g which gives a bound on the size of the smallest y for which $f(x_1, \ldots, x_n, y) = 0$; suppose, that is, that for each n-tuple $x_1, \ldots, x_n$ there is a y no greater than $g(x_1, \ldots, x_n)$ for which $f(x_1, \ldots, x_n, y) = 0$. In such a case, the function $\mathrm{Mn}[f]$ is necessarily primitive recursive, for we have

$$\mathrm{Mn}[f](x_1, \ldots, x_n) = \sum_{i=0}^{g(x_1, \ldots, x_n)} \mathrm{sg}\left(\prod_{j=0}^{i} f(x_1, \ldots, x_n, j)\right),$$

where the expression on the right is a composition out of functions known to be primitive recursive. To see what is going on, imagine that $x_1, \ldots, x_n$ are given fixed values for which the smallest y that makes $f(x_1, \ldots, x_n, y)$ vanish is, say, y_0. By our assumption, $y_0 \leqslant g(x_1, \ldots, x_n)$. Then for $i = 0, 1, \ldots, y_0 - 1$, each of the products $\prod_{j=0}^{i} f(x_1, \ldots, x_n, j)$ will be positive – for each of $f(x_1, \ldots, x_n, 0), \ldots, f(x_1, \ldots, x_n, y_0 - 1)$ is nonzero and thus positive. There are exactly y_0 such products, so the sum of their *signum* values will be y_0. Starting with $j = y_0$, however, all further products contain a factor of 0 and are thus 0: they do not contribute to the sum, and so the sum of the *signum* values of the products is y_0. By 7.8, 7.13, and 7.14, the function $\mathrm{Mn}[f]$ is thus primitive recursive if f and g are.

7.18 *Bounded minimization.* $\mathrm{Mn}_w[f]$ = the smallest y between 0 and w, inclusive, for which $f(x_1, \ldots, x_n, y) = 0$, and 0 if there is no such y. This function is primitive recursive if f is (whether or not f is regular), for we can define it by cases:

$$\mathrm{Mn}_w[f](x_1, \ldots, x_n) = \begin{cases} 0 & \text{if } \forall y(y \leqslant w \to f(x_1, \ldots, x_n, y) \neq 0), \\ \sum_{i=0}^{w} \mathrm{sg}\left(\prod_{j=0}^{i} f(x_1, \ldots, x_n, j)\right) & \text{otherwise.} \end{cases}$$

7.19 *Bounded maximization.* $\mathrm{Mx}_w[f]$ = the largest y between 0 and w, inclusive, for which $f(x_1, \ldots, x_n, y) = 0$, and 0 if there is no such y. Notice that if there is such a y, it can be characterized as the *smallest* y no greater than w for which we have $f(x_1, \ldots, x_n, y) = 0$, $f(x_1, \ldots, x_n, y+1) \neq 0$, $f(x_1, \ldots, x_n, y+2) \neq 0, \ldots, f(x_1, \ldots, x_n, w) \neq 0$. Thus we can define Mx_w in terms of Mn_w as follows:

$$\mathrm{Mx}_w[f] = \mathrm{Mn}_w[g],$$

where g is the characteristic function of the condition

$$f(x_1, \ldots, x_n, y) = 0 \,\&\, \forall u(u \leqslant w \rightarrow (u > y \rightarrow f(x_1, \ldots, x_n, u) \neq 0)).$$

7.20 *Graphs.* If f is an n-place function, then the $(n+1)$-place relation R such that

$$R(x_1, \ldots, x_n, x_{n+1}) \quad \text{iff} \quad f(x_1, \ldots, x_n) = x_{n+1}$$

for all $x_1, \ldots, x_{n+1}$ is called the *graph* of f. If f is primitive recursive, then so is the characteristic function c of its graph R, for since

$$R(x_1, \ldots, x_n, x_{n+1}) \quad \text{iff} \quad |f(x_1, \ldots, x_n) - x_{n+1}| = 0,$$

we have

$$c(x_1, \ldots, x_n, x_{n+1}) = \overline{\mathrm{sg}}\,(|f(x_1, \ldots, x_n) - x_{n+1}|).$$

(See 7.7 and 7.9.)

7.21 *Eliminating constants.* We have regarded such schemes as

$$h(0) = m$$
$$h(s(y)) = g(y, h(y)), \text{ where } m \text{ is a } \textit{natural number},$$

as admissible definitions of one-place primitive recursive functions h from two-place primitive recursive functions g. But we have gained no generality by doing so. If we let $g^* = \mathrm{Cn}[g, \mathrm{id}^3_2, \mathrm{id}^3_3]$, f_m = the one-place function whose value is always $= m$, and $h^* = \mathrm{Pr}[f_m, g^*]$, then $h = \mathrm{Cn}[h^*, z, \mathrm{id}^1_1]$, and h is thus definable from *the zero function* and the successor and identity functions by repeated applications of composition and primitive recursion provided that g is so definable. For f_m is certainly so definable, $h^*(x, 0) = f_m(x) = m$, and $h^*(x, s(y)) = g^*(x, y, h^*(x, y))$ $= g(\mathrm{id}^3_2(x, y, h^*(x, y)), \mathrm{id}^3_3(x, y, h^*(x, y))) = g(y, h^*(x, y))$. Thus $h(0) = m = h^*(0, 0)$; and if $h(y) = h^*(0, y)$, then $h(s(y)) = g(y, h(y)) = g(y, h^*(0, y)) = h^*(0, s(y))$ by substituting '0' for 'x'. By induction then, for all y, $h(y) = h^*(0, y) = h^*(z(y), \mathrm{id}^1_1(y))$, and thus $h = \mathrm{Cn}[h^*, z, \mathrm{id}^1_1]$.

8

Turing computable functions are recursive

We have seen that all recursive functions are abacus computable ($R \subseteq A$) and that all abacus computable functions are Turing computable ($A \subseteq T$). We shall now prove that all Turing computable functions are recursive ($T \subseteq R$). This will close the circle of inclusions

$$R \underset{7}{\subseteq} A \underset{6}{\subseteq} T \underset{8}{\subseteq} R$$

and complete the demonstration that the three notions of computability are equivalent:

$$R = A = T.$$

Let us suppose then that f is a function that is computed by a Turing machine M that meets conditions (1)–(5) of Chapter 6. We must show that f is recursive. For definiteness, we suppose f to be a function of two arguments; a construction similar to the present one can be given for any other number of arguments.

Let x_1, x_2 be arbitrary natural numbers. At the beginning of its computation of $f(x_1, x_2)$, M's tape will be completely blank except for two blocks of 1s, which are separated by a single blank square. The left block contains x_1+1 1s; the right block, x_2+1 1s. At the outset M is scanning the leftmost 1 in the left block. When it halts, it is scanning the leftmost 1 in a block of $f(x_1, x_2)+1$ 1s on an otherwise completely blank tape. And throughout the computation there are finitely many 1s to the left of the scanned square, finitely many 1s to the right, and at most one 1 in the scanned square.

Thus at any time during the computation, if there is a 1 to the left of the scanned square, there is a leftmost 1 to its left, and similarly for the right. We can therefore use *binary notation* to represent the contents of the tape and the content of the scanned square by means of *a pair of natural numbers*, in the following manner:†

† We are here indebted to Marvin L. Minsky, *Computation: Finite and Infinite Machines*, Prentice-Hall, Englewood Cliffs, N.J., 1967.

The infinite portion of the tape to the left of the scanned square can be thought of as containing a binary numeral (e.g. 1011, 0, 1) prefixed by an infinite sequence of superfluous 0s. ($0 = B$.) We call this numeral *the left numeral* and the number it denotes in binary notation *the left number*. The rest of the tape, consisting of the scanned square and the portion to its right, can be thought of as containing a binary numeral, WRITTEN BACKWARDS, to which an infinite sequence of superfluous 0s is also attached. We call this numeral, which appears backwards on the tape, *the right numeral* and the number it denotes *the right number*. Thus the scanned square contains the digit in the 'one's place' of the right numeral. We take the right numeral to be written backwards to insure that changes on the tape will always take place in the vicinity of the one's place of both numerals.

Example 8.1

The tape looks like this:

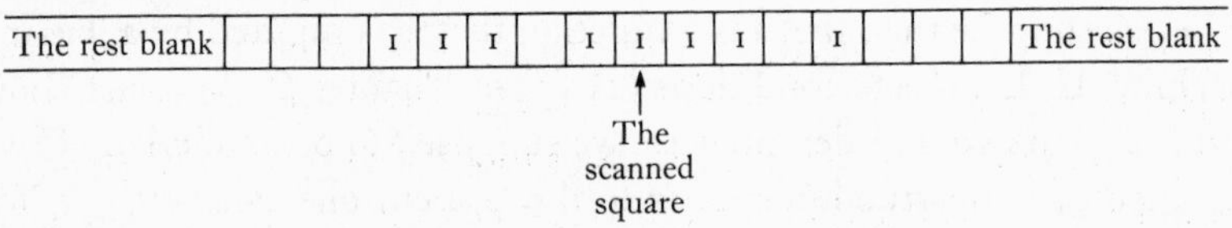

Then the left numeral is 11101, the right numeral is 10111, the left number is 29, and the right number is 23.

We use '0' and '1' as names both for numerals and for the numbers they denote and rely upon context to resolve ambiguity.

Example 8.2

The tape is completely blank. Then the left numeral = the right numeral = 0, and the left number = the right number = 0.

When the symbol in the scanned square changes from 0 to 1 (1 to 0) the right number increases (decreases) by one and the left number does not change. What happens to the numbers when M moves left or moves right one square?

Let l (r) be the old, i.e. pre-move, left (right) number and let l' (r') be the new left (right) number. We want to see how l' and r' depend upon l, r, and the direction of the move. There are four cases to consider, all quite similar, and we shall examine in detail only the one in which M moves left and l is odd.

Since l is odd, the old left numeral ends in a 1. If $r = 0$, then the new right numeral is 1, and $r' = 1 = 2r+1$. And, if $r > 0$, then the new right numeral comes from the old by appending a 1 to it at its one's place end; again $r' = 2r+1$. As for l', if $l = 1$, then the old left numeral is just 1, the new left numeral is 0, and $l' = 0 = (l-1)/2$. And if l is any odd number greater than 1, then the new left numeral comes from the old by deleting the 1 in its one's place (thus shortening the numeral). This new numeral denotes $(l-1)/2$, and again $l' = (l-1)/2$. To sum up: if M moves left and l is odd, then $l' = (l-1)/2$ and $r' = 2r+1$.

Example 8.1 continued

Suppose M moves left. Then the new left numeral is 1110, $l' = 14 = (29-1)/2$, the new right numeral is 101111, and $r' = 47 = 2 \cdot 23+1$.

Similar arguments show that

if M moves left and l is even, then $l' = l/2$ and $r' = 2r$;
if M moves right and r is odd, then $l' = 2l+1$ and $r' = (r-1)/2$; and
if M moves right and r is even, then $l' = 2l$ and $r' = r/2$.

What are the left and right numbers when M begins its computation? The tape is then completely blank to the left of the scanned square, and so the left numeral is 0 and the left number is 0. The right numeral is

$$\underbrace{1 \ldots 1}_{x_2+1 \text{ 1s}} \; 0 \; \underbrace{1 \ldots 1}_{x_1+1 \text{ 1s}}$$

A sequence of m 1s denotes $2^m - 1$ in binary notation; a sequence of m 1s followed by n 0s denotes $(2^m-1)2^n$; and a sequence of m 1s followed by one 0 followed by p 1s denotes $(2^m-1)2^{p+1}+(2^p-1)$. Thus the right number at the beginning of M's computation of $f(x_1, x_2)$ is $(2^{(x_2+1)} \dot{-} 1)2^{(x_1+2)}+(2^{(x_1+1)} \dot{-} 1)$. Let $s(x_1, x_2) = (2^{(x_2+1)} \dot{-} 1)2^{(x_1+2)}+(2^{(x_1+1)} \dot{-} 1)$. Then s is a primitive recursive function.

And what are the left and right numbers when M halts? When it halts, M is scanning the leftmost 1 of a block of $f(x_1, x_2)+1$ 1s on an otherwise blank tape. The left number is again 0, and the right number is $2^{f(x_1, x_2)+1}-1$, which is the number denoted in binary by a string of $f(x_1, x_2)+1$ 1s.

Let $\text{lo}(x)$ = the greatest $w \leqslant x$ such that $2^w \leqslant x$. By 7.19, lo is a primitive recursive function. (In fact, if $d(x, y) = 0$ if $2^y \leqslant x$, and

1 otherwise, then $\mathrm{lo}(x) = \mathrm{Mx}_x[d](x)$.) $\mathrm{lo}(2^{z+1}-1) = z$. Thus $\mathrm{lo}(2^{f(x_1,x_2)+1}-1) = f(x_1, x_2)$. It follows that if M halts its computation of $f(x_1, x_2)$ at t and r is the right number at t, then $\mathrm{lo}(r) = f(x_1, x_2)$.

We turn now to the task of representing the flow graph or machine table of M by two primitive recursive functions, a and q. We interpret the machine states q_1, etc. as natural numbers 1 etc., the symbols B, 1 as the natural numbers 0, 1, and the acts B, 1, L, R, as the natural numbers 0, 1, 2, 3. If in state q_i while scanning 0, M performs act $j(= 0, 1, 2, 3)$ and goes into state q_k, then we set $a(i, 0) = j$ and $q(i, 0) = k$; if in state q_i while scanning 1, M performs act j and goes into state q_k, we likewise set $a(i, 1) = j$ and $q(i, 1) = k$. For all pairs of numbers x, y not covered by this stipulation we set $a(x, y) = y$ and $q(x, y) = 0$. To halt, then, is to enter notional state 0. Since there are only finitely many quadruples in M's machine table, it is clear that a and q are primitive recursive functions, definable by cases.

Example 8.3

Consider the graph in Figure 8-1, of a machine which computes the successor function.

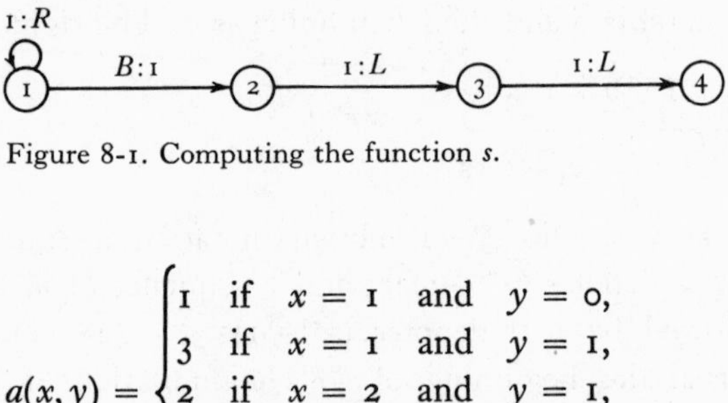

Figure 8-1. Computing the function s.

$$a(x,y) = \begin{cases} 1 & \text{if } x = 1 \text{ and } y = 0, \\ 3 & \text{if } x = 1 \text{ and } y = 1, \\ 2 & \text{if } x = 2 \text{ and } y = 1, \\ 2 & \text{if } x = 3 \text{ and } y = 1, \\ y & \text{otherwise} \end{cases}$$

$$q(x,y) = \begin{cases} 2 & \text{if } x = 1 \text{ and } y = 0, \\ 1 & \text{if } x = 1 \text{ and } y = 1, \\ 3 & \text{if } x = 2 \text{ and } y = 1, \\ 4 & \text{if } x = 3 \text{ and } y = 1, \\ 0 & \text{otherwise.} \end{cases}$$

In order to show that f is recursive, we need a device to code ordered triples of natural numbers as single natural numbers, and devices for recovering the first, second, and third components of a triple from the single number that codes the triple. The need for these devices arises from the fact that what the left number is at stage $t+1$ of the computation depends not only upon what the left number is at t but also upon what the right number is at t and upon the state of M at t. Similarly for the right number and the state of M at t. But once we have coded triples as single numbers, we shall be able to show that the triple l', c', r' consisting of the left number at $t+1$, the state at $t+1$, and the right number at $t+1$, depends (primitive recursively) on the triple l, c, r of the left number at t, the state at t, and the right number at t.

We thus define $\text{tpl}(x, y, z) = 2^x 3^y 5^z$. tpl is primitive recursive; if $\text{tpl}(x, y, z) = \text{tpl}(a, b, c)$, then $x = a$, $y = b$, and $z = c$. We shall henceforth write '$\langle x, y, z\rangle$' instead of '$\text{tpl}(x, y, z)$'.

Let $\text{lft}(w)$ be the greatest $x \leqslant w$ such that 2^x divides w (without remainder), let $\text{ctr}(w)$ be the greatest $x \leqslant w$ such that 3^x divides w, and let $\text{rgt}(w)$ be the greatest $x \leqslant w$ such that 5^x divides w. By 7.19, lft, ctr, and rgt are primitive recursive. The functions lft, ctr, and rgt are inverses to tpl in the sense that if $w = \langle x, y, z\rangle$, then $\text{lft}(w) = x$, $\text{ctr}(w) = y$, and $\text{rgt}(w) = z$.

We shall now define a primitive recursive function g of three arguments with the following property: *If t is a stage not later than the stage at which M halts when computing $f(x_1, x_2)$, then $g(x_1, x_2, t) = \langle$the left number at t, the state of M at t, the right number at $t\rangle$.* Since M is in state 1 when it begins its computation, we let

$$g(x_1, x_2, 0) = \langle 0, 1, s(x_1, x_2)\rangle.$$

We let $e(x) = 0$ if x is even, and 1 if x is odd. The function e is of course primitive recursive. A number is even if and only if there is a 0 in the one's place of its binary representation. Thus if r is the right number at t, then $e(r) = 0$ if the symbol in the square scanned at t is $0 (= B)$, and $e(r) = 1$ if this symbol is 1.

We can now complete our definition of g. In what follows we use 'l', 'c', and 'r' to abbreviate '$\text{lft}(g(x_1, x_2, t))$', '$\text{ctr}(g(x_1, x_2, t))$', and '$\text{rgt}(g(x_1, x_2, t))$', respectively. We also abbreviate '$q(c, e(r))$', i.e. '$q(\text{ctr}(g(x_1, x_2, t)), e(\text{rgt}(g(x_1, x_2, t))))$' by '$q$'. Let

$$g(x_1, x_2, t+1) = \begin{cases} \langle l, q, r \rangle & \text{if} \quad a(c, e(r)) = 0 \quad \text{and} \quad e(r) = 0, \\ \langle l, q, r \dot{-} 1 \rangle & \text{if} \quad a(c, e(r)) = 0 \quad \text{and} \quad e(r) = 1, \\ \text{(write a } B \text{ in the scanned square)} & \\ \langle l, q, r+1 \rangle & \text{if} \quad a(c, e(r)) = 1 \quad \text{and} \quad e(r) = 0, \\ \langle l, q, r \rangle & \text{if} \quad a(c, e(r)) = 1 \quad \text{and} \quad e(r) = 1, \\ \text{(write a 1 in the scanned square)} & \\ \langle l/2, q, 2r \rangle & \text{if} \quad a(c, e(r)) = 2 \quad \text{and} \quad e(l) = 0, \\ \langle (l \dot{-} 1)/2, q, 2r+1 \rangle & \text{if} \quad a(c, e(r)) = 2 \quad \text{and} \quad e(l) = 1, \\ \text{(move left one square)} & \\ \langle 2l, q, r/2 \rangle & \text{if} \quad a(c, e(r)) = 3 \quad \text{and} \quad e(r) = 0, \\ \langle 2l+1, q, (r \dot{-} 1)/2 \rangle & \text{if} \quad a(c, e(r)) = 3 \quad \text{and} \quad e(r) = 1, \\ \text{(move right one square)} & \\ 0 & \text{otherwise.} \end{cases}$$

(We take / to be defined primitive recursively so that x/y is the quotient on dividing x by y if there is a natural number that is the quotient, and to be (say) 0 if there is no such natural number.)

Since tpl, lft, ctr, rgt, s, q, e, a, $\dot{-}$, $+$, $\cdot$, and / are all primitive recursive, g is primitive recursive too.

Clause 2 of the definition of g can be interpreted: if M is in state c scanning a 1 and its machine table tells it to print a 0 on the scanned square, then the left number remains the same, the next state of M is the one prescribed by its table, and the right number decreases by 1. The other clauses have similar interpretations.

Our definitions guarantee that if t is the stage at which M halts, then M goes into state 0 for the first time at $t+1$. Thus $\text{ctr}(g(x_1, x_2, y)) \neq 0$ for all $y \leqslant t$. And if l, c, and r are the left number at t, the state of M at t, and the right number at t, respectively, then $a(c, e(r)) = e(r)(= 0, 1)$ and $q(c, e(r)) = 0$, and therefore $g(x_1, x_2, t+1) = \langle l, 0, r \rangle$ and $\text{ctr}(g(x_1, x_2, t+1)) = 0$. Thus M halts at t when computing $f(x_1, x_2)$ if and only if t is the least y such that $\text{ctr}(g(x_1, x_2, y+1)) = 0$. Let $h(x_1, x_2, y) = \text{ctr}(g(x_1, x_2, y+1))$. h is primitive recursive, and M halts at t when computing $f(x_1, x_2)$ iff $\text{Mn}[h](x_1, x_2) = t$. $\text{Mn}[h]$ is a recursive function. (It is not possible in general to conclude that it is a *primitive* recursive function.) And if M halts at t, and r is the right number at t, then $f(x_1, x_2) = \text{lo}(r) = \text{lo}(\text{rgt}(g(x_1, x_2, t)))$. Thus

$$f(x_1, x_2) = \text{lo}(\text{rgt}(g(x_1, x_2, \text{Mn}[h](x_1, x_2)))),$$

for all x_1, x_2. f is therefore a recursive function. The circle is closed.

We have also proved, by the way,

The Kleene normal form theorem for recursive functions
In obtaining a recursive function from basic functions via composition, primitive recursion, and minimization, the operation of minimization need not be used more than once.

Our usage of the term 'recursive' has been somewhat non-standard. What we have been calling the recursive functions are ordinarily called the *partial recursive* functions, while the functions that *are* standardly called 'recursive' are the partial recursive functions that are total, i.e. that assign values to *all* n-tuples of natural numbers (for the appropriate n). As we shall have no further dealings with partial recursive functions that are not total, we adopt standard usage henceforth: from now on, a recursive function is always *total*.

9
First-order logic revisited

We now take time off from computability to review some logic, after which we'll apply what we've learned about computability to logic and show, among other things, that there can be no perfectly satisfactory uniform mechanical procedure for determining whether or not inferences in the notation of first-order logic (= elementary logic) are valid.

We assume that you are already familiar with logical notation: that you've seen the *connectives*

$-$	$\&$	$\vee$	$\rightarrow$	$\leftrightarrow$
not	*and*	*or*	*if ... then*	*if and only if*

and the quantifiers

$\forall x \quad \forall y \quad \forall z \quad \ldots$	$\exists x \quad \exists y \quad \exists z \quad \ldots$
universal quantifiers	*existential quantifiers*

either in the notation illustrated above or in some other. We assume that once you are informed that the universe of discourse is mankind, that 'L' means 'loves', and that 'a' names Alma, you can read such formulas as

$$\forall x(xLx \rightarrow \exists y\, xLy). \tag{9.1}$$

('Anyone who loves himself loves someone') and

$$\forall x[\exists y(xLy \,\&\, yLa) \rightarrow -xLa]. \tag{9.2}$$

('None of Alma's lovers' lovers love her'). We also assume that with a bit of thought you can recognize (9.1) as a logical truth (= as valid) and that you can recognize that (9.2) implies '$-aLa$' (= the inference from (9.2) as premise to '$-aLa$' as conclusion is a *valid* one).

You are probably familiar, too, with the use of the sign

$$=$$

of identity (= the equals-sign) in making such statements as

$$\forall x\, \forall y\, \forall z\, [(xFz \,\&\, yFz) \rightarrow x = y] \tag{9.3}$$

which (reading 'F' as *is a father of*) says that nobody has more than one father. You may be familiar, too, with the use of function symbols such as 'f' for *the father of* in making such statements as

$$aLf(f(a)) \tag{9.4}$$

('Alma loves her paternal grandfather') which can also be made, more awkwardly, without function symbols.

In this chapter we will provide a framework within which such notions as validity can be discussed with the degree of clarity that will be needed when we show that the valid inferences are precisely those that pass a certain mechanical test. (We'll also show that there can be no such mechanical test for invalidity.)

The first notion we shall need is a division of the symbols that may occur in formulas in the notation of elementary logic into two sorts: *logical* and *non-logical.* The logical symbols are the variables and these eleven: $-$ $\&$ $\vee$ $\rightarrow$ $\leftrightarrow$ $\exists$ $\forall$ $=$ ( ) ,

We stress that the equals-sign counts as a *logical* symbol: we shall give it special treatment when we come to state the conditions under which sentences in logical notation are true or false. We suppose that there are enumerably infinitely many variables.

The non-logical symbols are of four disjoint sorts: *names, function symbols, sentence letters,* and *predicate letters.* Both function symbols and predicate letters are of various (unique) *numbers of places*: thus, ordinarily, + is a *two*-place function symbol and $<$ is a *two*-place predicate letter. Any positive integer can be the number of places of a predicate letter or a function symbol. (Occasionally names are regarded as zero-place function symbols and sentence letters as zero-place predicate letters. But though this is sometimes convenient, we shall not usually regard them in this way.)

The equals-sign, even though it is a logical symbol, is to count as a two-place predicate letter. But it is the only exception: the non-logical symbols are precisely the names, function symbols, sentence letters, and predicate letters other than =.

We shall appropriate the word 'language' to mean *an enumerable set of non-logical symbols.* A sentence or a formula *of* or *in* a language is one whose non-logical symbols all belong to the language. We suppose that sentences and formulas are formed in the ordinary, familiar ways, and that a sentence is, as usual, a formula in which there are no *free* occurrences of any variables. As always, an occurrence of a variable in a formula is free if it is governed by no quantifier (in that formula) containing that variable.

One language will be of great interest to us in later chapters. It is called *the language of arithmetic* – its nickname is 'L' – and is the set $\{\mathbf{o}, ', +, \cdot\}$. $\mathbf{o}$ is a name, $'$ is a one-place function symbol, and $+$ and $\cdot$ are two-place function symbols. There are no sentence letters or predicate letters in L. Two of the enumerably infinitely many sentences of L are

$$\mathbf{o}''+\mathbf{o}''' = \mathbf{o}''''' \tag{9.5}$$

and $$\forall x \forall y\, x+y = y+x. \tag{9.6}$$

((9.5) and (9.6) are really a sort of slang for what the official conventions about the positioning of function symbols and predicate letters would have us write, namely,

$$= (+('('(\mathbf{o})), '('('(\mathbf{o}))), '('('('('(\mathbf{o})))))$$

and $$\forall x \forall y = (+(x, y), +(y, x)).$$

We shall continue to speak with the vulgar.)

The empty set, $\varnothing$, is certainly an enumerable set of non-logical symbols, and so, by our lights, it is a language. One might be tempted to suppose that there weren't any sentences of this language, but there are: as $=$ is a predicate letter

$$\forall x\, x = x$$

is one of its sentences, and indeed is a sentence of *every* language.

Any formula in a language is a finite sequence drawn from an enumerable set of symbols. So there are at most enumerably many formulas in any language. (Cf. Exercise 2.4(*d*).) But as

$$\forall x\, x = x, \quad \forall y\, y = y, \quad \forall z\, z = z, \ldots$$

are sentences in every language, the set of sentences and hence the set of formulas of any language are actually enumerably infinite.

As yet we haven't said anything about when sentences in various languages are either true or false. *Interpretations* (or *models* or *structures*, as they are sometimes called) remedy this lack. An *interpretation of a language* specifies these things:

(1) A *domain*, *viz*. a nonempty set. The domain of an interpretation is the range (according to the interpretation) of any variables that occur in any sentences in the language. ('Universe' and 'universe of discourse' are often used as synonyms of 'domain'.)

(2) For each name in the language, and no others, a *designation* (or *bearer*, or *denotation*, or *reference*), i.e. an object in the domain specified in (1).

(3) For each function symbol in the language and no others, a *function* $\mathfrak{f}$ which assigns a value in the domain for any sequence of arguments in the domain; in the case of an n-place function symbol, f, the function $\mathfrak{f}$ has n argument places. Thus if f is a one-place function symbol, the domain of $\mathfrak{f}$ (the set of arguments of $\mathfrak{f}$) will be just the domain of the interpretation, and the range of $\mathfrak{f}$ (the set of values of $\mathfrak{f}$) will be a subset of the domain of the interpretation.

(4) For each sentence letter in the language and no others, a *truth-value*, 1 (truth) or 0 (falsity). (Truth is one.)

(5) For each predicate letter in the language and no others, a *characteristic function*. In the case of an n-place predicate letter R, the characteristic function ϕ has n argument places. For any objects $o_1, \ldots, o_n$ in the domain specified in (1), the value $\phi(o_1, \ldots, o_n)$ is to be 1 or 0 depending on whether or not R is supposed to be true in the interpretation of the sequence $o_1, \ldots, o_n$ of objects. We shall sometimes indicate what characteristic function a certain interpretation assigns to a predicate letter by stating a necessary and sufficient condition for the predicate letter to be true of a sequence of objects in the domain of the interpretation.

An *interpretation* is, of course, something that is an interpretation of some language. Our definition of 'interpretation of a language' should be taken as implying that each interpretation is the interpretation of only one language; so any interpretation assigns objects of the appropriate sort to at most enumerably many non-logical symbols. An interpretation of a *sentence* is, as one might guess, an interpretation of a language in which that sentence is a sentence; an interpretation of a *set of sentences* is an interpretation of all the sentences in the set.

A particularly important interpretation is *the standard interpretation of the language of arithmetic*, or '$\mathcal{N}$', for short. The domain of $\mathcal{N}$ is the set $\{0, 1, 2, \ldots\}$ of all natural numbers, and $\mathcal{N}$ assigns an appropriate sort of object to each of the symbols in L, the language of arithmetic. In fact, $\mathcal{N}$ assigns them just what you would expect (which is why it is the 'standard' interpretation): To $\mathbf{0}$ $\mathcal{N}$ assigns 0, the least natural number; to $'$ $\mathcal{N}$ assigns the successor function, whose value for each natural number n is $n+1$, the successor of n; to $\mathbf{+}$ $\mathcal{N}$ assigns the addition function, whose value for any natural numbers m, n is $m+n$; and to $\cdot$ $\mathcal{N}$ assigns the multiplication function, whose value for any natural numbers is $m \cdot n$.

Example 9.1. An interpretation of the sentence (9.4) '*aLf*(*f*(*a*))'

(1) *Domain*: the set {Alma, Max, Dan}.

(2) *Characteristic function* ϕ *of 'L'*: see the table.

		2nd argument		
	ϕ	Alma	Max	Dan
1st argument	Alma	1	1	1
	Max	1	0	1
	Dan	0	0	0

(3) *Denotation of 'a'*: Alma.

(4) *Function for 'f'*: f (Alma) = Max, f (Max) = Dan, f (Dan) = Dan. Clearly '*f*' can't be read as *the father of* in this interpretation: Dan can't be his own father. The facts of life being what they are, '*f*' can only be read in that way when the domain is infinite. Intuitively, sentence (9.4) is true in this interpretation: $f(f(\text{Alma})) = f(\text{Max}) = \text{Dan}$, and indeed ϕ (Alma, Dan) = 1 – Alma does love Dan in this interpretation.

Example 9.2. An interpretation of sentence (9.2)

(1) *Domain*: {Alma, Bert, Clara}.

(2) *Characteristic function* ϕ *of 'L'*: see the table.

ϕ	Alma	Bert	Clara
Alma	1	0	0
Bert	1	0	0
Clara	0	1	0

(3) *Denotation of 'a'*: Alma.

Intuitively, sentence (9.2) is false in this interpretation, for since ϕ (Alma, Alma) = 1, Alma is one of her own lovers; loving herself, she loves one of her lovers; and so, one of Alma's lovers' lovers does love her, contrary to what (9.2) asserts.

Example 9.3. An interpretation of sentence (9.5)

(1) *Domain*: {0, 1, 2, ...}.

(2) *Function assigned to* **′**: the successor function.

(3) *Function assigned to* **+** : the *multiplication* function.

(4) Denotation of **0**: zero.

Intuitively, sentence (9.5) is false in this interpretation, for according to the interpretation it asserts that two *times* three is five. (9.5) is true in $\mathcal{N}$, however, for according to $\mathcal{N}$ it says that two plus three is five.

Now let's make the notion of truth in an interpretation explicit and independent of intuition. If S is any sentence, and $\mathcal{I}$ is one of its interpretations, we want to define $\mathcal{I}(S)$, where

$$\mathcal{I}(S) = \begin{cases} 1 & \text{if } S \text{ is true in interpretation } \mathcal{I}, \\ 0 & \text{if } S \text{ is false in interpretation } \mathcal{I}. \end{cases}$$

We'll do this by successively reducing the question, 'What value does $\mathcal{I}$ assign to S?' to similar questions about shorter sentences and about interpretations closely related to $\mathcal{I}$, until finally we are concerned only with atomic sentences, for which such questions are answered separately.

A sentence that isn't atomic must have one of the seven forms:

$$-S,\ (S_1 \,\&\, S_2),\ (S_1 \vee S_2),\ (S_1 \rightarrow S_2),\ (S_1 \leftrightarrow S_2),\ \forall vF, \exists vF,$$

where the Ss are sentences, v is a variable, and F is a formula that contains no free occurrences of any variable other than v. (Remember that a sentence is a formula in which no variable has free occurrences; if $S = \forall vF$, the only variable that could possibly have free occurrences in F is v.) Let's consider these cases in turn.

Case 1. $\mathcal{I}(-S) = 1$ if $\mathcal{I}(S) = 0$;
$\mathcal{I}(-S) = 0$ if $\mathcal{I}(S) = 1$.

Case 2. $\mathcal{I}(S_1 \& S_2) = 1$ if $\mathcal{I}(S_1) = \mathcal{I}(S_2) = 1$;
$\mathcal{I}(S_1 \& S_2) = 0$ if either $\mathcal{I}(S_1) = 0$ or $\mathcal{I}(S_2) = 0$ or both.

Case 3. $\mathcal{I}(S_1 \vee S_2) = 1$ if either $\mathcal{I}(S_1) = 1$ or $\mathcal{I}(S_2) = 1$ or both;
$\mathcal{I}(S_1 \vee S_2) = 0$ if $\mathcal{I}(S_1) = \mathcal{I}(S_2) = 0$.

Case 4. $\mathcal{I}(S_1 \rightarrow S_2) = 1$ if either $\mathcal{I}(S_1) = 0$ or $\mathcal{I}(S_2) = 1$ or both;
$\mathcal{I}(S_1 \rightarrow S_2) = 0$ if $\mathcal{I}(S_1) = 1$ and $\mathcal{I}(S_2) = 0$.

Case 5. $\mathcal{I}(S_1 \leftrightarrow S_2) = 1$ if $\mathcal{I}(S_1) = \mathcal{I}(S_2)$;
$\mathcal{I}(S_1 \leftrightarrow S_2) = 0$ if $\mathcal{I}(S_1) \neq (S_2)$.

Here we have simply described the truth tables for the connectives.

Case 6. $\mathcal{I}(\forall v F) = 1$ if $\mathcal{I}^a_o(F_v a) = 1$ for every o in the domain of $\mathcal{I}$;
$\mathcal{I}(\forall v F) = 0$ if $\mathcal{I}^a_o(F_v a) = 0$ for even one o in that domain.

Explanation. $F_v a$ is the sentence obtained by writing a in place of all free occurrences of v in F; a is required to be a name that does not occur in F (it does not matter which), and $\mathscr{I}^a_o$ is the interpretation which is just like $\mathscr{I}$ except that in it, the name a is assigned the designation o. $\mathscr{I}^a_o$ therefore always has the same domain as $\mathscr{I}$. ($\mathscr{I}$ may or may not assign a any designation at all; if by chance $\mathscr{I}$ already assigns a the designation o, then $\mathscr{I}^a_o = \mathscr{I}$; $\mathscr{I}^a_o$ is 'not defined' if o is not in the domain of $\mathscr{I}$.)

Case 7. $\mathscr{I}(\exists v F) = 1$ if $\mathscr{I}^a_o(F_v a) = 1$ for even one o in the domain of $\mathscr{I}$;
$\mathscr{I}(\exists v F) = 0$ if $\mathscr{I}^a_o(F_v a) = 0$ for every o in that domain.

Note. Again, a must not occur in F, and $\mathscr{I}^a_o$ must assign designation o to a, and otherwise be just like $\mathscr{I}$.

This enumeration of cases may dazzle the eye, but all that's being done, here, is to restate the familiar interpretations of the connectives and quantifiers in a form that allows us to write out reasonably neat-looking calculations.

If a sentence has none of the foregoing seven forms, it must be atomic, i.e. must either be a sentence letter or consist of the equals-sign flanked by a pair of *terms* (case 8) or consist of an n-place predicate letter followed by a string of n *terms* (case 9).

What is a term? A term is either a name or something obtained by writing n (shorter) terms in the blanks of an n-place function symbol.

Examples. a, $f(a)$, $f(f(a))$, $g(a, f(a))$, $g(g(f(f(a)), a), f(a))$. (Here we assume that f is a one-place, and g a two-place, function symbol.)

If t is a term, what's $\mathscr{I}(t)$, the denotation of t in interpretation $\mathscr{I}$? If t is a name, $\mathscr{I}(t)$ is given in part (2) of the definition of interpretation. Otherwise t is of the form $f(t_1, \ldots, t_n)$ and $\mathscr{I}(t) = \mathfrak{f}(\mathscr{I}(t_1), \ldots, \mathscr{I}(t_n))$, where $\mathfrak{f}$ is the function that $\mathscr{I}$ assigns to f – see part (3).

If the sentence S is a sentence letter, then $\mathscr{I}(S)$ is explicitly given in part (4) of the definition of interpretation. Otherwise we have

Case 8. $\mathscr{I}(t_1 = t_2) = 1$ if $\mathscr{I}(t_1) = \mathscr{I}(t_2)$;
$\mathscr{I}(t_1 = t_2) = 0$ if $\mathscr{I}(t_1) \neq \mathscr{I}(t_2)$.

Case 9. $\mathscr{I}(Rt_1 \ldots t_n) = \phi(\mathscr{I}(t_1), \ldots, \mathscr{I}(t_n))$, where ϕ is the characteristic function of the n-place predicate letter R in interpretation $\mathscr{I}$, and the ts are terms whose denotations are $\mathscr{I}(t_1), \ldots, \mathscr{I}(t_n)$ in interpretation $\mathscr{I}$.

Now let's see how this gadgetry works on our three examples.

Example 9.1 again

With $\mathscr{I}$ as in Example 9.1 we can evaluate the sentence '*aLffa*'. (Some parentheses omitted.)

$$\begin{aligned}
\mathscr{I}(aLffa) &= \phi(\mathscr{I}a, \mathscr{I}ffa) && \text{by case 9,}\\
&= \phi(\mathscr{I}a, f\mathscr{I}fa) && \text{by the definition of } \mathscr{I}(t),\\
&= \phi(\mathscr{I}a, ff\mathscr{I}a) && \text{by the definition of } \mathscr{I}(t) \text{ again,}\\
&= \phi(\text{Alma}, ff\,\text{Alma}) && \text{since } \mathscr{I}a = \text{Alma},\\
&= \phi(\text{Alma}, f\,\text{Max}) && \text{since } f\,\text{Alma} = \text{Max},\\
&= \phi(\text{Alma}, \text{Dan}) && \text{since } f\,\text{Max} = \text{Dan},\\
&= 1 && \text{by the table for } \phi.
\end{aligned}$$

Example 9.2 again

With $\mathscr{I}$ as in Example 9.2 we can show that sentence (9.2),

$$\forall x\,[\exists y\,(xLy \,\&\, yLa) \rightarrow -xLa],$$

is false in $\mathscr{I}$: $\mathscr{I}(9.2) = 0$ if $\mathscr{I}^b_o(\exists y\,(bLy \& yLa) \rightarrow -bLa) = 0$ for $o =$ Alma or $o =$ Bert or $o =$ Clara. (Case 6, with $v =$ 'x' and using 'b' instead of 'a' since 'a' occurs in 9.2.) Let's try $o =$ Alma:
$\mathscr{I}^b_{\text{Alma}}(\exists y\,(bLy \,\&\, yLa) \rightarrow -bLa) = 0$ if (see case 4) we have both of these:

(i) $\mathscr{I}^b_{\text{Alma}}(\exists y\,(bLy \,\&\, yLa)) = 1$. By case 7, (i) holds iff

$$\mathscr{I}^{b\ c}_{\text{Alma}\ o}(bLc \,\&\, cLa) = 1$$

for even one o, e.g., $o =$ Alma: $\mathscr{I}^{b\ \ \ \ c}_{\text{Alma}\ \text{Alma}}(bLc \,\&\, cLa) = 1$ iff (by cases 2 and 9) $\phi(\text{Alma, Alma}) = \phi(\text{Alma, Alma}) = 1$, which is true, by the table for ϕ.

(ii) $\mathscr{I}^b_{\text{Alma}}(-bLa) = 0$. By case 1, (ii) holds iff $\mathscr{I}^b_{\text{Alma}}(bLa) = 1$, and by case 9 that holds iff ϕ (Alma, Alma) $= 1$, which is true by the table for ϕ.

Then conditions (i) and (ii) both hold, and we have shown that $\mathscr{I}(9.2) = 0$.

Example 9.3 again

With $\mathscr{I}$ as in Example 9.3 we can show that sentence (9.5) is false in $\mathscr{I}$: Since $\mathscr{I}(\mathbf{0}) = 0, \mathscr{I}(\mathbf{0}') = 0 + 1 = 1, \mathscr{I}(\mathbf{0}'') = 1 + 1 = 2,$

$$\mathscr{I}(\mathbf{0}''') = 2 + 1 = 3, \quad \mathscr{I}(\mathbf{0}'''') = 3 + 1 = 4, \quad \text{and}$$

$$\mathscr{I}(\mathbf{0}''''') = 4 + 1 = 5.$$

But then

$$\mathscr{I}(\mathbf{o}''+\mathbf{o}''') = \mathscr{I}(\mathbf{o}'')\cdot\mathscr{I}(\mathbf{o}''') = 2.3 = 6,$$

and so $\mathscr{I}(\mathbf{o}''+\mathbf{o}''') \neq \mathscr{I}(\mathbf{o}'''''')$, from which it follows by case 8, that $\mathscr{I}(\mathbf{o}''+\mathbf{o}''' = \mathbf{o}'''''') = 0$.

In terms of the basic concept $\mathscr{I}(S)$ of the truth value (1 or 0) of a sentence S in one of its interpretations $\mathscr{I}$ we can define:

Satisfaction. $\mathscr{I}$ *satisfies* S (or, equivalently, S *is true in* $\mathscr{I}$) iff $\mathscr{I}(S) = 1$.

Satisfiability. S is *satisfiable* iff $\mathscr{I}(S) = 1$ for some $\mathscr{I}$.

Validity. S is *valid* iff $\mathscr{I}(S) = 1$ for every interpretation $\mathscr{I}$ of S.

Example 9.4. Sentence (9.1), '$\forall x(xLx \to \exists yxLy)$', is valid

To prove this, we deduce a contradiction from the assumption that

(i) $\mathscr{I}(\forall x(xLx \to \exists y\, xLy)) = 0$ and thus prove that there can be no such interpretation $\mathscr{I}$ of S. By case 6, (i) holds iff

(ii) $\mathscr{I}^{a}_{o}(aLa \to \exists y\, aLy) = 0$ for some o in the domain D of $\mathscr{I}$, and by case 4, (ii) holds iff we have both (iii) and (iv):

(iii) $\mathscr{I}^{a}_{o}(aLa) = 1$, i.e. by case 9, $\phi(o,o) = 1$.

(iv) $\mathscr{I}^{a}_{o}(\exists y\, aLy) = 0$, i.e. by case 7, $\mathscr{I}^{a\,b}_{o\,p}(aLb) = 0$ for each p in D, i.e. by case 9, $\phi(o,p) = 0$ for each p in D. Then in particular, with $p = o$, we must have $\phi(o,o) = 0$.

Then (iii) and (iv) contradict each other, and since (i) holds if and only if both (iii) and (iv) do, (i) is refuted, and (9.1) is seen to be valid.

We write '$\vdash S$' to indicate that S is valid, and write '$S_1 \vdash S_2$' to indicate that S_1 implies S_2, i.e., the inference from S_1 as premise to S_2 as conclusion is valid.

Implication: $S_1 \vdash S_2$ iff for every $\mathscr{I}$ which is an interpretation of both S_1 and S_2, $\mathscr{I}(S_2) = 1$ if $\mathscr{I}(S_1) = 1$.
It is easy to see that we have

$$S_1 \vdash S_2 \text{ iff } \vdash (S_1 \to S_2). \tag{9.7}$$

Proof. By the definition of validity and case 4, the second of these conditions holds iff $\mathscr{I}(S_2) = 1$ if $\mathscr{I}(S_1) = 1$ for every interpretation $\mathscr{I}$ of S_1 and S_2, and by definition of implication, this is precisely when the first condition holds.

Example 9.5. $a = b \vdash f(a) = f(b)$

To prove this we derive a contradiction from the assumption that $\mathscr{I}$ is an interpretation of both sentences for which we have both (i) and (ii):

(i) $\mathscr{I}(a=b) = 1$, i.e. by case 8, $\mathscr{I}(a) = \mathscr{I}(b) = o$, say.

(ii) $\mathscr{I}(f(a) = f(b)) = 0$, i.e. $\mathscr{I}(f(a)) \neq \mathscr{I}(f(b))$, i.e. by the definition of $\mathscr{I}(t)$, $\mathfrak{f}(\mathscr{I}(a)) \neq \mathfrak{f}(\mathscr{I}(b))$.

But by (i), $\mathscr{I}(a) = \mathscr{I}(b) = o$, and therefore by (ii), $\mathfrak{f}(o) \neq \mathfrak{f}(o)$, which is impossible since $\mathfrak{f}$, being a function, is single-valued.

We can generalize the definition of *implication* as follows:

$S_1, \ldots, S_n \vdash S_{n+1}$ iff for every $\mathscr{I}$ which is an interpretation of all $n+1$ of the Ss we have:

$$\text{If } \mathscr{I}(S_1) = \ldots = \mathscr{I}(S_n) = 1 \text{ then } \mathscr{I}(S_{n+1}) = 1. \tag{9.8}$$

The relationship (9.7) between implication and validity can then be generalized:

$$S_1, \ldots, S_n \vdash S_{n+1} \text{ iff } \vdash [(S_1 \,\&\, \ldots \,\&\, S_n) \rightarrow S_{n+1}]. \tag{9.9}$$

Proof. By (9.8), the first condition fails iff for some $\mathscr{I}$ we have $\mathscr{I}(S_1) = \ldots = \mathscr{I}(S_n) = 1$ and $\mathscr{I}(S_{n+1}) = 0$; and by the definition of validity and cases 2 and 4, this is precisely when the second condition fails.

A *model* of a sentence is an interpretation of that sentence which satisfies it. Then the following are different ways of saying the same thing:

$\mathscr{I}(S) = 1$, S is true in $\mathscr{I}$, $\mathscr{I}$ satisfies S, $\mathscr{I}$ is a model of S.

We'll also speak of interpretations as satisfying *sets* of sentences and as being models of sets of sentences. To make such talk quite general, we'll want to allow the case in which the set is empty. We'll use capital Greek gamma (Γ) and delta (Δ) to stand for sets of sentences, and we use the usual symbol ($\varnothing$) for the empty set. The union $\Gamma \cup \Delta$ of sets Γ and Δ is the set whose members are the members of Γ together with the members of Δ. The set whose only member is the sentence S is $\{S\}$; the set whose members are the sentences $S_1, \ldots, S_n$ is $\{S_1, \ldots, S_n\}$. We define:

Satisfaction. $\mathscr{I}$ satisfies Γ ($=$ *$\mathscr{I}$ is a model of* Γ) iff $\mathscr{I}$ is a model of every sentence in Γ.

Implication. $\Gamma \vdash S$ iff $\mathscr{I}(S) = 1$ whenever $\mathscr{I}$ is a model of Γ and an interpretation of S. (We may read '$\Gamma \vdash S$' as 'S follows from Γ', 'S is a (logical) consequence of Γ', or 'Γ implies S'.)

Satisfiability. Γ is satisfiable iff it has a model.

These definitions are contrived so as to make the empty set satisfiable: $\varnothing$ is satisfiable iff it has a model, and according to the definition of *$\mathscr{I}$ is a model of* Γ (definition of *satisfaction*, above) $\varnothing$ has a model iff for for some $\mathscr{I}$, $\mathscr{I}$ satisfies $\varnothing$, which in turn means that *for all S, if S is in* $\varnothing$ *then* $\mathscr{I}(S) = 1$. This quantified conditional is true because the antecedent is false for every S. Therefore,

$$\varnothing \text{ is satisfiable.} \tag{9.10}$$

Indeed, $\varnothing$ is satisfied by every interpretation. Things go most smoothly if we contrive the definition of satisfaction so as to make (9.10) true; and we have done so.

Note that

$$\vdash S \text{ iff } \varnothing \vdash S. \tag{9.11}$$

Proof. The right-hand side holds iff $\mathscr{I}$ is a model of S whenever it is a model of $\varnothing$ and an interpretation of S. By (9.10) this comes to the same thing as: *every interpretation of S is a model of S.* And by definition of validity, that comes to the same thing as $\vdash S$.

Further, note the following important connection between the concepts of implication and satisfiability:

$$\Gamma \vdash S \text{ iff } \Gamma \cup \{-S\} \text{ is unsatisfiable.} \tag{9.12}$$

Proof. $\Gamma \vdash S$ iff every interpretation of S which is a model of Γ is a model of S. By quantificational logic, this statement is equivalent to: for no $\mathscr{I}$ is $\mathscr{I}$ an interpretation of S *and* a model of Γ *and* not a model of S. But an interpetation of S which is not a model of S assigns value 0 to S and thus (case 1) assigns value 1 to $-S$; such an interpretation of S is a model of $-S$. Then the statement is equivalent to this: no $\mathscr{I}$ is a model of Γ *and* of $-S$, and by definition of *satisfiability*, *that* statement is equivalent to the claim that $\Gamma \cup \{-S\}$ is unsatisfiable.

In later chapters, we shall make use of the notion of a *theory* over and over again. This is the appropriate place to define it. A *theory* is a set whose members are just the sentences in some language that follow from the set. So if T is a theory, then for some language K, all members of T are sentences of K, and any sentence of K that follows from T is also a member of T. As '$\forall x\, x = x$' is a sentence of every language, and one that follows from any set whatsoever, every theory must contain '$\forall x\, x = x$'; thus the empty set is not a theory. The members of a theory T are ordinarily called its *theorems*, and we write: $\vdash_T A$ to mean that A is a theorem of T.

Q is a theory whose acquaintance we shall make in Chapter 14: it is the set of consequences in the language of arithmetic of the set

$$\{\forall x \forall y (x' = y' \rightarrow x = y),$$
$$\forall x\, \mathbf{o} \neq x',$$
$$\forall x (x \neq \mathbf{o} \rightarrow \exists y\, x = y'),$$
$$\forall x\, x+\mathbf{o} = x,$$
$$\forall x \forall y\, x+y' = (x+y)',$$
$$\forall x\, x \cdot \mathbf{o} = \mathbf{o},$$
$$\forall x \forall y\, x \cdot y' = (x \cdot y)+x\}.$$

All of the members of this set are true in $\mathcal{N}$, and hence all the theorems of Q are true in $\mathcal{N}$. As we shall see later, the converse is definitely not the case! There are sentences of L that are true in $\mathcal{N}$, but not consequences of Q. (In fact, '$\forall x \forall y\, x+y = y+x$' is one of them; see Exercise 14.2.)

Our definition of a theory T is quite general: we do not require that any sentences in T be singled out as 'axioms', we do not require that there be a finite or even an effectively specifiable set of sentences in T from which all the others follow, we do not require that there be any effective procedure for deciding whether any given sentence is a theorem of the theory, and we do not require that a theory be satisfiable. All that is required of a set of sentences T for it to be a theory is that there be a language with respect to which T is closed under logical consequence, i.e. that any consequence of the theory that is in the language also be a member of the theory.

The last definition in this series is

Logical equivalence. $S_1 \simeq S_2$ (S_1 is logically equivalent to S_2) iff for all $\mathcal{I}$, if $\mathcal{I}$ is an interpretation of S_1 and of S_2, then $\mathcal{I}(S_1) = \mathcal{I}(S_2)$.

It should be obvious that $\simeq$ is an *equivalence relation on the set of sentences* in the technical sense that $\simeq$ is:

Reflexive on the set of sentences: for all sentences S, $S \simeq S$.
Symmetrical. For all S_1 and S_2, if $S_1 \simeq S_2$, then $S_2 \simeq S_1$.
Transitive. For all S_1, S_2, S_3, if $S_1 \simeq S_2$ and $S_2 \simeq S_3$, then $S_1 \simeq S_3$.

Logical implication and equivalence are related as follows:

$$S_1 \simeq S_2 \quad \text{iff} \quad S_1 \vdash S_2 \quad \text{and} \quad S_2 \vdash S_1. \tag{9.13}$$

Proof. The definition of implication can be reformulated as: $S \vdash T$ iff for every $\mathscr{I}$ which is an interpretation both of S and T, $\mathscr{I}(S) \leqslant \mathscr{I}(T)$. Since we have both $\mathscr{I}(S_1) \leqslant \mathscr{I}(S_2)$ and $\mathscr{I}(S_2) \leqslant \mathscr{I}(S_1)$ iff $\mathscr{I}(S_1) = \mathscr{I}(S_2)$, (9.13) then follows from the definition of $\simeq$.

It will be useful to generalize our definitions of the relations $\vdash$ and $\simeq$ so that they can be asserted to hold between formulas that aren't sentences, e.g.

$$(\exists x\, Px \rightarrow Py), \quad \text{and} \quad \forall x\, (Px \rightarrow Py),$$

where the variable y has a free occurrence in each formula. In particular, it would be good to have a sense of $\simeq$ in which it would be true that

> *when logical equivalents are substituted for each other, the results are logically equivalent*; i.e. if F_1 is a subformula of G_1, and F_2 is substituted for an occurrence of F_1 in G_1 to get a new formula G_2, then $G_2 \simeq G_1$ if $F_2 \simeq F_1$. (9.14)

Thus the logical equivalence noted above would allow us to conclude that

$$\underbrace{\exists y\, (\overbrace{\exists x\, Px \rightarrow Py}^{F_1})}_{G_1} \quad \simeq \quad \underbrace{\exists y\, \overbrace{\forall x\, (Px \rightarrow Py)}^{F_2}}_{G_2}$$

The following generalized definitions do the job:

Implication. $F_1 \vdash F_2$ iff $\mathscr{I}(F_1^*) \leqslant \mathscr{I}(F_2^*)$ for every $\mathscr{I}$ which is an interpretation of both F_1^* and F_2^*. Here, F_1^* and F_2^* are the results of substituting names $a_1, \ldots, a_n$ for all variables $v_1, \ldots, v_n$ that have free occurrences in F_1 or F_2, with distinct names being substituted for distinct variables, and with all of $a_1, \ldots, a_n$ distinct from all names that occur in F_1 or F_2. Where a variable occurs free in both F_1 and F_2, the same name is to be substituted for it in both formulas.

Logical equivalence. $F_1 \simeq F_2$ iff $\mathscr{I}(F_1^*) = \mathscr{I}(F_2^*)$ for every $\mathscr{I}$ which is an interpretation of both F_1^* and F_2^*.

Example 9.6. $F_1 \simeq F_2$, where $F_1 = (\exists x Px \rightarrow Py)$, $F_2 = \forall x(Px \rightarrow Py)$

The claim is that $\mathscr{I}(F_1^*) = \mathscr{I}(F_2^*)$ for each $\mathscr{I}$ for which both sides of the equation are defined, where we may take $F_1^* = (\exists x\, Px \rightarrow Pa)$ and $F_2^* = \forall x\, (Px \rightarrow Pa)$. To prove the claim, consider the two sides separately:

$\mathscr{I}(F_1^*) = 1$ iff $\mathscr{I}(\exists x\, Px) = 0$ or $\mathscr{I}(Pa) = 1$ (case 4)
iff either for each $o\, \mathscr{I}_o^b(Pb) = 0$ or $\mathscr{I}(Pa) = 1$ (case 7).

$\mathscr{I}(F_2^*) = 1$ iff for each o, either $\mathscr{I}_o^b(Pb) = 0$ or $\mathscr{I}_o^b(Pa) = 1$ (cases 6, 4).

As the names a and b are distinct, $\mathscr{I}_o^b(Pa) = \mathscr{I}(Pa)$ for each o, and so the two conditions are equivalent.

It should be fairly clear, now, why (9.14) holds. Thus, with F_1 and F_2 as above and $G_1 = \exists y\, F_1$ and $G_2 = \exists y\, F_2$, we have $G_1 \simeq G_2$ because in applying case 7 to $\mathscr{I}(G_1)$ and $\mathscr{I}(G_2)$, $F_{1\,v}a$ and $F_{2\,v}a$ are respectively F_1^* and F_2^*. It should also be evident that (9.13) holds with formulas F_1 and F_2 in place of sentences S_1 and S_2.

Before we conclude this chapter, we shall use (9.14) to prove that any formula can be put into *prenex normal form.* A formula is in *prenex form* iff all quantifiers (if any) occur at the extreme left, without intervening parentheses: The form is

$$Q_1 v_1 \ldots Q_n v_n F,$$

where F is a quantifier-free formula, and each Q_i is either $\forall$ or $\exists$. Thus in Example 9.6, F_2 is in prenex form but F_1 is not; and since F_2 is a prenex formula logically equivalent to F_1, F_2 is a prenex form of F_1. Then the claim is,

Corresponding to each formula F_1 there is a formula F_2 where F_2 is prenex and $F_2 \simeq F_1$. (9.15)

To prove (9.15) it is sufficient to note that the following are really logical equivalences, and that with their use one can successively move quantifiers to the left (in a sequence of logically equivalent formulas) until finally a prenex formula is obtained.

$$-\forall v\, F \simeq \exists v - F; \quad -\exists v\, F \simeq \forall v - F. \tag{9.16}$$

Here the general form is

$$-Qv\, F \simeq Q'v - F,$$

where Q' is $\exists$ if Q is $\forall$, and Q' is $\forall$ if Q is $\exists$.

Provided that v does not occur free in G,

$$\left.\begin{array}{ll} (a) & (Qv\, F \,\&\, G) \simeq Qv(F \,\&\, G), \\ (b) & (G \,\&\, Qv\, F) \simeq Qv(G \,\&\, F), \\ (c) & (Qv\, F \vee G) \simeq Qv(F \vee G), \\ (d) & (G \vee Qv\, F) \simeq Qv(G \vee F), \\ (e) & (G \to Qv\, F) \simeq Qv(G \to F), \\ (f) & (Qv\, F \to G) \simeq Q'v(F \to G). \end{array}\right\} \tag{9.17}$$

$$Qv\, F \simeq Qw\, F_v w, \tag{9.18}$$

where w does not occur in F and $F_v w$ is the result of replacing all free occurrences of the variable v in F by occurrences of the variable w.

Example 9.7. A prenex normal form of $(Qv\, F \leftrightarrow G)$, where v does not occur free in G, and F and G contain no quantifiers

We first put the biconditional into a logically equivalent form,

$$(Qv\, F \rightarrow G) \,\&\, (G \rightarrow Qv\, F).$$

By (9.17*e*) and (9.17*f*) this becomes $Q'v\,(F \rightarrow G)$ & $Qv\,(G \rightarrow F)$, and by (9.17*a*) it becomes $Q'v((F \rightarrow G) \,\&\, Qv\,(G \rightarrow F))$. Now v will presumably occur free in the first conjunct $(F \rightarrow G)$, so we must use (9.18) to get $Q'v\,((F \rightarrow G) \,\&\, Qw\,(G \rightarrow F_v w))$, where w is new to the entire formula. We can then apply (9.17*b*) to get the prenex form

$$Q'v\, Qw\,((F \rightarrow G) \,\&\, (G \rightarrow F_v w)).$$

If we had brought the quantifiers out in the other order, a different (but logically equivalent) prenex form would have been obtained.

In general, biconditionals must be replaced in formulas containing them by appropriate equivalents (such as conjunctions of suitable conditionals) before the prenexing operations can be applied.

The following further equivalences are sometimes useful in obtaining prenex forms with as few quantifiers as possible.

$$Qv\, F \simeq F, \text{ if } v \text{ does not occur free in } F. \tag{9.19}$$

$$(\forall v\, F \,\&\, \forall v\, G) \simeq \forall v\,(F \,\&\, G). \tag{9.20}$$

$$(\exists v\, F \vee \exists v\, G) \simeq \exists v\,(F \vee G). \tag{9.21}$$

$$(\forall v\, F \rightarrow \exists v\, G) \simeq \exists v\,(F \rightarrow G). \tag{9.22}$$

Exercises

9.1 Equivalences (9.20)–(9.22) fail when the quantifiers are all changed: in each case, implication then holds in one direction but not in the other. Prove that

$$(\exists v\, F \,\&\, \exists v\, G) \nvdash \exists v\,(F \,\&\, G), \quad \text{but} \quad \exists v\,(F \,\&\, G) \vdash (\exists v\, F \,\&\, \exists v\, G). \tag{9.23}$$

$$(\forall v\, F \vee \forall v\, G) \vdash \forall v\,(F \vee G), \quad \text{but} \quad \forall v\,(F \vee G) \nvdash (\forall v\, F \vee \forall v\, G). \tag{9.24}$$

$$(\exists v\, F \rightarrow \forall v\, G) \vdash \forall v\,(F \rightarrow G), \quad \text{but} \quad \forall v\,(F \rightarrow G) \nvdash (\exists v\, F \rightarrow \forall v\, G). \tag{9.25}$$

9.2 Test for validity and put into prenex form

(*a*) $\exists x(Px \rightarrow \forall x\, Px)$,

(*b*) $\exists x(\exists x\, Px \rightarrow Px)$.

9.3 Put into prenex form

$$\{\forall x[Sx \rightarrow \exists y(Py \& yOx)] \leftrightarrow \exists x[Px \& \forall y(Sy \rightarrow xOy)]\}.$$

Partial solutions

9.1 Counterexamples to the invalid ones, i.e. interpretations that make premise true and conclusion false: In each case the domain is $\{o_1, o_2\}$; characteristic functions of F and G are ϕ_1 and ϕ_2, respectively. For (9.23) and (9.24), let $\phi_1(o_1) = \phi_2(o_2) = 0$ and $\phi_1(o_2) = \phi_2(o_1) = 1$. For (9.25) let $\phi_1(o_1) = \phi_2(o_1) = 0$ and $\phi_2(o_2) = \phi_1(o_2) = 1$.

9.2 (*a*) $\mathscr{I}(\exists x(Px \rightarrow \forall x\, Px)) = 0$ *iff* $\mathscr{I}^a_o(Pa \rightarrow \forall x\, Px) = 0$ for each o, *iff* for each o, $\phi(o) = 1$ and $\mathscr{I}^{a\ b}_{o\ p}(Pb) = 0$ for some p, *iff* for each o, $\phi(o) = 1$ and $\phi(p) = 0$ for some p. With $o = p$ this is a contradiction. Then (*a*) is valid. Prenex form: $\exists x \forall y(Px \rightarrow Py)$. (*b*) Valid. Prenex form:

$$\exists x \forall y(Py \rightarrow Px).$$

9.3 One answer is

$$\begin{aligned}&\exists x_1 \forall y_1 \exists x_2 \forall y_2 \forall x_3 \exists y_3 \forall x_4 \exists y_4(\{[Sx_1 \rightarrow (Py_1 \& y_1 Ox_1)]\\ &\quad \rightarrow [Px_2 \& (Sy_2 \rightarrow x_2 Oy_2)]\} \& \{[Px_3 \& (Sy_3 \rightarrow x_3 Oy_3)]\\ &\quad \rightarrow [Sx_4 \rightarrow (Py_4 \& y_4 Ox_4)]\}).\end{aligned}$$

10
First-order logic is undecidable

In Chapters 11 and 12 we shall demonstrate the existence of a mechanical positive test for first-order unsatisfiability or, what comes to the same thing, a mechanical positive test for first-order validity. We shall now demonstrate the non-existence of any corresponding negative tests. We shall thereby have proved the unsolvability of the decision problem for first-order satisfiability and validity: we shall have proved first-order logic to be *undecidable*.

In general, the *decision problem* for a property is solvable if there is a mechanical test (= a computational routine, an effective procedure) which, applied to *any* object of the appropriate sort, *eventually* classifies that object *correctly* as a positive or a negative instance of that property. ('Eventually' here means 'after some finite number of steps'.) A positive test for a property is a mechanical test which eventually classifies as positive all and only its positive instances. A negative test is one which eventually classifies as negative all and only the negative instances. If both a positive and a negative test for a property exist, then, and only then, is the decision problem for that property solvable; for since any appropriate object will be either a positive or a negative instance, if one is equipped with both sorts of test, one can apply both to the object – dividing one's time so as to alternate steps of the two tests – and thus eventually discover which sort of instance the object is. (Conversely, any test which eventually classifies correctly any object as either a positive or a negative instance counts as both a positive and a negative test.) Here, the properties that interest us are validity and satisfiability, and the 'objects of the appropriate sort' are sentences in the notation of first-order logic. The derivations to be introduced in Chapter 11 will give us a positive test for validity of sentences (and therewith a negative test for satisfiability). In this chapter we prove that those tests cannot be supplemented – by a positive test for satisfiability or a negative test for validity – so as to solve the decision problem for satisfiability or validity of sentences of first-order logic.

Our proof that there is no solution to the decision problem for first-order validity will take the form of a *reductio ad absurdum*: we shall prove that if there were such a test, then the halting problem would be solvable, i.e., there would be a computational routine for discovering whether or not Turing machines eventually halt, when started in state q_1 scanning the leftmost of a string of 1's on an otherwise blank tape. But at the very end of Chapter 5 we saw that the halting problem is unsolvable, if Church's thesis is correct. In particular, we proved that no Turing machine can carry out a computational routine which solves the halting problem, and we noted that by Church's thesis this means that there is no computational routine of any sort which solves the halting problem.

The *reductio* will go in this way: we show how, given the machine table or flow graph or other suitable description of a Turing machine, and any n, we can effectively write down a *finite* set Δ of sentences and a sentence H such that $\Delta \vdash H$ if and only if the machine in question does eventually halt when given input n, i.e. when started in state q_1 scanning the leftmost of an unbroken string of n 1s on an otherwise blank tape. For each machine and input, we also specify an interpretation $\mathscr{I}$. Under $\mathscr{I}$, the sentence H will say that the machine eventually halts, and the sentences in Δ will describe the operation of the machine, will say that its input is n, and will also say something about the successor function $'$ (where for any integer i, $i' = i+1$). Thus, if we could solve the decision problem for validity of sentences we could effectively determine whether or not the machine eventually halts, for we have $\Delta \vdash H$ if and only if a certain sentence is valid, *viz*., the conditional whose antecedent is the conjunction of all the sentences in Δ and whose consequent is H.

The thing is really quite simple.† We shall have no need to suppose that only the symbols S_0 and S_1 are used, nor need we suppose that the tape is infinite only to the right. We imagine that the squares of the tape are numbered:

...	−3	−2	−1	0	1	2	3	...

We imagine that time is broken up into a series of moments t at which machines perform exactly one of their operations, and that there is a moment 0 at which our machine starts, scanning square 0. The moments of time are supposed to extend endlessly into the future and the past, just as the tape is supposed to extend endlessly to the right and the left.

† Simple, anyway, now that J. R. Büchi has shown us how to do it simply: see his 'Turing machines and the *Entscheidungsproblem*', *Math. Annalen* **148** (1962), 201–13. That there is no decision procedure for first-order validity was first shown by Alonzo Church.

We assume that the machine is 'plugged in' at moment 0 and 'unplugged' at the first moment (if any) after the one at which it halts; we assume that at all negative times and at all times later than the first one (if any) at which it halts, the machine is in none of its states, scanning none of its squares, and that no symbol (not even the blank) occurs anywhere on its tape.

For each state q_i which the machine can be in, we pick a two-place predicate letter Q_i, and for each symbol S_j which the machine can read or print, we pick a two-place predicate letter, which we shall also designate by 'S_j'. In addition to the Q_is and the S_js (and '=' and the usual logical apparatus) the only symbols that occur in the sentences in $\Delta \cup \{H\}$ are the name $\mathbf{o}$, the one-place function symbol $'$, and the two-place predicate letter $<$.

In the intended interpretation $\mathcal{I}$ of the sentences in $\Delta \cup \{H\}$, the variables range over the integers – positive, negative, and zero. $\mathcal{I}$ assigns zero to $\mathbf{o}$ and the successor function to $'$. The Q_is, the S_js and $<$ are interpreted as follows:

$\mathcal{I}$ stipulates that Q_i is to be true of t, x *iff* at time t, the machine is in state q_i, scanning square number x;

$\mathcal{I}$ stipulates that S_j is to be true of t, x *iff* at time t, the symbol S_j is in square number x; and

$\mathcal{I}$ stipulates that $<$ is to be true of x, y *iff* x is less than y.

We now say what the sentences of Δ are. (We shall use 't' as a variable when a time is intended, and 'x' and 'y' as variables when tape squares are intended, in order to remind the reader of the intended interpretation. Formally, the function of a variable is signalled by its position at the left (time) or right (tape square) of the symbols Q_i and S_j.) Suppose the machine can read or print the symbols $S_0, \ldots, S_r$. Corresponding to the three sorts of expression that can appear to the right of the colon over an arrow of a flow graph of the machine, we have three sorts of sentences which, under interpretation $\mathcal{I}$, describe features of the machine's operation.

For each label in the flow graph of form: $\textcircled{i} \xrightarrow{S_j:S_k} \textcircled{m}$ we have in Δ the sentence

$$\forall t \forall x \forall y \{[tQ_i x \,\&\, tS_j x] \rightarrow [t'Q_m x \,\&\, t'S_k x \,\&\, (y \neq x \rightarrow (tS_0 y \rightarrow t'S_0 y) \,\&\, \ldots \,\&\, (tS_r y \rightarrow t'S_r y))]\}. \quad (10.1)$$

Here is an English translation of (10.1) under the intended interpretation $\mathcal{I}$:

> If the machine is in state q_i at time t and is then scanning square number x on which symbol S_j occurs, then at time $t+1$ the machine is in state q_m scanning square number x, where the symbol S_k occurs, and in all squares other than x, the same symbols appear at time $t+1$ as appeared at time t (for all t and x).

The case is not excluded in which $i = m$; the diagram would then be:

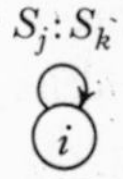

Corresponding remarks apply to the diagrams below.

For each label in the flow graph of form: $(i) \xrightarrow{S_j:R} (m)$ we have in Δ the sentence

$$\forall t\,\forall x\,\forall y\,\{[tQ_ix \,\&\, tS_jx] \rightarrow [t'Q_mx' \,\&\, (tS_0y \rightarrow t'S_0y) \,\&\, \ldots \,\&\, (tS_ry \rightarrow t'S_ry)]\}. \quad (10.2)$$

And for each label in the graph of form: $(i) \xrightarrow{S_j:L} (m)$ we have in Δ the sentence

$$\forall t\,\forall x\,\forall y\,\{[tQ_ix' \,\&\, tS_jx'] \rightarrow [t'Q_mx \,\&\, (tS_0y \rightarrow t'S_0y) \,\&\, \ldots \,\&\, (tS_ry \rightarrow t'S_ry)]\}. \quad (10.3)$$

One sentence in Δ says that initially the machine is in state q_1 scanning the leftmost of an unbroken string of n 1s on an otherwise blank tape:

$$\mathbf{o}Q_1\mathbf{o} \,\&\, \mathbf{o}S_1\mathbf{o} \,\&\, \mathbf{o}S_1\mathbf{o}' \,\&\, \ldots \,\&\, \mathbf{o}S_1\mathbf{o}^{(n-1)} \,\&\, \forall y\,[(y \neq \mathbf{o} \,\&\, y \neq \mathbf{o}' \,\&\, \ldots \,\&\, y \neq \mathbf{o}^{(n-1)}) \rightarrow \mathbf{o}S_0y]. \quad (10.4)$$

'$\mathbf{o}^{(n-1)}$' here abbreviates the result of attaching n successor symbols to the symbol **o**. Note that if there are n 1s on the tape at time 0, the leftmost of them is in square 0, so the rightmost must be in square $n-1$, not square n. If $n = 0$, (10.4) is

$$\mathbf{o}Q_1\mathbf{o} \,\&\, \forall y\,\mathbf{o}S_0y.$$

One sentence in Δ says that each integer is the successor of exactly one integer:

$$\forall z\,\exists x\, z = x' \,\&\, \forall z\,\forall x\,\forall y\,(z = x' \,\&\, z = y' \rightarrow x = y). \quad (10.5)$$

We require a guarantee that if p and q are different natural numbers, then the sentence $\forall x\, x^{(p)} \neq x^{(q)}$ is implied by Δ. All such sentences are consequences of the following:

$$\forall x\,\forall y\,\forall z\,(x<y \,\&\, y<z \rightarrow x<z) \,\&\, \forall x\,\forall y\,(x' = y \rightarrow x<y) \,\&\, \forall x\,\forall y\,(x<y \rightarrow x \neq y). \quad (10.6)$$

Example: from (10.6) we can infer $x'' < x'''$ and $x''' < x''''$, whence $x'' < x''''$, whence $x'' \neq x''''$, i.e. $x^{(2)} \neq x^{(4)}$.

We take Δ to be the set of all the sentences (10.1), (10.2) and (10.3) which correspond to the various arrows in the machine's flow graph, together with the three additional sentences (10.4), (10.5) and (10.6). It is clear that $\mathscr{I}$ is a model of Δ.

As for H, we note that a machine halts at time t if it is then in a state q_i scanning a symbol S_j and there is no entry for q_i, S_j in its machine table. (Otherwise put: there is no arrow in the machine's flow chart leaving node i before whose colon S_j occurs.) So for our sentence H, we take the disjunction of all sentences

$$\exists t \exists x (tQ_i x \,\&\, tS_j x) \tag{10.7}$$

such that there is no entry for q_i, S_j in the table of our machine. If there is always an entry for every q_i, S_j, then the machine never halts and we take H to be some sentence that is false in $\mathscr{I}$, e.g. $\mathbf{o} \neq \mathbf{o}$.

And there we have it: given a machine and an input n, we have shown how to find a finite set Δ of sentences and a sentence H such that (so we claim) we have $\Delta \vdash H$ *if and only if the machine eventually halts when given n as an input*. Of course we now have to *verify* that claim. Our proof of this fact will appeal freely to various facts about first-order logical entailment (to facts about $\vdash$) with which the reader is presumed to be familiar.

They can be verified by the method of Chapter 11, or by more direct arguments in the terms of Chapter 9, or by use of other methods of proof (e.g. natural deduction or trees).

The verification

The 'only if' part is trivial. All of the sentences in Δ are true in the intended interpretation $\mathscr{I}$. Therefore if $\Delta \vdash H$, H is true in $\mathscr{I}$. But H is true in $\mathscr{I}$ if and only if the machine eventually halts with input n.

The 'if' part is harder.

First, we need a convention about 'negative numerals': if p is a negative integer, and $p = -q$, then the formulas

$$xQ_i\mathbf{o}^{(p)},$$
$$xS_j\mathbf{o}^{(p)},$$
$$y \neq \mathbf{o}^{(p)}$$

are to be regarded as abbreviations of the formulas

$$\exists z(xQ_iz \,\&\, z^{(q)} = \mathbf{o}),$$
$$\exists z(xS_jz \,\&\, z^{(q)} = \mathbf{o}),$$
$$\exists z(y \neq z \,\&\, z^{(q)} = \mathbf{o}),$$

respectively. (Similar formulas are to be disabbreviated in a similar way.)

We now introduce a special kind of sentence, called a *description of time s*. A description of time s is a sentence which says, in the obvious way, what state the machine is in at time s, what square it is then scanning, and what symbols are on what squares of the tape, and does so by using the language of the sentences in $\Delta \cup \{H\}$. More precisely, a sentence is a description of time s if it has this form:

$$\mathbf{o}^{(s)}Q_i\mathbf{o}^{(p)} \,\&\, \mathbf{o}^{(s)}S_{j_1}\mathbf{o}^{(p_1)} \,\&\, \ldots \,\&\, \mathbf{o}^{(s)}S_j\mathbf{o}^{(p)} \,\&\, \ldots \,\&\, \mathbf{o}^{(s)}S_{j_v}\mathbf{o}^{(p_v)} \,\&$$
$$\forall y[(y \neq \mathbf{o}^{(p_1)} \,\&\, \ldots \,\&\, y \neq \mathbf{o}^{(p)} \,\&\, \ldots \,\&\, y \neq \mathbf{o}^{(p_v)}) \rightarrow \mathbf{o}^{(s)}S_0y]. \quad (10.8)$$

Here, we require that $p_1, \ldots, p, \ldots, p_v$ be an increasing sequence of integers; p *may* be p_1 or p_v. *Example*: (10.4) is a description of time 0.

Suppose now that the machine eventually halts with input n. Then for some s, i, p, and j, at time s the machine is in state q_i scanning square number p on which the symbol S_j occurs, but there is no entry for q_i, S_j in its machine table.

Suppose further that Δ implies some description G of time s. Since $\mathscr{I}$ is a model of Δ, G will be true in $\mathscr{I}$. Therefore two of the conjuncts of G will be $\mathbf{o}^{(s)}Q_i\mathbf{o}^{(p)}$ and $\mathbf{o}^{(s)}S_j\mathbf{o}^{(p)}$, and therefore G will imply

$$\exists t \exists x(tQ_ix \,\&\, tS_jx),$$

which is one of the disjuncts of H. Therefore Δ will imply H.

Therefore we need only show that for each s which is not negative, *if the machine has not halted before time s, then* Δ *implies some description of time s*. We prove this by mathematical induction on s.

Basis step. $s = 0$. Δ contains, and hence implies (10.4), which is a description of time 0.

Induction step. Suppose the italicized statement is true (for s). Suppose further that the machine has not halted before time $s+1$. Then the machine has not halted before time s and does not halt at time s. Then Δ implies some description (10.8) of time s. We must show that Δ implies some description of time $s+1$.

Since $\mathscr{I}$ is a model of Δ, (10.8) is true in $\mathscr{I}$. Therefore at time s, the machine is in state q_i, scanning some square (number p) on which the symbol S_j occurs. Since the machine does not halt at s, there must appear in its flow graph an arrow of one of the three forms

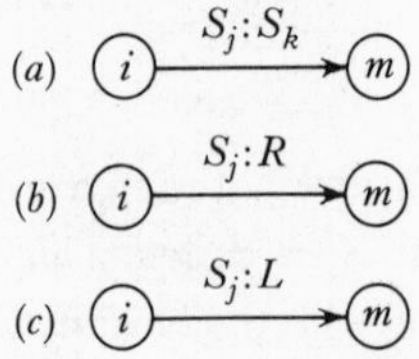

If (a) holds, then one of the sentences of Δ is

$$\forall t\,\forall x\,\forall y\,\{[tQ_i x \,\&\, tS_j x] \rightarrow [t'Q_m x \,\&\, t'S_k x \\ \&\,(y \neq x \rightarrow ((tS_0 y \rightarrow t'S_0 y)\,\&\,\ldots\,\&\,(tS_r y \rightarrow t'S_r y)))]\}.$$

This, together with (10.5), (10.6) and (10.8), implies

$$\mathbf{o}^{(s+1)}Q_m\mathbf{o}^{(p)}\,\&\,\mathbf{o}^{(s+1)}S_{j_1}\mathbf{o}^{(p_1)}\,\&\ldots\&\,\mathbf{o}^{(s+1)}S_k\mathbf{o}^{(p)}\,\&\ldots\&\,\mathbf{o}^{(s+1)}S_{j_v}\mathbf{o}^{(p_v)} \\ \&\,\forall y\,[(y \neq \mathbf{o}^{(p_1)}\,\&\ldots\&\,y \neq \mathbf{o}^{(p)}\,\&\ldots\&\,y \neq \mathbf{o}^{(p_v)}) \rightarrow \mathbf{o}^{(s+1)}S_0 y]$$

which is a description of time $s+1$.

If (b) holds, then one of the sentences of Δ is

$$\forall t\,\forall x\,\forall y\,\{[tQ_i x \,\&\, tS_j x] \rightarrow [t'Q_m x' \,\&\,(tS_0 y \rightarrow t'S_0 y)\,\&\ldots \\ \&\,(tS_r y \rightarrow t'S_r y)]\}.$$

There is some symbol S_q such that this, together with (10.5), (10.6) and (10.8), implies

$$\mathbf{o}^{(s+1)}Q_m\mathbf{o}^{(p+1)}\,\&\,\mathbf{o}^{(s+1)}S_{j_1}\mathbf{o}^{(p_1)}\,\&\ldots\&\,\mathbf{o}^{(s+1)}S_j\mathbf{o}^{(p)}\,\&\,\mathbf{o}^{(s+1)}S_q\mathbf{o}^{(p+1)}\,\&\ldots \\ \&\,\mathbf{o}^{(s+1)}S_{j_v}\mathbf{o}^{(p_v)}\,\&\,\forall y[(y \neq \mathbf{o}^{(p_1)}\,\&\ldots \\ \&\,y \neq \mathbf{o}^{(p)}\,\&\,y \neq \mathbf{o}^{(p+1)}\,\&\ldots\&\,y \neq \mathbf{o}^{(p_v)}) \rightarrow \mathbf{o}^{(s+1)}S_0 y]$$

which is a description of time $s+1$.

If (c) holds, then one of the sentences of Δ is

$$\forall t\,\forall x\,\forall y\,\{[tQ_i x' \,\&\, tS_j x'] \rightarrow [t'Q_m x \,\&\,(tS_0 y \rightarrow t'S_0 y)\,\&\ldots \\ \&\,(tS_r y \rightarrow t'S_r y)]\}.$$

There is some symbol S_q such that this, together with (10.5), (10.6) and (10.8), implies

$$\mathbf{o}^{(s+1)}Q_m\mathbf{o}^{(p-1)}\,\&\,\mathbf{o}^{(s+1)}S_{j_1}\mathbf{o}^{(p_1)}\,\&\ldots\&\,\mathbf{o}^{(s+1)}S_q\mathbf{o}^{(p-1)}\,\&\,\mathbf{o}^{(s+1)}S_j\mathbf{o}^{(p)}\,\&\ldots \\ \&\,\mathbf{o}^{(s+1)}S_{j_v}\mathbf{o}^{(p_v)}\,\&\,\forall y\,[(y \neq \mathbf{o}^{(p_1)}\,\&\ldots \\ \&\,y \neq \mathbf{o}^{(p-1)}\,\&\,y \neq \mathbf{o}^{(p)}\,\&\ldots\&\,y \neq \mathbf{o}^{(p_v)}) \rightarrow \mathbf{o}^{(s+1)}S_0 y]$$

which is a description of time $s+1$.

In all three cases Δ implies a description of time $s+1$ and therewith the undecidability of first-order logic is proved.

Exercises

10.1 Write out the sentences in Δ and the sentence H for the following machine with $n = 2$, $n = 1$, and $n = 0$, and verify that $\Delta \vdash H$ in case $n = 0$.

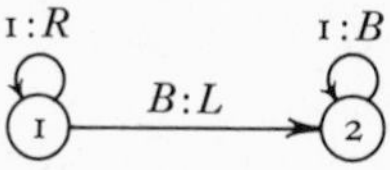

10.2 In each of cases (*a*), (*b*), and (*c*), verify that Δ implies the description of time $s+1$.

Solutions

10.1 Members of Δ corresponding to the arrows:

$$tQ_2x \,\&\, tS_1x \to t'Q_2x \,\&\, t'S_0x \,\&\, (y \neq x \to (tS_0y \to t'S_0y) \,\&\, (tS_1y \to t'S_1y)).$$

$$tQ_1x \,\&\, tS_1x \to t'Q_1x' \,\&\, (tS_0y \to t'S_0y) \,\&\, (tS_1y \to t'S_1y).$$

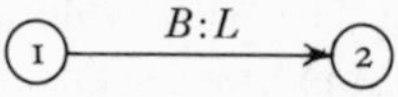

$$tQ_1x' \,\&\, tS_0x' \to t'Q_2x \,\&\, (tS_0y \to t'S_0y) \,\&\, (tS_1y \to t'S_1y).$$

(Here, universal quantifiers $\forall t, \forall x, \forall y$, braces { } and brackets [ ] have been suppressed, following the convention that free variables are to be read as universally quantified and that '&' has more binding force than '$\to$'.) The member of Δ which describes time 0 will be one thing or another, depending on the value of n.

If $n = 0$: $\mathbf{0}Q_1\mathbf{0} \,\&\, \forall y\, \mathbf{0}S_0y$.

If $n = 1$: $\mathbf{0}Q_1\mathbf{0} \,\&\, \mathbf{0}S_1\mathbf{0} \,\&\, \forall y(y \neq \mathbf{0} \to \mathbf{0}S_0y)$.

If $n = 2$: $\mathbf{0}Q_1\mathbf{0} \,\&\, \mathbf{0}S_1\mathbf{0} \,\&\, \mathbf{0}S_1\mathbf{0}' \,\&\, \forall y(y \neq \mathbf{0} \,\&\, y \neq \mathbf{0}' \to \mathbf{0}S_0y)$.

Finally, we have (10.5) and (10.6), which are always the same no matter what the particular machine may be, and no matter what the value of n may be.

Since the only 'missing' arrow is of form 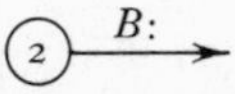

the sentence H will be $\exists t \exists x (tQ_2x \,\&\, tS_0x)$. Now in case $n = 0$, the machine halts at time 1 (i.e. time 0′), for it will then be in state q_2, scanning a blank.

Proof that $\Delta \vdash \boldsymbol{H}$. The operative member of Δ is the sentence corresponding to the arrow $(1) \xrightarrow{B:L} (2)$. From that sentence we derive

$$[\mathbf{o} = x' \,\&\, \mathbf{o}Q_1\mathbf{o} \,\&\, \mathbf{o}S_0\mathbf{o}] \rightarrow [\mathbf{o}'Q_2x \,\&\, (\mathbf{o}S_0x \rightarrow \mathbf{o}'S_0x)].$$

From the description of time 0 (with $n = 0$) we derive

$$\mathbf{o}Q_1\mathbf{o} \,\&\, \mathbf{o}S_0\mathbf{o} \,\&\, (\mathbf{o} = x' \rightarrow \mathbf{o}S_0x).$$

Then we have

$$\mathbf{o} = x' \rightarrow \mathbf{o}'Q_2x \,\&\, \mathbf{o}'S_0x$$

(where the whole is understood to be governed by '$\forall x$'). By (10.5) we have

$$\exists x\, \mathbf{o} = x'.$$

These last two sentences imply

$$\exists x(\mathbf{o}'Q_2x \,\&\, \mathbf{o}'S_0x)$$

from which H follows by existential generalization.

A further exercise

10.3 A sentence S is called *finitely* satisfiable if $\mathscr{I}(S) = 1$ for some interpretation $\mathscr{I}$ whose domain is finite. Show that for each machine M and input n a sentence S can effectively be found which is finitely satisfiable iff M halts with input n. Hint: modify Δ to obtain a suitable finite set of sentences Δ' such that the conjunction of sentences in Δ' and H is finitely satisfiable iff M halts with input n. You will probably want to discard $'$ and replace it with a new two-place predicate letter ('P', say for 'predecessor of'). (10.3) might then be revised to read

$$\forall t\,\forall x\,\forall y\{[tQ_ix \,\&\, tS_jx] \rightarrow \exists z\,\exists w\,[tPz \,\&\, wPx \,\&\, zQ_mw \,\&\, (tS_0y \rightarrow zS_0y) \,\&\, \ldots \,\&\, (tS_ry \rightarrow zS_ry)]\}.$$

(10.1) and (10.2) would be similarly revised. (10.4) would contain lots of existential quantifiers. (10.5) might read

$$\forall x\,\forall y\,\forall z\,(xPy \,\&\, xPz \rightarrow y = z) \,\&\, \forall x\,\forall y\,\forall z\,(yPx \,\&\, zPx \rightarrow y = z).$$

The middle conjunct of (10.6) might be replaced by

$$\forall x\,\forall y\,(xPy \rightarrow x < y).$$

Complete the proof.

11
First-order logic formalized: derivations and soundness

In Chapter 9 we saw that every sentence is equivalent to one in prenex normal form, and, indeed, saw how to find a prenex equivalent of any given sentence in an effective manner. We now use this fact to *formalize* first-order logic: to provide a sound, complete mechanical procedure for demonstrating the validity of valid sentences (= a mechanical positive test for validity). *Soundness* means that if the procedure classifies a sentence as valid, then the sentence really is valid. *Completeness* is defined conversely: if a sentence is valid, the procedure will so classify it. In this chapter we shall begin describing the procedure and prove a theorem (the soundness theorem) of which a consequence will be that the procedure is sound. In the next chapter we shall finish the description and prove completeness. Some important consequences of completeness are drawn in the next chapter.

A test for validity of sentences is readily modified so as to yield a test for the unsatisfiability of finite sets of sentences, for as is clear from the definitions of Chapter 9, a finite set Δ is unsatisfiable if and only if the negation of the conjunction of its members is valid. And conversely, a test for unsatisfiability is readily modified so as to yield a test for validity, for S is valid if and only if $-S$ is unsatisfiable, if and only if $\{-S\}$ is unsatisfiable. Here we shall treat the notion of unsatisfiability as basic; we will provide a positive test for unsatisfiability of finite sets of sentences which is sound (the set is unsatisfiable if the test so classifies it) and complete (the test does so classify it if it is unsatisfiable).

Note that we are now in a position to conclude that there can't be a sound, complete positive test for satisfiability (= a sound, complete negative test for unsatisfiability). For by the main result of the last chapter, a given machine fails to halt on a given input if and only if a certain sentence (the conjunction of the negation of H and the members of Δ) is satisfiable. So a positive test for satisfiability would yield a negative test for halting. But of course a mechanical *positive* test for halting does exist, *viz.*, *imitate the operations of the machine when supplied with the input*! So if there were a mechanical positive test for satisfiability, there would be both a positive and a negative test for halting, from which it would follow that the halting problem was solvable.

The positive test for unsatisfiability of Δ will be a systematic search for a finite *refutation* of Δ, *viz.* a certain sort of finite list of sentences. Soundness of the test will come to this: if there is a finite refutation of Δ, then Δ is unsatisfiable. And completeness will come to this: if Δ is unsatisfiable, then there is a finite refutation of Δ. After we define 'refutation of Δ' (soon!) it will become apparent that if Δ is finite, or even if Δ only has the weaker property that membership in it is effectively decidable, then the property of being a finite refutation of Δ is also effectively decidable. Then a straightforward (but inefficient) way of systematically searching for a refutation of (effectively decidable) Δ would be to choose (*ad lib.*) some effective enumeration of the finite lists of sentences, and test those lists in turn to see whether they are refutations of Δ. The proof of soundness will assure us that if we do find a refutation of Δ in this way, we can be sure that Δ is unsatisfiable. And the proof of completeness will assure us that if Δ is unsatisfiable, then for some finite n, we shall discover that the nth list in our enumeration will prove to be a refutation of Δ. But if the set Δ is satisfiable, there may be no end to our search for a refutation. Zeus, testing lists of sentences faster and faster for the property of being a refutation of Δ, could get through the whole enumeration in a finite period of time, and thus establish the satisfiability of Δ by an exhaustive search which fails; but that technique is not open to us or our machines, and in Chapter 10 we saw that no other technique can work for us either – if Church's thesis is true. Of course, even where a refutation exists, it may occur so late in the chosen enumeration of all finite lists of sentences as to be inaccessible in practice: the completeness of the test guarantees that where Δ is unsatisfiable, list number n will be a refutation for some finite n, but it guarantees absolutely nothing about the size of n. We shall have more to say about that later.

Meanwhile, let us begin describing the test, and prove its soundness. In what follows, we don't assume that Δ is finite, or even that membership in Δ is effectively decidable. To clarify the description it will be well to have an example in mind, e.g., the inference

$$\exists x \forall y\, xLy \vdash \forall y \exists x\, xLy$$

which is valid if and only if the set consisting of the sentences

$$\exists x \forall y\, xLy, \quad -\forall y \exists x\, xLy$$

is unsatisfiable. Putting the second of these into prenex normal form, we have, as the set Δ which is to be tested for unsatisfiability, the set consisting of the sentences in lines 1 and 2 of the following annotated list.

1	$\exists x \forall y\, xLy$	Δ
2	$\exists y \forall x - xLy$	Δ
3	$\forall y\, aLy$	1
4	$\forall x - xLb$	2
5	$-aLb$	4
6	aLb	3

This list is a refutation of Δ. A refutation of Δ is a special sort of *derivation from* Δ, *viz.*, one in which

some finite set of quantifier-free lines is unsatisfiable.

(In our example, lines 5 and 6 make up an unsatisfiable set.) A derivation from Δ is defined as

> a finite or infinite list of sentences which can be annotated by writing either 'Δ' or the number of an earlier line at the right of each line, in such a way that where the annotation is 'Δ', the annotated sentence is a member of Δ, while if the annotation is *m*, the annotated sentence is obtainable from the *m*th sentence by one of the rules UI and EI which are defined below.

UI ('Universal instantiation'). The earlier line (the *premise*) is of form

$$m \quad \forall v\, F$$

with some annotation, and the later line (the *conclusion*) is of form

$$n \quad F_v t \quad m,$$

where t (the '*instantial term*') may be any term, e.g. a or $f(a, g(b, f(b)))$ or b.

EI ('Existential instantiation'). The two lines are as in the statement of UI except that we have '$\exists$' in place of '$\forall$' in line *m*, and the instantial term *t* in line *n* must be *a name which appears in no sentence of* Δ *and in no line earlier than n in the derivation.*

In our example, line 3 is annotated '1' and since line 1 begins with an existential quantifier, the operative rule must be EI, with $v = x$, $F = \forall y\, xLy$, and $t = a$. The instantial term appears nowhere in Δ and (therefore) nowhere earlier than line 3 of this derivation. Similarly, line 2 must be the premise of an application of EI of which line 4 is the conclusion, with the instantial term b (a name) appearing nowhere earlier than line 4 of this derivation (and thus, nowhere in Δ, since in this case, the sentences of Δ are the first two lines of the derivation). Finally, lines 5 and 6 are conclusions by UI of lines 4 and 3 respectively, with a and b respectively as instantial terms.

Here is an example of a list of sentences that is *not* a derivation from $\{\forall x \exists y\, xLy, \forall y \exists x - xLy\}$:

1	$\forall x \exists y\ xLy$
2	$\forall y \exists x - xLy$
3	$\exists y\ aLy$
4	aLb
5	$\exists x - xLb$
6	$-aLb$

This list is not a derivation because the sixth sentence in it, '$-aLb$' could have been inferred only by an application of EI from the fifth sentence, '$\exists x\ -xLb$'. The instantial term in this application of EI would then have been 'a'. But 'a' occurs in the earlier sentence 'aLb'. Lines 1 through 5 form a perfectly admissible derivation, however.

UI is a perfectly ordinary rule of inference, for the conclusion $F_v t$ always follows from the premise $\forall v\, F$, no matter what the instantial term t may be. Thus, if $\mathscr{I}$ is an interpretation of both the premise and the conclusion, the conclusion will be true in $\mathscr{I}$ if the premise is. But EI appears to have been stated backwards. Not only does the conclusion $F_v t$ of an application of EI not necessarily follow from the premise, the premise always follows from the conclusion! But although EI is not sound, it serves admirably as an ingredient in our sound, complete test for unsatisfiability, as long as the restriction is observed, that the instantial term t be a 'new' name – a name which appears nowhere in Δ and nowhere earlier than line n in the derivation. The thought is that when we infer (say) '$\forall x\ -xLb$' from '$\exists y \forall x\ -xLy$' we are 'giving a name' to an object y whose existence is asserted in the premise. In some interpretation $\mathscr{I}$ the premise assures us that there is someone whom no one loves. In the conclusion, we identify one such unfortunate. Since 'b' plays no role in Δ and appears nowhere earlier in the derivation (nowhere earlier than line 4 in our example), no assumptions about the bearer of the name 'b' have been formulated *except* for the assumption that he or she is totally unloved. The premise assures us that *that* assumption is true of someone, so that if the premise is true in some interpretation $\mathscr{I}$, we can be sure that somewhere in the domain of $\mathscr{I}$ there will be an object *o* which '*b*' can be made to denote in the near variant $\mathscr{I}_o^b$ of $\mathscr{I}$; and the conclusion, which may be assigned neither truth value by $\mathscr{I}$, will be assigned the truth value 1 by $\mathscr{I}_o^b$, provided only that the premise is assigned the value 1 by $\mathscr{I}$.

In general terms, the *basic property* of EI is this:

> Suppose that all members of a set Γ of sentences which includes the premise of an application of EI are true in an interpretation $\mathscr{I}$, and suppose that the instantial term of that application is a name *t* which does not occur in any sentence of Γ. Then there is an object *o* in the domain of $\mathscr{I}$ such that in the interpretation $\mathscr{I}_o^t$, every member of Γ is true, as is the conclusion of the application of EI.

(Recall that in general, $\mathscr{I}_o^t$ is the interpretation which differs from $\mathscr{I}$ – if at all – only in that it assigns the designation *o* to *t*.) In proving that EI has this basic property we shall use the following seemingly trivial fact about interpretations.

> *Continuity*. A sentence has the same truth value in any two interpretations which differ only in what (if anything) they assign to names, sentence letters, predicate letters, or function symbols which do *not* occur in the sentence.

Proof that EI has the basic property. Since all of the sentences in Γ are true in $\mathscr{I}$, and *t* occurs in none of them, continuity assures us that they are all true in an interpretation $\mathscr{J}$ which differs from $\mathscr{I}$ (if at all) in assigning *no* designation to *t*. So the premise, $\exists vF$, is true in $\mathscr{J}$. Then by case 7 in Chapter 9, there will be an object *o* in the domain of $\mathscr{J}$ (= the domain of $\mathscr{I}$) such that the conclusion $F_v t$ is true in $\mathscr{J}_o^t$; and by continuity, so are all the other sentences in Γ. But $\mathscr{J}_o^t$ is just the same interpretation as $\mathscr{I}_o^t$.

We are now ready to prove the

Soundness theorem

If there is a refutation of Δ then Δ is unsatisfiable.

Here, 'Δ' is a variable for sets of sentences in prenex normal form in which no vacuous quantifiers occur. ($\exists v$ or $\forall v$ is *vacuous* if and only if the variable v has no free occurrences in what follows the quantifier, e.g.

'$\exists x$' in '$\exists x p$' or '$\forall x$' in '$\forall x \exists x Gx$'.) Since every sentence has a prenex equivalent, and vacuous quantifiers can be dropped to get sentences equivalent to the original ones, these assumptions reflect no real restriction.

The soundness theorem follows from the

Strong soundness theorem

If $\mathscr{I}$ is a model of Δ and $\mathscr{D}$ is a derivation from Δ then the set of all sentences occurring in $\mathscr{D}$ has a model $\mathscr{L}$. Moreover, $\mathscr{L}$ can be chosen so that it differs from $\mathscr{I}$ (if at all) only in what designations or functions it assigns to those names or function symbols which occur in sentences of $\mathscr{D}$ that do not belong to Δ.

It is immediate from the first sentence of the strong soundness theorem that if Δ has a refutation, then Δ is unsatisfiable. (If Δ were satisfiable, it would have a model $\mathscr{I}$, and therefore the set of all sentences occurring in the refutation would have a model $\mathscr{L}$, which is impossible as some finite subset of them is unsatisfiable.) The strong soundness theorem thus implies the soundness theorem.

Proof of the strong soundness theorem. Suppose that $\mathscr{I}$ is a model of Δ. We define $\Delta_0 = \Delta$ and in general, if $\mathscr{D}$ has an *n*th line, we define

$$\Delta_n = \Delta \cup \{S_1, ..., S_n\}$$

where the *S*s are the sentences in the first *n* lines of $\mathscr{D}$. (Note that the *S*s need not be distinct.) We consider $\mathscr{D}$ together with an arbitrary annotation which meets the requirements laid down in the definition of 'derivation from Δ':

$$\begin{array}{ccc} 1 & S_1 & A_1 \\ \vdots & \vdots & \vdots \\ n & S_n & A_n \\ \vdots & \vdots & \vdots \end{array}$$

Thus, A_1 must be 'Δ'; A_2 may be 'Δ' or '1', depending on what Δ, S_1, and S_2 are; and so on. (We speak of *an* annotation because there are derivations which can be annotated in more than one way.) Now for each *n* for which there is an *n*th line in $\mathscr{D}$ we define an interpretation $\mathscr{I}_n$ which is a model of Δ_n. Our definition will be recursive. To begin we set

$$\mathscr{I}_0 = \mathscr{I}.$$

To say that $\mathcal{I}_0$ is a model of Δ_0 is to repeat, in different words, our assumption that $\mathcal{I}$ is a model of Δ. Suppose now that we have already defined a model $\mathcal{I}_k$ of Δ_k, and that there is a $(k+1)$st line in $\mathcal{D}$. We define a model $\mathcal{I}_{k+1}$ of Δ_{k+1} in one way or another, depending on the annotation A_{k+1} and perhaps also on the particulars of the instantial term.

Case 1. A_{k+1} = 'Δ'. Here we define $\mathcal{I}_{k+1} = \mathcal{I}_k$. Since $\Delta_{k+1} = \Delta_k$ in this case, we can be sure that $\mathcal{I}_{k+1}$ is a model of Δ_{k+1} for (to say the same thing in different words) we know that $\mathcal{I}_k$ is a model of Δ_k.

Case 2. A_{k+1} refers to an earlier line which begins with a universal quantifier (so that the operative rule must have been UI) and the instantial term in S_{k+1} contains only names or function symbols which occur in sentences of Δ_k. Then the model $\mathcal{I}_k$ of Δ_k is an interpretation of S_{k+1} and thus of Δ_{k+1}; and since the conclusion of an application of UI is implied by its premise (= is true in each of its interpretations in which its premise is true) then if we set $\mathcal{I}_{k+1} = \mathcal{I}_k$ we can be sure that $\mathcal{I}_{k+1}$ is a model of Δ_{k+1}.

Case 3. Like case 2 except that the instantial term in S_{k+1} contains one or more names or function symbols which occur in no sentences of Δ_k. Here we cannot rely on $\mathcal{I}_k$ to be an interpretation of S_{k+1}, although it may be. (It *will* be if and only if it happens to assign denotations and functions to the 'new' names and function symbols in the instantial term.) Now any interpretation which differs from $\mathcal{I}_k$ (if at all) only in the denotations and functions which it assigns to the 'new' names and function symbols in S_{k+1} will (by continuity) be a model of Δ_k and will (since the premise of an application of UI implies the conclusion) be a model of S_{k+1}. For definiteness, we define $\mathcal{I}_{k+1}$ as follows, where *d* is some element of the domain of $\mathcal{I}$ which we think of as having been chosen, once and for all instances of case 3 at the beginning of the construction. To each 'new' name, $\mathcal{I}_{k+1}$ assigns *d* as denotation, and to each 'new' function symbol it assigns the constant function which assumes the value *d* for every argument in the domain of $\mathcal{I}_k$; but $\mathcal{I}_{k+1}$ differs from $\mathcal{I}_k$ only in these assignments, if at all. (It may just happen that $\mathcal{I}_k$ itself makes those very assignments.)

Case 4. A_{k+1} refers to an earlier line which begins with an existential quantifier (so that the operative rule must have been EI). In this case the instantial term *t* is a name which appears in no sentence of Δ_k. As the premise of this application of EI belongs to Δ_k, it must be true in $\mathcal{I}_k$. Then by the basic property of EI, there will be at least one object *o* in the domain of $\mathcal{I}_k$ for which the interpretation $\mathcal{I}_k{}^t_o$ is a model of S_{k+1} and (by continuity) of Δ_k as well. We therefore define $\mathscr{I}_{k+1} = \mathscr{I}^t_{ko}$ for some such o.†

In all four cases, then, $\mathscr{I}_{k+1}$ is a model of Δ_{k+1}. Note that all of the $\mathscr{I}_k$s have the same domain, *viz*., the domain of $\mathscr{I}_0$ ($= \mathscr{I}$).

Now define $\mathscr{L}$ as the interpretation which is just like $\mathscr{I}$ except that to each name or function symbol which appears in $\mathscr{D}$ but not in Δ, $\mathscr{L}$ assigns whatever $\mathscr{I}_n$ assigns it – where S_n is the earliest sentence in $\mathscr{D}$ in which the name or function symbol occurs.† Each sentence S_k occurring in $\mathscr{D}$ is then true in an interpretation $\mathscr{I}_k$ from which $\mathscr{L}$ differs only in what (if anything) it assigns to names and function symbols not in Δ_k, and so not in S_k. By continuity, therefore, each sentence in $\mathscr{D}$ is true in $\mathscr{L}$. And $\mathscr{L}$ differs from $\mathscr{I}$ (if at all) only in what it assigns to names and function symbols that appear in $\mathscr{D}$ but not in Δ.

This completes our proof of the strong soundness theorem. We have as an easy consequence that

> An inference is valid if there is a refutation of a set each of whose members is a prenex equivalent of a premise or of the denial of the conclusion of the inference.

The converse follows from the completeness theorem of Chapter 12.

A last note: To prove that it is always a mechanical matter to answer the question, 'Is $\mathscr{D}$ a refutation of Δ?' where Δ is assumed to be effectively decidable and $\mathscr{D}$ to be finite we must prove that there is a decision procedure for determining whether a set of quantifier-free sentences is satisfiable. The truth-table test alone will not do the job, since (as Exercises 11.2, 11.3, and 11.4 show) unsatisfiability of such a set may depend essentially on laws of identity. A proof that there is such a decision procedure is given at the end of the following chapter.

Exercises

Verify that each of the following inferences is valid by finding a refutation of an appropriate set of sentences.

11.1 $\forall x\, xLf(x) \vdash \forall x \exists y\, xLy$.

11.2 $\exists x \forall y (Py \leftrightarrow y = x) \vdash \exists x\, Px$.

11.3 $\exists x \forall y (Py \leftrightarrow y = x) \vdash \forall x \forall y [(Px \,\&\, Py) \rightarrow x = y]$.

11.4 $\exists x\, Px, \forall x \forall y [(Px \,\&\, Py) \rightarrow x = y] \vdash \exists x \forall y (Py \leftrightarrow y = x)$.

† Cognoscenti will recognize that we have tacitly appealed to the axiom of (dependent) choice at these points in the proof. (Cf. Exercise 11.5.)

Solutions

11.1	1	$\exists x\forall y\, x\text{Ł}y$	Δ
	2	$\forall y\, a\text{Ł}y$	1
	3	$a\text{Ł}f(a)$	2
	4	$\forall x\, xLf(x)$	Δ
	5	$aLf(a)$	4

$$\Delta = \{\forall x\, xLf(x), \exists x\forall y\, x\text{Ł}y\}.$$

The derivation above is a refutation of Δ since the set consisting of the quantifier-free sentences in lines 3 and 5 is unsatisfiable. Since the members of Δ are prenex equivalents of the premise or of the denial of the conclusion of the inference, this refutation of Δ establishes the validity of the inference.

11.2	1	$\exists x\forall y(Py \leftrightarrow y = x)$	Δ
	2	$\forall x -Px$	Δ
	3	$\forall y(Py \leftrightarrow y = a)$	1
	4	$Pa \leftrightarrow a = a$	3
	5	$-Pa$	2

$\Delta = \{S_1, S_2\}$ in the derivation above. The derivation is a refutation because the set $\{S_4, S_5\}$ is unsatisfiable (because '$a = a$' is valid).

11.3	1	$\exists x\forall y(Py \leftrightarrow y = x)$	Δ
	2	$\exists x\exists y(Px \,\&\, Py \,\&\, x \neq y)$	Δ
	3	$\forall y(Py \leftrightarrow y = a)$	1
	4	$\exists y(Pb \,\&\, Py \,\&\, b \neq y)$	2
	5	$Pb \,\&\, Pc \,\&\, b \neq c$	4
	6	$Pb \leftrightarrow b = a$	3
	7	$Pc \leftrightarrow c = a$	3

This derivation is a refutation of Δ ($= \{S_1, S_2\}$) because the last three (quantifier-free) lines form an unsatisfiable set.

11.4	1	$\exists x\, Px$	Δ
	2	$\forall x\forall y[(Px \,\&\, Py) \to x = y]$	Δ
	3	$\forall x\exists y(Py \leftrightarrow y \neq x)$	Δ
	4	Pa	1
	5	$\exists y(Py \leftrightarrow y \neq a)$	3
	6	$Pb \leftrightarrow b \neq a$	5
	7	$\forall y[(Pb \,\&\, Py) \to b = y]$	2
	8	$(Pb \,\&\, Pa) \to b = a$	7

The above is a refutation of $\{S_1, S_2, S_3\}$.

To see that $\{S_4, S_6, S_8\}$ is unsatisfiable, note that by sentential logic it implies '$-Pb$' and '$b = a$', whence (substituting equals for equals) '$-Pa$', which contradicts S_4.

Exercise 11.5

(For those worried about the use of the axiom of dependent choice in the soundness proof.)

The axiom of dependent choice asserts that if X is a non-empty set, and for any x in X there is a y in X such that x bears relation R to y, then there is a function f such that for any natural number n, $f(n)$ is in X and $f(n)$ bears R to $f(n+1)$. This assertion is not to be confused with the weaker assertion whose antecedent is the same, but whose consequent is 'for any natural number n, there is a function f such that for any natural number $i < n$, $f(i)$ is in X and $f(i)$ bears R to $f(i+1)$. Unlike the axiom of dependent choice, this second assertion can be proved (in set theory) without any extra assumptions. Since the strong soundness theorem is equivalent in set theory to the axiom of dependent choice, uses of that axiom can only be disguised, not essentially avoided, in any proof of the theorem.

Show that the soundness theorem (as opposed to the strong soundness theorem) can be proved without appeal to the axiom of dependent choice. *Hint*: use the fact that if there is a refutation of Δ, there is one which contains only finitely many sentences. A proof of the result of modifying the strong soundness theorem by inserting the words 'finitely long' before the word 'derivation' in its statement may be given in which appeal need be made only to the weaker statement mentioned above instead of the axiom of dependent choice. In the original proof we needed the axiom of dependent choice to guarantee us of the existence of $\mathscr{L}$ if $\mathscr{D}$ was infinite; if $\mathscr{D}$ is assumed to be finite, we need only the truth of the weaker statement.

12
Completeness of the formalization; compactness

We now prove the

Completeness theorem

If Δ† is unsatisfiable, there is a refutation of Δ.

An analysis of the completeness proof will reveal two important facts. First, the

Compactness theorem

If Δ is unsatisfiable, some finite subset of Δ must be unsatisfiable.

Second, the

Skolem–Löwenheim theorem

If Δ has a model, it has a model with an enumerable domain.

The ramifications of this second fact are sufficiently striking to warrant extensive treatment: see Chapter 13.

We prove the completeness theorem by showing how to generate special sorts of derivation – *canonical* derivations – which have the characteristic that

If Δ is unsatisfiable, any canonical derivation from Δ will be a refutation of Δ.

Definition

$\mathscr{D}$ is a *canonical derivation* from Δ if and only if $\mathscr{D}$ is a derivation from Δ which has these five characteristics:

(1) Every sentence in Δ occurs in $\mathscr{D}$.

(2) If a sentence $\exists v\, F$ occurs in $\mathscr{D}$, then for some term t, the sentence $F_v t$ occurs in $\mathscr{D}$.

(3) If a sentence $\forall v\, F$ occurs in $\mathscr{D}$, then for some term t, the sentence $F_v t$ occurs in $\mathscr{D}$.

(4) If a sentence $\forall v\, F$ occurs in $\mathscr{D}$, then for every term t that can be

† Here and henceforth, sets of sentences are always assumed to be enumerable.

formed from names and function symbols appearing in $\mathscr{D}$, the sentence $F_v t$ occurs in $\mathscr{D}$.

(5) All function symbols appearing in $\mathscr{D}$ appear in Δ.

To generate a canonical derivation from Δ, follow the instructions given in the proof of the following

Lemma I

For any set Δ, there is a canonical derivation $\mathscr{D}$ from Δ.

Proof. Let F_1, F_2, ... be an enumeration of all the sentences of Δ, if any. (Δ may be empty. If it is not, the enumeration is to be a gapless list – finite or infinite.) If Δ is empty, $\mathscr{D}$ will be the vacuous list. Otherwise, we form $\mathscr{D}$ in a series of stages. At each stage, a finite number of lines will be added to $\mathscr{D}$. There will be infinitely many stages, even if Δ is finite. Each stage will have three parts.

Stage 1*a*. Enter

$$1 \quad F_1 \quad \Delta$$

as the first line of $\mathscr{D}$.

Stage 1*b*. Extend $\mathscr{D}$ by adding to it as many lines as can possibly be inferred by EI under certain restrictions (stated below).

Stage 1*c*. Extend $\mathscr{D}$ further by adding to it as many lines as can possibly be inferred by UI (under certain restrictions).

Stage 2*a*. Enter

$$n \quad F_2 \quad \Delta$$

as the next line (where n is the appropriate number) if there is a second entry in the enumeration of Δ.

Stage 2*b*. Extend $\mathscr{D}$ by adding to it as many lines as can possibly be inferred by EI (under restrictions).

Stage 2*c*. Extend $\mathscr{D}$ further by adding to it as many lines as can possibly be inferred by UI (under restrictions).

And so on.

The restrictions are these:

We never enter the same sentence in two different lines of $\mathscr{D}$.
No sentence may be the premise of more than one application of EI.
Whenever, during part c of stage N, we apply UI, the instantial term is always one which *contains fewer than N occurrences of function symbols* and which can be formed from names and function symbols

already occurring in sentences of $\mathscr{D}$ (*except* in the one case where $N = 1$ and F_1 begins with a universal quantifier and contains no names. In that case we apply UI to F_1 using as an instantial term a name which does not occur in F_1).

Of course, in part *b* of stage N, new sentences may be added to $\mathscr{D}$ which can themselves be the premises of applications of EI; and any such sentences are supposed to be used *in part b of stage N* as the premises of applications of EI. Similarly for part *c* of stage N, and UI. During each part of each stage, only finitely many sentences are added to the derivation: we always eventually run out of terms to substitute or sentences to use as premises or both. (That is precisely the point of the third restriction, e.g. in part *c* of stage 3 we could use only the first three members of the series '*a*', '*f*(*a*)', '*f*(*f*(*a*))', '*f*(*f*(*f*(*a*)))', ... as instantial terms.)

Example

$\Delta = \{F_1, F_2\}$, where $F_1 =$ '$\forall x\, xLfx$' and $F_2 =$ '$\exists x \forall y\, x\bar{L}y$', as in the solution to Exercise 11.1. Here is the part of a canonical derivation from Δ which is yielded by the first two stages of our procedure.

1	$\forall x\, xLfx$	Δ	(1*a*)
2	$aLfa$	1	(1*c*)
3	$\exists x \forall y\, x\bar{L}y$	Δ	(2*a*)
4	$\forall y\, b\bar{L}y$	3	(2*b*)
5	$bLfb$	1	(2*c*)
6	$faLffa$	1	(2*c*)
7	$fbLffb$	1	(2*c*)
8	$b\bar{L}a$	4	(2*c*)
9	$b\bar{L}b$	4	(2*c*)
10	$b\bar{L}fa$	4	(2*c*)
11	$b\bar{L}fb$	4	(2*c*)

Note that at stage 2*c*, the allowable instantial terms are those with fewer than two occurrences of '*f*': '*a*', '*b*', '*fa*' and '*fb*'. All of these were used in lines 8–11, and all but '*a*' were used in lines 5–7. (Use of '*a*' there would have produced a replica of line 2, in violation of the first restriction.) At stage 3, only part *c* will be applicable: UI will be applied to lines 1 and 4, using as instantial terms in each case whichever of '*a*', '*b*' '*fa*', '*fb*', '*ffa*', and '*ffb*' do not produce violations of the first restriction. Although these first 11 lines make up a refutation of Δ (since lines 5 and 11 form an unsatisfiable set), they do not constitute a canonical derivation. Any canonical derivation from Δ in this case will be unending, for at each stage, application of UI to line 1 will generate new terms which provide fuel for the next stage.

The next concept we shall need has to do with interpretations $\mathscr{I}$ and sets Γ of *quantifier-free* sentences:

$\mathscr{I}$ *matches* Γ if and only if
  $\mathscr{I}$ is a model of Γ, and
  if any terms at all occur in Γ† then each object in the domain of $\mathscr{I}$ is the denotation of some such term.

The case in which no terms occur in Γ is simply the case in which Γ is a set of sentences built out of sentence letters '*p*' etc., so that the domain of $\mathscr{I}$ is irrelevant to the truth-values which $\mathscr{I}$ assigns to members of Γ. Of course, if $\mathscr{I}$ is any model of Γ, every term occurring in a sentence of Γ must denote something or other in the domain of $\mathscr{I}$. But where $\mathscr{I}$ matches Γ, the converse holds as well: every object in the domain of $\mathscr{I}$ is denoted by one or another term which appears in a sentence of Γ.

Where Γ is the set of all quantifier-free sentences in a canonical derivation from Δ, we can rely on any interpretation which matches Γ to be a model of Δ:

Lemma II

Suppose that $\mathscr{D}$ is a canonical derivation from Δ, that Γ is the set of all quantifier-free sentences in $\mathscr{D}$, and that $\mathscr{I}$ matches Γ. Then $\mathscr{I}$ is a model of the set of *all* sentences in $\mathscr{D}$, and hence is a model of Δ.

Proof. Since every non-logical symbol, i.e. name, function symbol, sentence letter or predicate letter that appears in $\mathscr{D}$ appears in Γ, $\mathscr{I}$ will assign a truth-value to each sentence in $\mathscr{D}$. To prove the lemma it suffices to prove that in no case is the truth-value 0 (falsity). To prove this by *reductio ad absurdum*, suppose $\mathscr{I}$ assigns the value 0 to one or more sentences in $\mathscr{D}$. Then among the lengths of all such sentences, there must be a minimum, say, *m*. (The length of a sentence is the number of symbols in it, *counting terms as single symbols*.) Let *M* be some sentence of length *m* in $\mathscr{D}$, to which $\mathscr{I}$ assigns the value 0. Since *M* cannot be quantifier-free, *M* must begin with a quantifier. *If the quantifier is existential*, then by

† A term is understood to occur in Γ even if it occurs only as a part of some other term that occurs in Γ.

clause (2) of the definition of 'canonical derivation', some instance of M, shorter (of length $m-2$) and hence true in $\mathscr{I}$, occurs in $\mathscr{D}$; but then M would be true in $\mathscr{I}$, as any instance implies it. *If the quantifier is universal*, then by clause (4) and the fact that every object in the domain of $\mathscr{I}$ is denoted by some term appearing in Γ, some instance of M of which the instantial term appears in Γ must occur in $\mathscr{D}$ and be false in $\mathscr{I}$ – which is impossible, since instances of M are shorter than M and hence are true in $\mathscr{I}$ if they occur in $\mathscr{D}$.

Let us pause for a moment to see how far we have come. We are trying to prove that if Δ is unsatisfiable, then there is a canonical derivation from Δ in which some finite number of quantifier-free sentences form an unsatisfiable set. So let us suppose that Δ is unsatisfiable. By Lemma I, there is a canonical derivation $\mathscr{D}$ from Δ. Since Δ is unsatisfiable and is a subset of the set of all sentences appearing in $\mathscr{D}$, there is no model of the set of all such sentences. By Lemma II, then, there is no model of the set Γ of quantifier-free sentences in $\mathscr{D}$ that matches Γ. If we could prove a proposition to the effect that

> if every finite subset of Γ is satisfiable then some interpretation matches Γ

then we should have proved the completeness theorem, for we should know that some finite set of the quantifier-free sentences in $\mathscr{D}$ is unsatisfiable.

In order to prove this proposition we shall introduce the concept of an *O.K.* set of sentences, and shall briefly discuss the concepts of an *equivalence relation* and an *equivalence class.*

Definition

A set θ of sentences is *O.K.* if and only if every finite subset of θ is satisfiable.
(Then the antecedent of the proposition which we seek to prove is that Γ is O.K.) An important fact about O.K.-ness is that

> If θ is O.K. and S is any sentence, then either $\theta \cup \{S\}$ is O.K. or $\theta \cup \{-S\}$ is.

Proof. Note that if each of the sets

$$\{A_1, \ldots, A_m, S\}, \quad \{B_1, \ldots, B_n, -S\}$$

is unsatisfiable, so is the set

$$\{A_1, \ldots, A_m, B_1, \ldots, B_n\}.$$

(For if this last set is satisfied by some interpretation, it is satisfied by an interpretation in which one of the sentences S, $-S$, is true, and hence in which all members of one of the former two sets is true.) So if $\theta \cup \{S\}$ is not O.K., some subset $\{A_1, \ldots, A_m, S\}$ is unsatisfiable where, we may assume, all of the As belong to θ. Similarly if $\theta \cup \{-S\}$ is not O.K., one of *its* subsets $\{B_1, \ldots, B_n, -S\}$ is unsatisfiable, where the Bs all belong to θ. Then if both fail to be O.K., a subset $\{A_1, \ldots, A_m, B_1, \ldots, B_n\}$ of θ is unsatisfiable, and thus θ is not O.K.

Equivalence relations

Suppose that X is a nonempty set and that R is a relation on X, i.e. suppose that whenever we have xRy, then x and y both belong to X.

Definition. R is an equivalence relation on X if and only if,
R is *reflexive* on X (xRx whenever x is in X),
R is *transitive* (xRz whenever xRy and yRz), and
R is *symmetrical* (xRy whenever yRx).

If R is an *equivalence relation* on X, and x is in X, then the set of those members of X to which x bears R is called *the equivalence class of x under the relation R*. A customary designation for the equivalence class of x under R is '$[x]_R$'. Usually it is clear from the context what relation R is in question, and then the subscript 'R' is generally omitted. Thus we define

$$z \in [x] \text{ if and only if } xRz.$$

We shall need the following facts about equivalence classes and equivalence relations:

If R is an equivalence relation on X and x and y are in X, then
(1) x is in $[x]$, and
(2) xRy if and only if $[x] = [y]$.

Proof of (1). R is reflexive.

Proof of (2). For the 'only if' part, suppose that xRy. If z is in $[x]$ then xRz and thus, by symmetry, zRx, so that by transitivity, zRy, and by symmetry again, yRz, so that z is in $[y]$: and if z is in $[y]$ then by completely parallel reasoning we have z in $[x]$. Therefore $[x] = [y]$. For the 'if' part, suppose that $[x] = [y]$. Since by (1), y is in $[y]$, we have y in $[x]$ and hence xRy. This completes the proof of (2).

Now as the sentences, and hence the quantifier-free sentences, in any derivation form an enumerable set, we shall have established the completeness theorem when we have proved the following proposition:

Lemma III

If Γ is an enumerable, O.K. set of quantifier-free sentences, then there is an interpretation $\mathscr{I}$ which matches Γ.

Since the proof is long, let us first outline it. We shall define a sequence of sentences $A_1, A_2, \ldots$; then, a sequence $\Gamma_1, \Gamma_2, \ldots$ of O.K. sets; and then a sequence $B_1, B_2, \ldots$, of sentences. The sequence $B_1, B_2, \ldots$ will be used to define an equivalence relation on the set of terms appearing in Γ, and this equivalence relation is then used to define $\mathscr{I}$. The Bs are then shown to be all true in $\mathscr{I}$. By the time this has been shown, it will have become clear that if $\mathscr{I}$ is a model of Γ, then $\mathscr{I}$ matches Γ. Finally, all members of Γ are shown to be true in $\mathscr{I}$.

Proof of lemma III. Suppose that Γ is an enumerable, O.K. set of quantifier-free sentences. We may assume that Γ is nonempty, for otherwise every interpretation $\mathscr{I}$ matches Γ.

Let $A_1, A_2, \ldots$ be an enumeration of all atomic sentences which are either sentence letters appearing in sentences of Γ or sentences which can be formed by filling the blanks of the equals-sign and predicate letters appearing in Γ with terms that occur in sentences in Γ. (Include sentences formed by filling the blanks of the equals-sign even when that sign does not appear in Γ.)

We now define the sequence $\Gamma_1, \Gamma_2, \ldots$ and verify that all members are O.K. Let $\Gamma_1 = \Gamma$, which is O.K. by hypothesis. Now suppose that Γ_n has been defined so as to be O.K. Then as we have noted, at least one of the sets $\Gamma_n \cup \{A_n\}$, $\Gamma_n \cup \{-A_n\}$ is O.K. We define Γ_{n+1} to be the O.K. one of the two if exactly one is O.K., and we define $\Gamma_{n+1} = \Gamma_n \cup \{A_n\}$ if both are O.K. Then Γ_{n+1} is O.K. if Γ_n is, and all members of the sequence are O.K.

It is clear from the definition of the Γ's that if $i \leqslant j$ then $\Gamma_i \subseteq \Gamma_j$, and also that $\Gamma \subseteq \Gamma_i$ for all i. Since all the Γs are O.K., it never happens that both A_i and $-A_i$ are in some Γ_j. But since at least one of A_i, $-A_i$ is in

Γ_{i+1}, *exactly one* of them is in Γ_{i+1}, and hence exactly one of them is in Γ_m, for all $m \geqslant i+1$. *We define B_i to be whichever of A_i, $-A_i$ is in Γ_{i+1}.*

We now examine the sequence of Bs rather carefully. If r and s are terms (*viz*., terms which occur in Γ – henceforth we shall usually omit this qualification), exactly one of the two sentences $r = s$, $r \neq s$ is in the sequence of Bs. (*Proof*. If both $r = s$ and $r \neq s$ were in the sequence of Bs, then for some i and j, $r = s$ would be B_i and $r \neq s$ would be B_j, and so both would be in Γ_{m+1}, where m is the larger of i and j. But then Γ_{m+1} would not be O.K., as it would have $\{r = s, r \neq s\}$ as an unsatisfiable subset.) We now define

$$r \sim s \quad \textit{if and only if the sentence } r = s \textit{ is one of the } B\text{s.}$$

Note that for any term r we have $r \sim r$. (*Proof*. If B_i were $r \neq r$ then Γ_{i+1} would not be O.K., for it would have $\{r \neq r\}$ as an unsatisfiable finite subset. Then $r \neq r$ is not one of the Bs, and hence $r = r$ must be one, and so $r \sim r$.) Then the relation $\sim$ is reflexive on the set of terms. It is also transitive. (*Proof*. For *reductio ad absurdum*, suppose $r \sim s$ and $s \sim t$ but $r \nsim t$, i.e. suppose that $r = s$ is B_i, $s = t$ is B_j, but $r \neq t$ is B_k, for some i, j, k. Then all of $r = s$, $s = t$, $r \neq t$ would be in Γ_{m+1}, where m is the largest of i, j, k, and Γ_{m+1} would then not be O.K. Thus, if B_i is $r = s$ and B_j is $s = t$ then for no k is B_k $r \neq t$, and hence for some k, B_k is $r = t$, and so $r \sim t$.) Finally, the relation $\sim$ is symmetrical. (*Proof*. If $r \sim s$ then for some i, B_i is $r = s$. Then for no j is B_j $s \neq r$ and thus for some j, B_j is $s = r$, so that $s \sim r$.) We have now established that $\sim$ is an equivalence relation on the set of terms. Every term t thus belongs to a unique equivalence class $[t]$ under the relation $\sim$, a class of which each member bears $\sim$ to every other member, and which contains every term which bears $\sim$ to any one of its members.

We can now begin to define the interpretation $\mathscr{I}$ which matches Γ:

(A) If no names and predicate letters (and hence no terms) appear in Γ, the domain of $\mathscr{I}$ may be any nonempty set at all; otherwise, the domain is to be the set of all equivalence classes $[t]$.

We wish to define $\mathscr{I}$ so that $\mathscr{I}$ assigns to each term t its own equivalence class as denotation, and so that $\mathscr{I}$ makes each of the Bs true: we want to have $\mathscr{I}(t) = [t]$ and $\mathscr{I}(B_i) = 1$ for each t and i. There is essentially only one way to finish the definition of $\mathscr{I}$ so as to satisfy this wish:

(B) $\mathscr{I}$ assigns to each *name* t the equivalence class $[t]$ as its designation.
(C) $\mathscr{I}$ assigns to each n-place function symbol f the function $\mathbf{f}$ which is

determined by the condition that for all $[t_1], \dots, [t_n]$ in the domain of $\mathscr{I}$, $\not{f}([t_1], \dots, [t_n]) = [f(s_1, \dots, s_n)]$ if there are terms $s_1, \dots, s_n$ in $[t_1], \dots, [t_n]$ respectively such that $f(s_1, \dots, s_n)$ is a term appearing in Γ; and otherwise we set $\not{f}([t_1], \dots, [t_n]) = [t]$ where t is any term we please appearing in Γ.

(D) $\mathscr{I}$ specifies that a sentence letter is to be true if and only if that letter is one of the Bs.

(E) $\mathscr{I}$ specifies that an n-place predicate letter R is to be true of $[t_1], \dots, [t_n]$ (in that order) if and only if $Rt_1, \dots, t_n$ is one of the Bs.

Now it may appear that there is something improper about clauses (C) and (E) of the definition of $\mathscr{I}$. Thus, in (C) we have specified that if there are terms $s_1, \dots, s_n$ in $[t_1], \dots, [t_n]$ such that $f(s_1, \dots, s_n)$ appears in Γ, then $\not{f}([t_1], \dots, [t_n]) = [f(s_1, \dots, s_n)]$. But it is entirely possible that there are also terms $r_1, \dots, r_n$ in $[t_1], \dots, [t_n]$ such that $f(r_1, \dots, r_n)$ appears in $\mathscr{I}$ (without, say, r_1 being identical with s_1), and then $\not{f}([t_1], \dots, [t_n])$ must also be $[f(r_1, \dots, r_n)]$. Then unless we have some guarantee that in this case $[f(s_1, \dots, s_n)] = [f(r_1, \dots, r_n)]$, we cannot claim to have determined a *unique* value of the function $\not{f}$ for the arguments $[t_1], \dots, [t_n]$, and hence we cannot claim to have defined a *function* $\not{f}$ at all.

We have such a guarantee. Because the set

$$\{r_1 = s_1, \dots, r_n = s_n, f(r_1, \dots, r_n) \neq f(s_1, \dots, s_n)\}$$

is unsatisfiable, we can conclude that if r_1 and s_1 are both in $[t_1], \dots,$ and r_n and s_n are both in $[t_n]$, then $r_1 \sim s_1, \dots,$ and $r_n \sim s_n$, and hence

$$f(r_1, \dots, r_n) \sim f(s_1, \dots, s_n)$$

and hence $[f(r_1, \dots, r_n)] = [f(s_1, \dots, s_n)]$.

A similar question arises in connection with clause (E): What guarantee have we that we have determined a unique truth-value for the predicate letter R with respect to $[t_1], \dots, [t_n]$? That we have such a guarantee is shown by the observation that since the sets

$$\{s_1 = t_1, \dots, s_n = t_n, Rs_1, \dots s_n, -Rt_1, \dots, t_n\}$$

and

$$\{s_1 = t_1, \dots, s_n = t_n, -Rs_1, \dots, s_n, Rt_1, \dots, t_n\}$$

are both unsatisfiable, if all of the conditions $s_1 \sim t_1, \dots, s_n \sim t_n$ hold then $Rs_1, \dots, s_n$ is one of the Bs if and only if $Rt_1, \dots, t_n$ is.

It follows inductively from clauses (B) and (C) that each term t appearing in Γ denotes $[t]$. For by (B), each name t denotes $[t]$; and if t appears in Γ with $t = f(t_1, \dots, t_n)$, and if $\not{f}$ is as in (C) and t_1 denotes $[t_1], \dots$ and t_n denotes $[t_n]$, then t denotes $\not{f}([t_1], \dots, [t_n]) = [f(t_1, \dots, t_n)] = [t]$.

We now want to see that each of the Bs is true in $\mathscr{I}$. For each i, either $B_i = A_i$ or $B_i \neq A_i$. If $B_i \neq A_i$ then B_i is true in $\mathscr{I}$ if and only if A_i is not true in $\mathscr{I}$. We thus want to see that A_i is true in $\mathscr{I}$ if and only if $B_i = A_i$. Now A_i can be either a sentence letter, a sentence $Rt_1, \ldots, t_n$, or a sentence $s = t$. If A_i is a sentence letter then by clause (D), A_i is true in $\mathscr{I}$ if and only if $A_i = B_i$. If $A_i = Rt_1, \ldots, t_n$ then by clause (E) and the fact that each t denotes $[t]$, A_i is true if and only if $A_i = B_i$. Finally, if A_i is $s = t$ then A_i is true in $\mathscr{I}$ if and only if the denotation of s is the same as the denotation of t, i.e. $[s] = [t]$, i.e. $s \sim t$, i.e. $s = t$ is one of the Bs, i.e. $s = t$ is B_i, i.e. $A_i = B_i$.

In order to see that $\mathscr{I}$ matches Γ we have to see that in case the members of Γ are not simply built out of sentence letters and connectives, every object in the domain of $\mathscr{I}$ is the denotation (according to $\mathscr{I}$) of some term appearing in Γ, and also that $\mathscr{I}$ is a model of Γ. But where the members of Γ are not simply built out of sentence letters and connectives, the objects in the domain of $\mathscr{I}$ are just the equivalence classes of terms appearing in Γ, and we have already seen that each of these is denoted by any one of its members, which are terms that appear in Γ. So we need only show that $\mathscr{I}$ is a model of Γ.

Suppose then that S is a member of Γ. For any i, B_i is true in $\mathscr{I}$. Therefore, if B_i is true in an interpretation $\mathscr{J}$, A_i will have the same truth-value in $\mathscr{J}$ that it has in $\mathscr{I}$. S is a truth-functional compound of some finite number of the As, and therefore there is a positive integer k such that all of the As of which S is a compound are members of $\{A_1, \ldots, A_k\}$. Therefore in any interpretation $\mathscr{J}$ in which all of $B_1, \ldots, B_k$ are true, each of $A_1, \ldots, A_k$ has the same truth-value that it has in $\mathscr{I}$, and hence S has the same truth-value that it has in $\mathscr{I}$. All of $B_1, \ldots, B_k$ are in Γ_{k+1}, as is S, which is in Γ_1. Then $\{B_1, \ldots, B_k, S\}$ is a subset of Γ_{k+1}. Since Γ_{k+1} is O.K. and this subset is finite, it is satisfiable, and hence true in some interpretation $\mathscr{J}$. As all of $B_1, \ldots, B_k$ and S are true in $\mathscr{J}$, S is true in $\mathscr{I}$. Thus $\mathscr{I}$ is a model of Γ, and the completeness theorem is proved.

Let us now ponder three of the more significant consequences of the completeness theorem. First:

The compactness theorem for first-order logic

A set θ of sentences is unsatisfiable if and only if some finite subset θ_0 of θ is unsatisfiable.

Proof. The 'if' part is trivial: if *some* subset (finite or not) of θ is unsatisfiable, so is θ. For the 'only if' part, suppose θ unsatisfiable. We may assume that each member of θ is in prenex normal form. By the completeness theorem, there is a derivation $\mathscr{D}$ from θ such that some finite set $\{A_1, \ldots, A_m\}$ of quantifier-free sentences that occur in $\mathscr{D}$ is unsatisfiable. We may suppose that these sentences first occur in $\mathscr{D}$ in the order $A_1, \ldots, A_m$. Consider the derivation $\mathscr{D}_0$ that is obtained from $\mathscr{D}$ by deleting all sentences that occur in $\mathscr{D}$ after A_m. $\mathscr{D}_0$ contains only finitely many sentences, and therefore there are only finitely many members $F_1, \ldots, F_n$ of θ that occur in $\mathscr{D}_0$. Let $\theta_0 = \{F_1, \ldots, F_n\}$. Since all of $A_1, \ldots, A_m$ occur in $\mathscr{D}_0$, $\mathscr{D}_0$ is a derivation from θ_0 such that some set of quantifier-free sentences that occur in $\mathscr{D}_0$ is unsatisfiable. Therefore, by the soundness theorem, θ_0 is unsatisfiable, and is thus a finite unsatisfiable subset of θ. Upshot:

> A sentence is implied by a set of sentences if and only if it is implied by some finite subset of it.

Proof. If Γ implies S then $\Gamma \cup \{-S\}$ is unsatisfiable, whence, by the compactness theorem, some finite subset θ_0 of $\Gamma \cup \{-S\}$ is unsatisfiable. If $\Gamma_0 = \theta_0 - \{-S\}$,† Γ_0 is a finite subset of Γ and implies S, for $\Gamma_0 \cup \{-S\}$ is unsatisfiable. Second:

The Skolem–Löwenheim theorem

If a set Δ of sentences has a model, then it has a model with an enumerable domain.

Proof. Suppose that Δ has a model, i.e. is satisfiable. By Lemma I there is a canonical derivation $\mathscr{D}$ from Δ. By the soundness theorem, $\mathscr{D}$ is not a refutation of Δ and so no finite subset of the set Γ of quantifier-free sentences in $\mathscr{D}$ is unsatisfiable. So Γ is an enumerable O.K. set of sentences. Thus by Lemma III there is an interpretation $\mathscr{I}$ which matches Γ, and by Lemma II $\mathscr{I}$ is a model of Δ.

Now if Δ is a set of sentences in none of which are there any predicate letters, then all of the non-logical symbols appearing in Δ are sentence letters, and any interpretation $\mathscr{J}$, whose domain is some arbitrarily chosen enumerable set, and which assigns to the sentence letters appearing in Δ whatever truth-values $\mathscr{I}$ assigns to them will be a model of Δ with an enumerable domain.

† $A-B$ is the set of things in A that are not in B.

But if there is at least one predicate letter appearing in Δ, then there is at least one term appearing in Γ, and since $\mathscr{I}$ matches Γ, everything in the domain of $\mathscr{I}$ is the denotation of some term appearing in Γ. But as Γ is enumerable, there are at most enumerably many such terms, and hence the domain of $\mathscr{I}$ will be enumerable, and thus in this case $\mathscr{I}$ itself will be a model of Δ with an enumerable domain. Third:

There is an effective positive test for unsatisfiability (and hence, for validity)

We are now in a position to see that there is an effective positive test for unsatisfiability: an effective procedure which, when applied to an arbitrary sentence S of some first-order language, terminates with a 'yes' if and only if S is unsatisfiable. An effective positive test for unsatisfiability is not the same thing as a decision procedure for unsatisfiability, which we may take to be an effective method which terminates with a 'yes' if the sentence to which it is applied is unsatisfiable, and terminates with a 'no' if the sentence is satisfiable. (Not terminating with a 'yes' is not the same thing as terminating with a 'no', for one way of not terminating with a 'yes' is not terminating at all!) Our description of the procedure will be brief and intuitive, but it ought to be quite clear from it how one might go about actually writing down the table of a (quite large) Turing machine M^* which, when given an arbitrary sentence of some first-order language as input, halted after some finite number of steps with the words 'yes, unsatisfiable' ('yes, valid') appearing on its tape if and only if the sentence was unsatisfiable (valid).

The proof of Lemma I showed us that for any prenex sentence S there is a canonical derivation from $\{S\}$. In general, there are infinitely many canonical derivations from any set Δ. But if we make certain further decisions not already indicated in the proof of Lemma I (about which sentences and which terms are to be used in applications of UI and EI if there is a choice), we can associate with each prenex S a *unique* canonical derivation $\mathscr{D}_S$ from $\{S\}$, and can give effective instructions for writing down arbitrarily long finite initial segments of $\mathscr{D}_S$. That is to say, given any prenex sentence S and any number n, we can (effectively) find effective instructions for writing down the first n sentences of $\mathscr{D}_S$ – or the whole of $\mathscr{D}_S$, if $\mathscr{D}_S$ contains fewer than n sentences.

In detail, our effective positive test can be described as follows. *First*: find a prenex equivalent P of S. (Effective instructions can be given for finding a prenex equivalent of any given sentence.) *Second*: list longer and longer initial portions of $\mathscr{D}_P$, pausing after each sentence is written down to decide whether the set of all quantifier-free sentences so far listed is satisfiable or not (an effective procedure for doing this is given below). Stop altogether and write 'yes' if the set is unsatisfiable, and otherwise continue, writing down the next sentence in $\mathscr{D}_P$. The soundness and completeness theorems imply that this procedure eventually terminates with a 'yes' if and only if $\{P\}$ is unsatisfiable, and hence if and only if $\{S\}$ is unsatisfiable.

Our procedure will thus always give the right answer when that answer is 'yes, unsatisfiable'. But where the right answer is 'No, satisfiable', our procedure will not in general show that to be the right answer: we may continue to list sentences *ad infinitum* without ever being in a position to effectively calculate that in fact we will continue to list sentences *ad infinitum*. Moreover, this is not a defect of the particular procedure we have described, for as we saw in Chapter 10, there can be no effective negative test for unsatisfiability. But our *positive* test for unsatisfiability is premised on the existence of both positive and negative tests for unsatisfiability of finite sets of *quantifier-free* sentences. We conclude by proving that such tests exist.

How to decide whether or not a finite set of quantifier-free sentences is satisfiable. Lemma III guarantees us that there is a procedure for deciding whether a finite set Γ of quantifier-free sentences is satisfiable or not, for either (1) all members of Γ are built out of sentence letters with the aid of connectives and parentheses, in which case we can use truth tables or some other well known method to determine whether Γ is satisfiable, or (2) there is a positive integer n which is the number of distinct terms occurring in sentences of Γ. Now if Γ is satisfiable, Γ is O.K., whence by Lemma III there is a model of Γ whose domain contains no more than n members. Therefore, Γ is satisfiable if and only if it has a model whose domain contains no more than n members.

It follows that Γ is satisfiable if and only if satisfied by some interpretation $\mathscr{I}$ of which the domain is $\{1, \ldots, m\}$ (where m is some number between 1 and n, inclusive) and which specifies nothing about symbols which do not occur in Γ. There are only finitely many such interpretations and, supplied with Γ, we can effectively give instructions for writing down an explicit description – sometimes called a *diagram* – of each of them. In a diagram of $\mathscr{I}$ it is explicitly stated what number each name in Γ denotes, what the truth-value of each sentence letter in Γ is, what the value is for the function $\mathscr{I}$ assigns to each function symbol in Γ, for each sequence of arguments in the domain, and which truth-value $\mathscr{I}$ assigns to each predicate letter, for each sequence of arguments in the domain. Given such a diagram, we can effectively calculate the denotation in $\mathscr{I}$ of any term in Γ, then calculate the truth-value in $\mathscr{I}$ of any atomic sentence of which any sentence of Γ is compounded, and then, via the rules of interpretation for the propositional connectives (cases 1–5 of Chapter 9, i.e. essentially, via the usual truth tables) calculate the truth-values which $\mathscr{I}$ assigns to the several sentences which make up Γ.

Then our effective procedure is this: write down all diagrams of all interpretations of Γ with domain $\{1, ..., m\}$ $(1 \leqslant m \leqslant n)$, calculate the truth-values of the members of Γ in each of them, and see whether all members of Γ are true in at least one of these interpretations. If so, Γ is satisfiable; if not, not.

Exercises

12.1 Use the construction given in the proof of Lemma III to find interpretations which match the sets of quantifier-free sentences in canonical derivations from the following sets:

$$(a)\quad \{\forall x \exists y\, xLy\}; \qquad (b)\quad \{\exists x\, xLf(gx, x)\}.$$

12.2 A set Γ of sentences is said to have *arbitrarily large finite models* when, for each positive integer n, there is a model of Γ whose domain has at least n members. Show that

If Γ has arbitrarily large finite models, it has a model whose domain is infinite.

Conclude that there is no *axiom of finitude*, i.e., no sentence which is true in all and only those interpretations whose domains are finite.

12.3 Show that there is an effective procedure for deciding the validity of prenex sentences in which no function symbols occur, and in which no existential quantifier is to the left of any universal quantifier.

12.4 Show ('from scratch') that if Γ is a set of sentences which contain only sentence letters, connectives, and parentheses, then Γ is satisfiable if every finite subset is. (*Hint*: strip from the proof of Lemma III the numerous complexities which are required when terms, '=' and other predicate letters may occur in Γ.)

Solutions

12.1(*a*) $\Delta = \{\forall x \exists y\, x \not{L} y\}$. A canonical derivation $\mathscr{D}$ from Δ:

1	$\forall x \exists y\, x \not{L} y$	Δ
2	$\exists y\, a_1 \not{L} y$	1
3	$a_1 \not{L} a_2$	2
4	$\exists y\, a_2 \not{L} y$	1
5	$a_2 \not{L} a_3$	4
	$\vdots$	
$2n$	$\exists y\, a_n \not{L} y$	1
$2n+1$	$a_n \not{L} a_{n+1}$	$2n$

Here $\Gamma = \{a_1 \not{L} a_2, a_2 \not{L} a_3, \ldots, a_n \not{L} a_{n+1}, \ldots\}$. There is some arbitrariness in the choice of the sequence $A_1, A_2, \ldots$, but once that sequence is fixed (e.g. as at the left, below), the sequences $\Gamma_1, \Gamma_2, \ldots$, and $B_1, B_2, \ldots$ are fixed and thereby $\mathscr{I}$ is determined.

$A_1: a_1 L a_1$	$\Gamma_2 = \{a_1 L a_1\} \cup \Gamma$	$B_1: a_1 L a_1$
$A_2: a_1 = a_1$	$\Gamma_3 = \Gamma_2 \cup \{a_1 = a_1\}$	$B_2: a_1 = a_1$
$A_3: a_1 L a_2$	$\Gamma_4 = \Gamma_3$	$B_3: a_1 \not{L} a_2$
$A_4: a_1 = a_2$	$\Gamma_5 = \{a_1 \neq a_2\} \cup \Gamma_4$	$B_4: a_1 \neq a_2$
$A_5: a_2 L a_1$	$\Gamma_6 = \{a_2 L a_1\} \cup \Gamma_5$	$B_4: a_2 L a_1$
$A_6: a_2 = a_1$	$\Gamma_7 = \Gamma_6$	$B_5: a_2 \neq a_1$

If the subscripts on the pairs of names flanking 'L' and '$=$' in the sequence of As continue in the pattern

22, 13, 31, 23, 32, 33, 14, 41, 24, 42, 34, 43, 44, ...

then the sequence of Bs continues:

$$a_2 L a_2,\ a_2 = a_2,\ a_1 L a_3,\ a_1 \neq a_3,\ a_3 L a_1,\ a_3 \neq a_1,$$
$$a_2 \not{L} a_3,\ a_2 \neq a_3,\ a_3 L a_2,\ a_3 \neq a_2,\ a_3 L a_3,\ a_3 = a_3,$$
$$a_1 L a_4,\ a_1 \neq a_4, \ldots$$

where distinct names are always asserted to name distinct objects in the domain of $\mathscr{I}$ – i.e., $[a_m] \neq [a_n]$ if and only if $m \neq n$ – and where the relation assigned to 'L' by $\mathscr{I}$ is asserted to hold between all objects except those between which it is asserted not to hold in Γ, i.e., $\mathscr{I}(a_m L a_n) = 0$ if $n = m+1$ and otherwise $\mathscr{I}(a_m L a_n) = 1$. Had the sequence of As been chosen differently, the matching interpretation might have been very different.

12.2 For each $i \geq 2$, let A_i be a sentence which is true in an interpretation if and only if there are at least i members of the domain of the interpretation. E.g. A_3 might be

$$\exists x_1 \exists x_2 \exists x_3 (x_1 \neq x_2 \,\&\, x_2 \neq x_3 \,\&\, x_1 \neq x_3).$$

Let Γ' be the set of all such sentences, $\Gamma' = \{A_1, A_2, \ldots\}$. Every finite subset of $\Gamma \cup \Gamma'$ has a model. (Why?) So by the compactness theorem, $\Gamma \cup \Gamma'$ has a model. No model of Γ' can have a finite domain, so $\Gamma \cup \Gamma'$ and hence Γ have a model with infinite domain.

13
The Skolem–Löwenheim theorem

The cluster of theorems that is generally called 'the Skolem–Löwenheim theorem' is a group of results that concern the size of domains of interpretations of sentences in first-order logical languages, 'size' being understood in the sense of *(cardinal) number of members*. They typically have the form: if there is an interpretation with (semantical) property —, then there is an interpretation with (semantical) property —, whose domain has size – – –. For example, the best known Skolem–Löwenheim theorem, an easy consequence of the version proved in Chapter 12, reads: if there is an interpretation in which a sentence S is true, then there is an interpretation in which S is true, whose domain is enumerable (Löwenheim, 1915). In the present chapter we shall prove a strong form of the Skolem–Löwenheim theorem from which this and other versions follow.

In Chapter 25 we shall prove that if S is a sentence that contains k *one-place* predicate letters and r variables, and possibly also the equals-sign '=', but no names, function symbols, or two or more place predicate letters (a so-called 'monadic' sentence), then if S is true in any interpretation at all, it is true in one whose domain contains no more than $2^k \cdot r$ members. Thus if S is a monadic sentence true in some interpretation whose domain is infinite, S is also true in some other interpretation whose domain is finite. One might wonder whether the restriction to monadic sentences is essential. It is.

Let (a) = '$\forall x - xRx$', let (b) = '$\forall x \exists y\, xRy$', and let

$$(c) = \text{'}\forall x \forall y \forall z\, ((xRy \,\&\, yRz) \rightarrow xRz)\text{'}.$$

Let (1) be the conjunction of (a), (b), and (c). (1) is then a sentence which is true in no interpretation with a finite domain, but is true in some interpretation with an infinite domain.

To see that (1) is true in no interpretation with a finite domain, suppose that $\mathscr{I}((1)) = 1$, that D = the domain of $\mathscr{I}$, and that $\mathscr{I}$ specifies that 'R' is to be true of c, d iff cSd. Let's call a sequence $d_1, \ldots, d_n$ of elements of D a *good* sequence if $d_i S d_j$ whenever $i < j \leqslant n$. Observe that if $d_1, \ldots, d_n$ is a good sequence, then $d_1, \ldots,$ and d_n are all *distinct*, for if $i < j$, but $d_i = d_j$, then $d_i S d_i$, which is impossible, as $\mathscr{I}((a)) = 1$. We shall show that D is infinite by showing inductively that for each positive n, there is a good sequence $d_1, \ldots, d_n$. Since D is nonempty, there is, trivially, a good sequence containing just one member d_1 of D. Suppose that $d_1, \ldots, d_n$ is a good sequence. Since $\mathscr{I}((b)) = 1$, for some d_{n+1} in D, $d_n S d_{n+1}$. But then $d_1, \ldots, d_n, d_{n+1}$ is also a good sequence, for if $i < n$, then, since $d_i S d_n$, $d_n S d_{n+1}$, and $\mathscr{I}((c)) = 1$, we have that $d_i S d_{n+1}$, and therefore we have that if either $i < j \leqslant n$ or $i < j = n+1$, then $d_i S d_j$, and thus have that $d_i S d_j$ whenever $i < j \leqslant n+1$.

On the other hand, (*a*) no real number is less than itself; (*b*) every real number is less than some other real number; and (*c*) if one real is less than a second, which is less than a third, the first is less than the third. (1) is therefore true in the interpretation $\mathscr{I}$ whose domain is the set of all real numbers and which specifies that '*R*' is to be true of c, d iff $c < d$. The domain of this interpretation is unnecessarily large, however, for (1) is also true in the interpretation $\mathscr{J}$ whose domain is the set of all natural numbers, and which, like $\mathscr{I}$, specifies that '*R*' is to be true of c, d iff $c < d$. The domain of $\mathscr{I}$ is non-enumerable, that of $\mathscr{J}$, enumerable. Another question thus suggests itself: is there any sentence of a first-order logical language which, though true in some interpretation, is true only in interpretations whose domains are non-enumerable?

Löwenheim's 1915 theorem, stated five paragraphs back, answers this question negatively. In 1919 Skolem extended Löwenheim's result in a far-reaching way by proving a theorem that immediately implies not only Löwenheim's theorem, but the stronger statement proved in Chapter 12 that if an enumerable collection of sentences is satisfiable, then there is a single interpretation with an enumerable domain in which all members of the collection are true. Skolem's 1919 theorem was that *any interpretation*† *has an elementarily equivalent subinterpretation with an enumerable domain.*

What are 'subinterpretations' and what's 'elementary equivalence'? An interpretation $\mathscr{J}$ is called a *subinterpretation* of an interpretation $\mathscr{I}$ if $\mathscr{J}$ and $\mathscr{I}$ are interpretations of the same languages and

(1) The domain E of $\mathscr{J}$ is a subset of the domain of $\mathscr{I}$;

(2) $\mathscr{J}$ assigns names the same designations as $\mathscr{I}$;

(3) $\mathscr{J}$ assigns an n-place function symbol f the function g which takes

† We assumed in Chapter 9 that no interpretation assigns appropriate objects to non-enumerably many non-logical symbols, and hence that every interpretation interprets (defines truth-values for) at most enumerably many sentences. This assumption is vital to the proof of the Skolem–Löwenheim theorem (in Skolem's 1919 version) (cf. Exercise 13.1), as is the restriction to first-order logic (cf. Chapter 17).

values only on sequences $c_1, \ldots, c_n$ of objects in E and for which

$$g(c_1, \ldots, c_n) = \mathscr{f}(c_1, \ldots, c_n),$$

where $\mathscr{f}$ is the function $\mathscr{I}$ assigns f;

(4) $\mathscr{J}$ assigns sentence-letters the same truth-values as $\mathscr{I}$; and

(5) $\mathscr{J}$ specifies that an n-place predicate letter R is to be true of a sequence $c_1, \ldots, c_n$ of objects in E iff $\mathscr{I}$ specifies that R is to be true of $c_1, \ldots, c_n$.

Which sets E may be domains of subinterpretations of $\mathscr{I}$? First of all, E must be a nonempty subset of the domain of $\mathscr{I}$. Clause 2 imposes another requirement on E: that it contain all designations that $\mathscr{I}$ assigns names. Clause 3 imposes a further requirement: that if $c_1, \ldots, c_n$ are in E, and $\mathscr{I}$ assigns the function $\mathscr{f}$ to some n-place function symbol, then $\mathscr{f}(c_1, \ldots, c_n)$ must also be in E. If these three requirements are met, however, there is a unique subinterpretation of $\mathscr{I}$ whose domain is E. If nonempty, the set of denotations that $\mathscr{I}$ assigns to terms is a set that meets all three requirements.

If $\mathscr{J}$ is a subinterpretation of $\mathscr{I}$, $\mathscr{J}$ interprets the same sentences as $\mathscr{I}$: the same sentences have truth-values in both, though not necessarily the same truth-values. It follows from clauses 2 and 3 that $\mathscr{J}$ assigns each term the same denotation as $\mathscr{I}$. It then follows from clauses 4 and 5 that atomic sentences have the same truth-values in both, and therefore so do all *quantifier-free* sentences.

Two interpretations $\mathscr{I}$ and $\mathscr{J}$ are said to be *elementarily equivalent* if they interpret the same sentences and any sentence is true in $\mathscr{I}$ iff it is true in $\mathscr{J}$. This definition can be weakened to: they interpret the same sentences and all sentences true in $\mathscr{I}$ are true in $\mathscr{J}$; for if $\mathscr{I}$ and $\mathscr{J}$ interpret S and S is true in $\mathscr{J}$, then S's negation is not true in $\mathscr{J}$, hence not true in $\mathscr{I}$, and thus S is true in $\mathscr{I}$.

Skolem showed that for any interpretation $\mathscr{I}$, there is an interpretation $\mathscr{J}$ which (A) is a subinterpretation of $\mathscr{I}$, (B) has an enumerable domain, and (C) is elementarily equivalent to $\mathscr{I}$. It follows that given an interpretation of all the (enumerably many) sentences in a (first-order) language, no matter how large its domain, one can whittle down the domain in such a way that even though only an enumerable number of elements remains, the very same sentences are true in the reduced interpretation as were true in the original.

There are related results – 'upward' theorems – that speak of arbitrarily large models, and still stronger 'downward' theorems than the one we shall demonstrate, but we shall not discuss them (the reader is referred to Chang & Keisler's *Model Theory*). We shall now prove Skolem's 1919 theorem and derive another Skolem–Löwenheim theorem as a corollary. The theorem follows quite directly from the strong soundness theorem of Chapter 11, Lemmas I and II of Chapter 12, and the facts about subinterpretations just mentioned.

Theorem

Any interpretation $\mathscr{I}$ has an elementarily equivalent subinterpretation $\mathscr{J}$ whose domain is enumerable.

Proof. Let Δ be the set of all (prenex) sentences true in $\mathscr{I}$. By Lemma I of Chapter 12, there is a canonical derivation $\mathscr{D}$ from Δ. If a is a name or f is a function symbol assigned a denotation or a function by $\mathscr{I}$, then all of $a = a$, $\forall x f(x, \ldots, x) = f(x, \ldots, x)$, and $\forall x\, x = x$ occur in Δ and $\mathscr{D}$. By clause 5 of the definition of 'canonical derivation', all function symbols appearing in $\mathscr{D}$ appear in Δ, and so by the strong soundness theorem of Chapter 11, there is an interpretation $\mathscr{L}$, in which all sentences in $\mathscr{D}$ are true, and which differs from $\mathscr{I}$ only in what it assigns to those names that appear in $\mathscr{D}$ but are assigned no denotation by $\mathscr{I}$.

Let Γ = the set of quantifier-free sentences in $\mathscr{D}$. As $\mathscr{L}$ is a model of Γ, any term t appearing in Γ is assigned a denotation by $\mathscr{L}$; and if t is assigned a denotation by $\mathscr{L}$, t is formed from names and function symbols assigned things by $\mathscr{I}$, all of which appear in $\mathscr{D}$, and other names appearing in $\mathscr{D}$, and hence, as $\forall x\, x = x$ is in $\mathscr{D}$, $t = t$ is in Γ. Thus the terms assigned denotations by $\mathscr{L}$ are just those appearing in Γ.

Let E be the set of denotations $\mathscr{L}$ assigns to terms. E is non-empty, as $t = t$ is in Γ (for some term t). E is enumerable, because Γ is. Let $\mathscr{K}$ be the subinterpretation of $\mathscr{L}$ with domain E. $\mathscr{K}$ assigns a term the denotation d iff $\mathscr{L}$ assigns it d. Every object in $\mathscr{K}$'s domain is therefore denoted (according to $\mathscr{K}$) by a term appearing in Γ. All members of Γ, being quantifier-free and true in $\mathscr{L}$, are true in $\mathscr{L}$'s subinterpretation $\mathscr{K}$. So $\mathscr{K}$ matches Γ (cf. Chapter 12). So by Lemma II of Chapter 12, $\mathscr{K}$ is a model of the set of all sentences in $\mathscr{D}$, and hence of Δ.

If $\mathscr{J}$ is just like $\mathscr{K}$ except that it assigns denotations to the same names as $\mathscr{I}$, then $\mathscr{J}$ is a model of Δ with an enumerable domain, and thus an elementarily equivalent subinterpretation of $\mathscr{I}$ with an enumerable domain.

Corollary 1 (Skolem, 1919)

If θ is an enumerable collection of sentences that is satisfiable, there is an interpretation $\mathscr{J}$ with enumerable domain which is a model of θ.

Corollary 2 (Löwenheim, 1915)

If a sentence is satisfiable, then it is true in some interpretation with an enumerable domain.

Corollary 3

If a sentence that does not contain the equals-sign '=' is satisfiable, then it is true in some interpretation whose domain is the set of all natural numbers.

Proof. This corollary follows from Corollary 2 together with these two facts: (*a*) If a sentence not containing '=' is true in some interpretation $\mathscr{J}$ with finite domain, it is true in some interpretation $\mathscr{I}$ with enumerably infinite domain. (*b*) If a sentence is true in some interpretation $\mathscr{I}$ with enumerably infinite domain, it is true in some interpretation $\mathscr{K}$ whose domain is the set of natural numbers.

Proof of (*a*). Suppose that $\mathscr{J}$ is an interpretation with finite domain E. Let e be some element of E. Let $c_0, c_1, \ldots$ be an enumerably infinite sequence of objects *not* in E. Let D be the set that contains all members of E and all the c_is. If d is in D, define d^* as follows: $d^* = d$ if d is in E, and $d^* = e$ if d is not in E. Let $\mathscr{I}$ be the following interpretation, 'which makes all the c_is indistinguishable from e': The domain of $\mathscr{I}$ is D. $\mathscr{I}$ assigns names and sentence letters whatever $\mathscr{J}$ assigns them; $\mathscr{I}$ specifies that an n-place predicate letter R is to be true of $d_1, \ldots, d_n$ iff $\mathscr{J}$ specifies that R is to be true of $d_1^*, \ldots, d_n^*$; and $\mathscr{I}$ assigns an n-place function symbol f the function $\mathfrak{f}$ such that for any $d_1, \ldots, d_n$ in D, $\mathfrak{f}(d_1, \ldots, d_n) = \mathfrak{g}(d_1^*, \ldots, d_n^*)$, where $\mathfrak{g}$ is the function $\mathscr{J}$ assigns f. We shall show that for any sentences S not containing '=', $\mathscr{I}(S) = \mathscr{J}(S)$.

To that end, for each d in D, choose a *new name* a_d, i.e., one not assigned any designation by $\mathscr{J}$. (Choose different names for different members of D.) Let $\mathscr{I}_1$ be just like $\mathscr{I}$ except that for each a_d, $\mathscr{I}_1(a_d) = d$; let $\mathscr{J}_1$ be just like $\mathscr{J}$ except that for each a_d, $\mathscr{J}_1(a_d) = d^*$. Then for *any* name b, $\mathscr{I}_1(b) = d$ iff $\mathscr{J}_1(b) = d^*$. If now, for any i between 1 and n, $\mathscr{I}_1(t_i) = d_i$ and $\mathscr{J}_1(t_i) = d_i^*$, then

$$\mathscr{I}(f(t_1, \ldots, t_n)) = \mathfrak{f}(d_1, \ldots, d_n) = \mathfrak{g}(d_1^*, \ldots, d_n^*) = \mathscr{J}(f(t_1, \ldots, t_n)).$$

It follows inductively that for any term t, $\mathscr{I}_1(t) = d$ iff $\mathscr{J}_1(t) = d^*$. Hence $\mathscr{I}_1(Rt_1 \dots t_n) = 1$ iff $\mathscr{J}_1(Rt_1 \dots t_n) = 1$. Thus every atomic sentence not containing '=' is true in $\mathscr{I}_1$ iff true in $\mathscr{J}_1$. If we can conclude that *every* '='-free sentence has the same truth-value in $\mathscr{I}_1$ as in $\mathscr{J}_1$, we can conclude that the same holds good of $\mathscr{I}$ and $\mathscr{J}$; for a sentence not containing any of the a_ds will be true in $\mathscr{I}$ iff true in $\mathscr{I}_1$, and true in $\mathscr{J}$ iff true in $\mathscr{J}_1$. As a truth-functional compound of sentences having the same truth-values in $\mathscr{I}_1$ as in $\mathscr{J}_1$ clearly has the same truth-value in $\mathscr{I}_1$ as in $\mathscr{J}_1$, if there is a simplest sentence S having opposite truth-values in $\mathscr{I}_1$ and $\mathscr{J}_1$, S must be of one of the forms $\exists v\, H$ and $\forall v\, H$. Suppose $S = \exists v\, H$. (The argument is similar if $S = \forall v\, H$.) But then, since any member of D or E is the denotation according to $\mathscr{I}_1$ or $\mathscr{J}_1$, respectively, of some name, we have that $\mathscr{I}_1(S) = 1$ iff $\mathscr{I}_1(\exists v\, H) = 1$, iff for some name a, $\mathscr{I}_1(H_v\, a) = 1$, iff – as $H_v\, a$ is *simpler* than $\exists v\, H$ – for some name a, $\mathscr{J}_1(H_v\, a) = 1$, iff $\mathscr{J}_1(\exists v\, H) = 1$, iff $\mathscr{J}_1(S) = 1$. Thus no '='-free sentence S has opposite truth-values in $\mathscr{I}_1$ and $\mathscr{J}_1$, and therefore the same goes for $\mathscr{I}$ and $\mathscr{J}$. So (*a*) holds.

The assumption that S not contain '=' is indispensable: '$\exists x\, \forall y\, x = y$' is true in any interpretation whose domain contains exactly one member, but in no interpretation with a larger domain.

Proof of (*b*). Let $d_0, d_1, d_2, \dots$ be a repetition-free enumeration of all members of the domain of $\mathscr{I}$. Then the same sentences are true in $\mathscr{I}$ as in the interpretation $\mathscr{K}$ whose domain is the set of all natural numbers, and which assigns each sentence letter the same truth-value as $\mathscr{I}$, assigns a name the number i as designation iff $\mathscr{I}$ assigns it d_i, specifies that a predicate letter is to be true of a sequence of natural numbers $i_1, i_2, \dots, i_n$ iff $\mathscr{I}$ specifies that it is to be true of $d_{i_1}, d_{i_2}, \dots, d_{i_n}$, and assigns a function symbol f the function from the one $\mathscr{I}$ assigns f by similarly everywhere 'replacing' d_{i_j} by i_j.

At one time the Skolem–Löwenheim theorem was considered philosophically perplexing because some of its consequences were perceived as anomalous. The apparent anomaly, sometimes called 'Skolem's paradox', is that there exist certain interpretations in which a certain sentence, which seems to say that non-enumerably many sets of natural numbers exist, is true, even though the domains of these interpretations contain only enumerably many sets of natural numbers, and the predicate letter in the sentence we would be inclined to translate as 'set (of natural numbers)' is true of just the sets (of natural numbers) in the domains.

There is no denying that the state of affairs thought to be paradoxical does obtain. In order to see how it arises, we shall first need an alternative account of what it is for a set E of sets of natural numbers to be enumerable, and for this we shall need a few definitions.

We shall say that one (ordered) pair, $\langle x,y \rangle$ of natural numbers *precedes* another $\langle i,j \rangle$ *in order O* if either $x+y < i+j$ or ($x+y = i+j$ and $x < i$). We define the *pairing function J* by setting $J(x,y) = z$ if $\langle x,y \rangle$ is the $(z+1)$st pair in order O. J is then a one–one function from the sets of pairs of natural numbers onto the set of natural numbers, i.e. J assigns each pair of natural numbers as argument a unique natural number as value, assigns different pairs different numbers, and assigns every number to some pair or other. (J, incidentally, is recursive.)

We shall call a set w of natural numbers an *enumerator* of a set E of sets of natural numbers if

$$\forall z\,(z \text{ is a set of natural numbers } \& \ z \text{ is in } E \rightarrow$$
$$\exists x\,(x \text{ is a natural number } \& \ \forall y\,(y \text{ is a natural number} \rightarrow$$
$$(y \text{ is in } z \leftrightarrow J(x,y) \text{ is in } w)))).$$

The fact about enumerators and enumerability that we need is that *a set E of sets of natural numbers is enumerable iff E has an enumerator.*

(The reason: suppose E is enumerable. Let $e_0, e_1, e_2, \ldots$ be an enumeration of sets of natural numbers that contains all of the members of E, and possibly some other sets of natural numbers too. Then the set of numbers $J(x,y)$ such that y is in e_x is an enumerator of E. Conversely, if w is an enumerator of E, then, where $e_x =$ the set of those numbers y such that $J(x,y)$ is in w, $e_0, e_1, e_2, \ldots$ is an enumeration that contains all members of E, and therefore E is enumerable.)

We want now to look at a language and some of its interpretations. The language contains just the names '**0**', '**1**', '**2**', etc., one two-place function symbol '**J**', two one-place predicate letters '**N**' and '**S**', and one two-place predicate letter '**ϵ**'. The interpretations we are interested in are those whose domains contain all natural numbers and some (possibly all) sets of natural numbers, but nothing else; which assign each numeral its ordinary designation; which assign '**J**' the pairing function (extended so as to take some arbitrary value – say 17 – for the argument x, y if either x or y is a set); and which specify that '**N**' is to be true of any number, '**S**' is to be true of any set of numbers, and '**ϵ**' is to be true of y, z iff the number y is in the set z. One of these interpretations, $\mathscr{I}$, is called the *standard* interpretation. It is the one whose domain contains *all* sets of natural numbers.

In all of these interpretations the sentence

$$-\exists w(\mathbf{S}w \,\&\, \forall z(\mathbf{S}z \to \exists x(\mathbf{N}x \,\&\, \forall y(\mathbf{N}y \to (y \in z \leftrightarrow \mathbf{J}(x,y) \in w))))) \qquad (2)$$

will have a truth-value. It will be *true* in one of them iff there is no enumerator of the set of all sets of numbers in the domain of the interpretation *that is itself in the domain of the interpretation* (as we can see by checking back to the definition of 'enumerator'). We can't simply say that the sentence is true in an interpretation iff there is no enumerator of the set of all sets of numbers in the domain, because the quantifier '$\exists w$' is understood to range over, or 'refer to', members of the domain of the interpretation alone.

There is, as we know, *no* enumerator of the set of all sets of numbers in the domain of $\mathscr{I}$ since all sets of numbers are in $\mathscr{I}$'s domain. *A fortiori*, there is no such enumerator in the domain of $\mathscr{I}$, and sentence (2) is therefore true in $\mathscr{I}$, and can be said to mean 'Non-enumerably many sets exist', when interpreted 'over' $\mathscr{I}$, since it then denies that there is an enumerator of the set of all sets of numbers. By the Skolem–Löwenheim theorem, $\mathscr{I}$ has an elementarily equivalent subinterpretation $\mathscr{J}$, whose domain is enumerable and thus contains only enumerably many sets of numbers. (All of 0, 1, 2, etc. are in $\mathscr{J}$'s domain since these are the designations $\mathscr{I}$, and hence $\mathscr{J}$, assigns to '**0**', '**1**', '**2**', etc.) Since $\mathscr{I}$ and $\mathscr{J}$ are elementarily equivalent, (2) is a sentence true in $\mathscr{J}$, and therefore in an interpretation in whose domain there are only enumerably many sets of numbers, and in which '**S**' is true of just the sets of numbers in its domain. This is Skolem's 'paradox'.

How is the paradox to be resolved? Well, although the set of all sets in the domain of $\mathscr{J}$ does indeed have an enumerator, since it is enumerable, none of its enumerators can be *in* the domain of $\mathscr{J}$ (for otherwise,

$$(\mathbf{S}w \,\&\, \forall z(\mathbf{S}z \to \exists x(\mathbf{N}x \,\&\, \forall y(\mathbf{N}y \to (y \in z \leftrightarrow \mathbf{J}(x,y) \in w)))))$$

would be true in $\mathscr{J}$ of one of them and thus (2) would be false in $\mathscr{J}$.) So part of the explanation of how (2) can be true in $\mathscr{J}$ is that those sets which 'verify' the claim that the set of sets in the domain of $\mathscr{J}$ is enumerable are not themselves members of the domain of $\mathscr{J}$.

A further part of the explanation is that what a sentence should be understood as saying or meaning or denying is at least as much a function of the interpretation over which the sentence is interpreted (and even of the way in which that interpretation is described or referred to) as of the symbols that constitute it. (2) can be understood as saying 'non-enumerably many sets exist' when its quantifiers are understood as ranging over a collection containing all numbers and all sets of numbers, such as the domain of the standard interpretation $\mathscr{I}$, but it cannot be so understood when its quantifiers range over other domains, in particular, not when they range over members of countable domains. The sentence (2) – that sequence of symbols – 'says' something only when supplied with an interpretation. It may be surprising and even amusing that (2) is true in all sorts of interpretations, including, perhaps, some subinterpretations $\mathscr{J}$ of $\mathscr{I}$ that have enumerable domains, but it should not *a priori* seem impossible that it be true in these. Interpreted over such a $\mathscr{J}$, it will only say 'the domain of $\mathscr{J}$ contains no enumerator of the set of sets of numbers in $\mathscr{J}$' which is, of course, true.

Exercises

13.1 Suppose that interpretations were allowed to assign objects to non-enumerably many one-place predicate letters. Show that the Skolem–Löwenheim theorem, as we stated it, would then be false.

13.2 A subinterpretation $\mathscr{J}$ of $\mathscr{I}$ is called an *elementary subinterpretation* of $\mathscr{I}$ if for every formula F of the language of $\mathscr{I}$ and every sequence $o_1, \ldots, o_n$ in the domain of $\mathscr{J}$

$$\mathscr{I}^{a_1 \ldots a_n}_{o_1 \ldots o_n}(F^*) = \mathscr{J}^{a_1 \ldots a_n}_{o_1 \ldots o_n}(F^*).$$

Here F is supposed to contain at most the n variables $v_1, \ldots, v_n$ free, $a_1, \ldots, a_n$ are n names to which $\mathscr{I}$ assigns no designation, and F^* is the result of substituting $a_1, \ldots, a_n$ for all free occurrences of $v_1, \ldots, v_n$ (respectively) in F. Show that any interpretation $\mathscr{I}$ has an elementary subinterpretation $\mathscr{J}$ whose domain is enumerable; deduce the version of the Skolem–Löwenheim theorem proved in the text from this statement.

13.3 Show that the version of the Skolem–Löwenheim theorem proved in the text implies the axiom of dependent choice. (Cf. Exercise 11.5.)

Solutions

13.1 Let $\mathbb{R}$ be the set of real numbers. For each r in $\mathbb{R}$ let A_r be a one-place predicate letter. Let $\mathscr{I}$ be the interpretation with domain $\mathbb{R}$ which specifies that each A_r is to be true of r and r alone. For each r, $\exists x\, A_r x$ is true in $\mathscr{I}$, and hence true in every elementarily equivalent subinterpretation. Each real number r must therefore belong to the domain of every elementarily equivalent subinterpretation $\mathcal{J}$. Each such $\mathcal{J}$ will therefore have a non-enumerable domain.

13.3 Suppose X is a nonempty set, and for any x in X there is a y in X such that xRy. Let $\mathcal{I}$ be the interpretation whose domain is X, and according to which '**R**' is true of x, y iff xRy. $\forall x \exists y\, x\mathbf{R}y$ is true in $\mathcal{I}$. Let $\mathcal{J}$ be an elementarily equivalent subinterpretation of $\mathcal{I}$ whose domain E is enumerable. Let $e_0, e_1, e_2, \ldots$ be an enumeration of E. $\forall x \exists y\, x\mathbf{R}y$ is true in $\mathcal{J}$. Therefore for every e_i in E there will be an e_j in E such that $e_i R e_j$. Define f by: $f(0) = e_0$; for each n, $f(n+1) = e_j$ iff $f(n)\ Re_j$ and for every $k < j$, not: $f(n)\ Re_k$. The axiom of dependent choice is not required to guarantee the existence of f.

14
Representability in Q

The present chapter falls into three parts. In the first part we introduce the notion of *representability of a function* (of natural numbers) *in a theory* and present a theory, called 'Q'. In the second part we give an alternative characterization of the recursive functions,† and in the third we use this new characterization to show that every recursive function is representable in the theory Q. In the next chapter several important results about undecidability, indefinability and incompleteness will be shown to follow from the latter result. The converse, that every function representable in Q is recursive, is also true, and we shall also indicate why at the end of the next chapter (Exercise 15.2).

Part I

We shall take a theory to be a set of sentences in some language that contains all of its logical consequences that are sentences in that language. If a sentence A is a member of theory T, it is called a *theorem* of T; to indicate that A is a theorem of T, we write: $\vdash_T A$.

From now through Chapter 21, we shall confine our attention to *numerical* theories: theories whose language contains the name $\mathbf{0}$ and the one-place function symbol $'$. (Q will be such a theory.) The *numeral* for n, $\mathbf{n}$, is the result of attaching n occurrences of $'$ to (the right of) $\mathbf{0}$. Thus $\mathbf{3} = \mathbf{0}'''$ and the numeral for $n+1$ is $\mathbf{n}'$. For any natural number n, $\mathbf{n}$ is an expression or sequence of symbols, a *term* of the sort described.

If A is a formula that contains free occurrences of the n (distinct) variables $x_1, \ldots, x_n$, we shall sometimes refer to A as $A(x_1, \ldots, x_n)$. For any natural numbers $p_1, \ldots, p_n$, $A(\mathbf{p_1}, \ldots, \mathbf{p_n})$ is the result of substituting an occurrence of $\mathbf{p_i}$ for each free occurrence of x_i in $A(x_1, \ldots, x_n)$ (for each i between 1 and n). In discussing a formula $A(x_1, \ldots, x_n)$ we may wish to consider a formula, which we refer to as '$A(y_1, \ldots, y_n)$'. This is to be understood to be the formula that results when any bound occurrence of y_i that may occur in $A(x_1, \ldots, x_n)$ is first replaced by an occurrence of a new

† Cf. the last paragraph of Chapter 8.

variable z_i (different z_is for different y_is), and then an occurrence of y_i is substituted for each free occurrence of x_i in the result.

To reduce clutter, we shall write 'p' instead of '$p_1, \ldots, p_n$', 'x' instead of '$x_1, \ldots, x_n$', and '$\mathbf{p}$' instead of '$\mathbf{p}_1, \ldots, \mathbf{p}_n$'.

We can now define representability. An n-place function f is *representable* in a theory T if there is a formula $A(\mathsf{x}, x_{n+1})$ such that for any natural numbers p, j, if $f(\mathsf{p}) = j$, then $\vdash_T \forall x_{n+1}(A(\mathbf{p}, x_{n+1}) \leftrightarrow x_{n+1} = \mathbf{j})$. In this case the formula $A(\mathsf{x}, x_{n+1})$ is said to *represent* f in T.

The requirement that $\vdash_T \forall x_{n+1}(A(\mathbf{p}, x_{n+1}) \leftrightarrow x_{n+1} = \mathbf{j})$ should hold whenever $f(\mathsf{p}) = j$ is equivalent to the requirement that both $\vdash_T A(\mathbf{p}, \mathbf{j})$ and $\vdash_T \forall x_{n+1}(A(\mathbf{p}, x_{n+1}) \rightarrow x_{n+1} = \mathbf{j})$ should hold whenever $f(\mathsf{p}) = j$. If the sentence $\mathbf{j} \neq \mathbf{k}$ is a theorem of T whenever $j \neq k$ (and we shall see that Q is a theory of which this is so), then if A represents f in T and $f(\mathsf{p}) \neq k$, then $\vdash_T -A(\mathbf{p}, \mathbf{k})$ (for $\vdash_T \mathbf{j} \neq \mathbf{k}$, where $j = f(\mathsf{p})$).

The language of theory Q is L, the *language of arithmetic*. L contains four non-logical symbols, the name $\mathbf{o}$, the one-place function symbol $'$, and two two-place function symbols, $\mathbf{+}$ and $\cdot$. Q is the set of sentences in L that are logical consequences of these seven sentences, the *axioms of Q*:

$Q1 \quad \forall x \forall y(x' = y' \rightarrow x = y)$,

$Q2 \quad \forall x\, \mathbf{o} \neq x'$,

$Q3 \quad \forall x\, (x \neq \mathbf{o} \rightarrow \exists y\, x = y')$,

$Q4 \quad \forall x\, x \mathbf{+} \mathbf{o} = x$,

$Q5 \quad \forall x \forall y\, x \mathbf{+} y' = (x \mathbf{+} y)'$,

$Q6 \quad \forall x\, x \cdot \mathbf{o} = \mathbf{o}$,

$Q7 \quad \forall x \forall y\, x \cdot y' = (x \cdot y) \mathbf{+} x$.

Q is a consistent theory, for all of its axioms are true in the *standard interpretation* $\mathcal{N}$ *for its language L*, in which the domain is the set of all natural numbers, $\mathbf{o}$ is assigned zero as denotation, and $'$, $\mathbf{+}$, and $\cdot$ are assigned the successor, addition, and multiplication functions. Q is a theory that is rather strong in certain ways (all recursive functions are representable in it), but rather weak in others (e.g. $\forall x \forall y\, x \mathbf{+} y = y \mathbf{+} x$ is not a theorem of Q, as an exercise at the end of the chapter shows). Tarski, Mostowski, and R. Robinson have written that it 'is distinguished by the simplicity and clear mathematical content of its axioms'. We shall devote the remainder of this chapter to showing that all recursive functions are representable in Q.

Part II

We recall from Chapters 7 and 8 that the recursive functions can be characterized as those functions obtainable from the zero function, the successor function and the identity functions by means of a finite number of applications of the operations of composition, primitive recursion, and minimization of those functions called *regular* functions.

The *zero* function z is the one-place function whose value for all arguments is zero.

The *successor* function $'$ ($= s$) is the one-place function whose value for any argument i is $i+1$ ($= i'$, the successor of i).

For each $m \geqslant 1$ and each $n \leqslant m$, there is an m-place *identity* function id_n^m. For any natural numbers $i_1, \ldots, i_m$, $\mathrm{id}_n^m(i_1, \ldots, i_m) = i_n$.

If f is an m-place function, and $g_1, \ldots, g_m$ are all n-place functions, then the n-place function h is said to be obtained from $f, g_1, \ldots, g_m$ by *composition* if for any natural numbers p, $h(\mathsf{p}) = f(g_1(\mathsf{p}), \ldots, g_m(\mathsf{p}))$.

If f is an n-place function and g is an $(n+2)$-place function, then the $(n+1)$-place function h is said to be obtained from f and g by *primitive recursion* if for any natural numbers p, k, $h(\mathsf{p}, 0) = f(\mathsf{p})$ and

$$h(\mathsf{p}, k+1) = g(\mathsf{p}, k, h(\mathsf{p}, k)).$$

An $(n+1)$-place function f is called *regular* if for any natural numbers p, there exists at least one natural number i such that $f(\mathsf{p}, i) = 0$. If f is a regular $(n+1)$-place function, then the n-place function g is said to be obtained from f by *minimization* if for any natural numbers p,

$$g(\mathsf{p}) = \mu i f(\mathsf{p}, i) = 0,$$

where 'μi' means 'the least natural number i such that'.

If R is an n-place relation of natural numbers (i.e. a set of ordered n-tuples of natural numbers), then the *characteristic function* of R is the n-place function f_R such that for any p,

$$f_R(\mathsf{p}) = \begin{cases} 1 & \text{if } R\mathsf{p} \text{ (i.e. if } \mathsf{p} \text{ is in } R), \\ 0 & \text{if not } R\mathsf{p}. \end{cases}$$

$f_=$ is thus the characteristic function of the identity relation. For any i, j, $f_=(i,j) = 1$ if $i = j$ and $f_=(i,j) = 0$ if $i \neq j$.

We shall call a function *Recursive* (capital 'R') if it can be obtained from the functions $+, \cdot, f_=$, and the various id_n^m by means of a finite number of applications of the two operations of composition and minimization of regular functions.

All Recursive functions are recursive, for $+$, $\cdot$, $f_=$ and the functions id^m_n are recursive, and the recursive functions are closed under composition and minimization of regular functions. On the other hand the zero function z is Recursive, for, as we saw in Chapter 7, z can be obtained from $\cdot$ by minimization. And s is obtainable by composition from Recursive functions and thus is Recursive too: for all i,

$$s(i) = i+1 = \mathrm{id}^1_1(i)+f_=(\mathrm{id}^1_1(i), \mathrm{id}^1_1(i)).$$

In the rest of Part II we show that all other recursive functions are also Recursive, and for this it suffices to show that if f and g are Recursive functions from which h is obtained by primitive recursion, then h is also Recursive.

We must first see that certain relations and functions are Recursive. A relation is Recursive iff its characteristic function is Recursive. (So $=$ is Recursive.)

Suppose, for example, that d is a two-place Recursive function and e is a three-place Recursive function. Let R be the 6-place relation defined by: Ri, j, k, m, n, q iff $d(j, n) = e(n, k, m)$. Then R is Recursive, for

$$f_R(i,j,k,m,n,q) = f_=(d(\mathrm{id}^6_2(i,j,k,m,n,q), \mathrm{id}^6_5(i,j,k,m,n,q)),$$
$$e(\mathrm{id}^6_5(i,j,k,m,n,q), \mathrm{id}^6_3(i,j,k,m,n,q), \mathrm{id}^6_4(i,j,k,m,n,q))).$$

Similarly, all other relations obtained by 'setting Recursive functions equal to each other' are Recursive.

Suppose that R and S are n-place Recursive relations. Then the intersection $(R\&S)$ of R and S and the complement $-R$ of R are Recursive, for

$$f_{(R\&S)}(\mathsf{p}) = f_R(\mathsf{p})\cdot f_S(\mathsf{p}), \text{ and } f_{-R}(\mathsf{p}) = f_=(f_R(\mathsf{p}), z(f_R(\mathsf{p}))) \; (=f_=(f_R(\mathsf{p}), 0)).$$

As & and $-$ suffice to define all truth-functional connectives, any relation obtained from Recursive relations by truth-functional, i.e., Boolean, operations is also Recursive. E.g. if Ri, j, k if and only if either $i = k$ or $k \neq j$, then R is Recursive.

If R is an $(n+1)$-place relation, then e will be said to be obtained from R by *minimization* if for any p, $e(\mathsf{p}) = \mu i R\mathsf{p}, i$. ($e$ may be undefined for some p.) An $(n+1)$-place relation will be called *regular* if for any p, there is an i such that $R\mathsf{p}, i$. The function obtained from a regular relation by minimization is everywhere defined.

If R is a regular, Recursive $(n+1)$-place relation, and e is obtained from R by minimization, then e is Recursive, for $e(\mathsf{p}) = \mu i f_{-R}(\mathsf{p}, i) = 0$. ($R\mathsf{p}, i$ iff $f_{-R}(\mathsf{p}, i) = 0$.)

Finally, if R is an $(n+1)$-place relation, then the $(n+1)$-place relation S will be said to be obtained from R by *bounded universal quantification*† if (for all p, j) $S\mathsf{p}, j$ iff $\forall i < j\, R\mathsf{p}, i$. If R is Recursive and S is obtained from R by bounded universal quantification, then S is Recursive. *Proof.* Let T be defined by: $T\mathsf{p}, j, i$ iff either not $R\mathsf{p}, i$ or $i = j$. T is regular (for all p, j, $T\mathsf{p}, j, j$) and Recursive (by the foregoing). Let d be defined by: $d(\mathsf{p}, j) = \mu i T\mathsf{p}, j, i$. d is Recursive. For any p, j, $d(\mathsf{p}, j) \leqslant j$. And $d(\mathsf{p}, j) = j$ iff for every $i < j$, $R\mathsf{p}, i$; iff $S\mathsf{p}, j$. So if e is defined by: $e(\mathsf{p}, j) = f_{=}(j, d(\mathsf{p}, j))$, then e is Recursive and the characteristic function of S.

S is said to be obtained from R by *bounded existential quantification*† if (for all p, j) $S\mathsf{p}, j$ iff $\exists i < j\, R\mathsf{p}, i$. Analogously, any relation obtained from a Recursive relation by bounded existential quantification is Recursive.

We'll now define J, the pairing function.

Definition

$$J(a, b) = \tfrac{1}{2}(a+b)(a+b+1)+a.$$

Lemma 14.1

J is a one–one function whose domain is the set of all ordered pairs $\langle a, b\rangle$ of natural numbers and whose range is the set of all natural numbers.

Proof. There are $n+1$ pairs $\langle a, b\rangle$ such that $a+b = n$ (*viz.*, $\langle 0, n\rangle$, $\langle 1, n-1\rangle, \ldots, \langle n, 0\rangle$). So there are $0+1+2+\ldots+n$, $= \tfrac{1}{2}n(n+1)$, pairs $\langle c, d\rangle$ such that $c+d < n$. We'll say that $\langle c, d\rangle$ *precedes* $\langle a, b\rangle$ *in order* O (cf. Chapter 13) if either $c+d < a+b$ or ($c+d = a+b$ and $c < a$). There are a natural numbers less than a. So if $a+b = n$, there are $\tfrac{1}{2}n(n+1)+a$ pairs that precede $\langle a, b\rangle$ in order O. But if $a+b = n$, then

$$\tfrac{1}{2}n(n+1)+a = J(a, b).$$

So $J(a, b)$ is precisely the number of pairs preceding $\langle a, b\rangle$ in order O.

$a, b \leqslant J(a, b)$. J is Recursive, for J is obtained from a regular Recursive function by minimization: $J(a, b) = \mu i[i+i = (a+b)(a+b+1)+2a]$.

Define K and L, the inverse pairing functions, by:

$K(i) = \mu a \exists b \leqslant i J(a, b) = i$ (i.e. $\mu a[\exists b < i J(a, b) = i \vee J(a, i) = i]$),
$L(i) = \mu b \exists a \leqslant i J(a, b) = i$. By Lemma 14.1, K and L are Recursive.

† These definitions differ slightly from those given in Chapter 7 in that '$<$' is used instead of '$\leqslant$'.

We now define some more relations and functions; it should be evident from their definitions that they are Recursive.

m *divides* $n \leftrightarrow \exists i \leqslant n\ i \cdot m = n$.

p *is prime* $\leftrightarrow \{p \neq 0\ \&\ p \neq 1\ \&\ \forall m \leqslant p[m \text{ divides } p \rightarrow (m = 1 \vee m = p)]\}$.

$m < n \leftrightarrow \exists i < n\ i = m$.

$m \dot{-} n = \mu i([n < m \rightarrow n+i = m]\ \&\ [-n < m \rightarrow i = 0])$.

n *is a power of the prime* $p \leftrightarrow \{n \neq 0\ \&\ p \text{ is prime}\ \&\ \forall m \leqslant n[m \text{ divides } n$
$\rightarrow (m = 1 \vee p \text{ divides } m)]\}$.

(Notice that we can't simply say that n is a power of k iff for every $m \leqslant n$, if m divides n, then $m = 1$ or k divides m; let $n = k = 6$, $m = 2$.)

$\eta(p, b) = \mu i[(p \text{ is prime}\ \&\ i \text{ is a power of the prime } p\ \&\ i > b\ \&\ i > 1)$
$\vee\ (p \text{ is not prime}\ \&\ i = 0)]$.

For prime p, $\eta(p, b)$ is the least number whose base p numeral is *longer* than the base p numeral for b. E.g. $\eta(7, 25) = 49$. (Note that $25 = 34_7$ and $49 = 100_7$.)

$a \underset{p}{*} b = a \cdot \eta(p, b) + b$.

If $a \neq 0$, $a \underset{p}{*} b$ is $\neq 0$ and is the number denoted in base p notation by the result of writing the base p numeral for b directly to the right of that for a. So, e.g. $4 \underset{7}{*} 25 = 4 \cdot 49 + 25 = 221$, and $4_7 = 4$, $34_7 = 25$, and $434_7 = 221$. In what follows, association is assumed to be to the left: '$a \underset{p}{*} b \underset{p}{*} c$' means '$(a \underset{p}{*} b) \underset{p}{*} c$', not '$a \underset{p}{*} (b \underset{p}{*} c)$'. Then if $a \neq 0$, $a \underset{p}{*} b \underset{p}{*} c \underset{p}{*} \ldots \underset{p}{*} z$ is the number denoted in base p notation by the result of writing down the base p numeral for b directly to the right of that for a, then that for c directly to the right of *that*, ... and then that for z directly to the right of *that*.

$a\, part_p\, b \leftrightarrow \exists c \leqslant b\, \exists d \leqslant b\, [c \underset{p}{*} a \underset{p}{*} d = b \vee c \underset{p}{*} a = b \vee a \underset{p}{*} d = b \vee a = b]$.

a part$_p$ b iff $a = 0$ or $a = b$ or b's base p numeral can be obtained by attaching base p numerals to the left and/or right of a's base p numeral.

$\alpha(p, q, j) = \mu i[(p \dot{-} 1) \underset{p}{*} j \underset{p}{*} i \text{ part}_p\, q \vee i = q]$. ('$i = q$' is for 'waste cases'.)

$\beta(i, j) = \alpha(K(i), L(i), j)$.

Lemma 14.2. (*The* β*-function lemma*)

For any k and any finite sequence of natural numbers $i_0, \ldots, i_k$, there exists a natural number i such that for every $j \leqslant k$, $\beta(i, j) = i_j$.

Proof. Let $i_0, \ldots, i_k$ be a finite sequence of natural numbers. Let p be a prime such that $p-1$ is greater than all of $i_0, \ldots, i_k, k$. (There are infinitely many primes.) Let $s=p-1$. $s \neq 0$. All of $s, 0, i_0, \ldots, k, i_k$ are represented by single digits in base p notation (!). Let

$$q = s \underset{p}{*} 0 \underset{p}{*} i_0 \underset{p}{*} s \underset{p}{*} 1 \underset{p}{*} i_1 \underset{p}{*} \ldots \underset{p}{*} s \underset{p}{*} k \underset{p}{*} i_k.$$

Then for every $j \leqslant k$, $\alpha(p,q,j) = i_j$. Let $i = J(p,q)$. Then for every $j \leqslant k$, $\beta(i,j) = i_j$.

Suppose now that f is an n-place function, that g is an $(n+2)$-place function, and that h is obtained from f and g by primitive recursion. Then $h(\mathsf{p}, 0)=f(\mathsf{p})$ and (for any k) for every $j<k, h(\mathsf{p},j')=g(\mathsf{p},j,h(\mathsf{p},j))$. By the β-function lemma, for any k there is an i such that for every $j \leqslant k$, $\beta(i,j) = h(\mathsf{p},j)$. These is are precisely those such that $\beta(i,0) = f(\mathsf{p})$ and for every $j < k$, $\beta(i,j') = g(\mathsf{p},j,\beta(i,j))$. Therefore, if R is the $(n+2)$-place relation defined by:

$$R\mathsf{p}, k, i \text{ iff } \beta(i,0) = f(\mathsf{p}) \;\&\; \forall j < k\; \beta(i,j') = g(\mathsf{p},j,\beta(i,j)),$$

then R is regular; and R is Recursive if f and g are. So if d is the $(n+1)$-place function defined by: $d(\mathsf{p},k) = \mu i R\mathsf{p},k,i$, then d is Recursive if f and g are. Moreover $d(\mathsf{p},k)$ is the least i such that for every $j \leqslant k$, $\beta(i,j) = h(\mathsf{p},j)$. For any such i, $\beta(i,k) = h(\mathsf{p},k)$. We may thus define h by composition from β, d, id^{n+1}_{n+1}: $h(\mathsf{p},k) = \beta(d(\mathsf{p},k), \mathrm{id}^{n+1}_{n+1}(\mathsf{p},k))$. As β and id^{n+1}_{n+1} are Recursive, h is Recursive if f and g are Recursive.

Thus any function obtained by primitive recursion from Recursive functions is itself Recursive.

We have therefore shown that a function is recursive if and only if it is Recursive.

Part III

We'll now show that all Recursive functions are representable in Q, from which we conclude that all recursive functions are representable in Q.

The identity functions id^m_n are all representable in Q: since for any $i_1, \ldots, i_m$, $\forall x_{m+1}((\mathbf{i_1} = \mathbf{i_1} \;\&\; \ldots \;\&\; \mathbf{i_m} = \mathbf{i_m} \;\&\; x_{m+1} = \mathbf{i_n}) \leftrightarrow x_{m+1} = \mathbf{i_n})$ is *valid*,

$$(x_1 = x_1 \;\&\; \ldots \;\&\; x_m = x_m \;\&\; x_{m+1} = x_n)$$

represents id^m_n in Q.

We now show that addition is represented in Q by the formula

$$x_1 + x_2 = x_3.$$

Lemma 14.3

Suppose that $i+j = k$. Then $\vdash_Q \mathbf{i}+\mathbf{j} = \mathbf{k}$.

Proof. The proof is an induction on j. Basis step: $j = 0$. We must show that $\vdash_Q \mathbf{i}+\mathbf{0} = \mathbf{i}$. But this follows from $Q4$. Induction step: $j = m'$. Then for some n, $k = n'$ and $i+m = n$, whence by the induction hypothesis, $\vdash_Q \mathbf{i}+\mathbf{m} = \mathbf{n}$, and therefore $\vdash_Q (\mathbf{i}+\mathbf{m})' = \mathbf{n}'$. Since

$$\vdash_Q (\mathbf{i}+\mathbf{m})' = \mathbf{i}+\mathbf{m}'$$

by $Q5$, it follows that $\vdash_Q \mathbf{i}+\mathbf{j} = \mathbf{k}$.

Lemma 14.4

$x_1 + x_2 = x_3$ represents addition in Q.

Proof. $\forall x_3(\mathbf{i}+\mathbf{j} = x_3 \leftrightarrow x_3 = \mathbf{k})$ is a logical consequence of $\mathbf{i}+\mathbf{j} = \mathbf{k}$, which, by 14.3, is a theorem of Q if $i+j = k$.

Multiplication:

Lemma 14.5

Suppose that $i \cdot j = k$. Then $\vdash_Q \mathbf{i} \cdot \mathbf{j} = \mathbf{k}$.

Proof. Induction on j. If $j = 0$, we must show that $\vdash_Q \mathbf{i} \cdot \mathbf{0} = \mathbf{0}$. But this follows from $Q6$. If $j = m'$, then $k = n+i$, where $n = i \cdot m$. By the hypothesis of the induction, $\vdash_Q \mathbf{i} \cdot \mathbf{m} = \mathbf{n}$. By 14.3, $\vdash_Q \mathbf{n}+\mathbf{i} = \mathbf{k}$. By $Q7$, $\vdash_Q \mathbf{i} \cdot \mathbf{m}' = \mathbf{i} \cdot \mathbf{m}+\mathbf{i}$. So $\vdash_Q \mathbf{i} \cdot \mathbf{m}' = \mathbf{k}$, i.e., $\vdash_Q \mathbf{i} \cdot \mathbf{j} = \mathbf{k}$.

Lemma 14.6

$x_1 \cdot x_2 = x_3$ represents multiplication in Q.

Proof. This follows from 14.5 just as 14.4 followed from 14.3.

So $+$ and $\cdot$ are representable in Q.

Let's now verify that if $i \neq j$, then $\mathbf{i} \neq \mathbf{j}$ is a theorem of Q.

Lemma 14.7

If $i \neq j$, then $\vdash_Q \mathbf{i} \neq \mathbf{j}$.

Proof. We may suppose without loss of generality that $i < j$. Induction on i. If $i = 0$, then $j > 0$, and so for some n, $j = n'$. We must show that $\vdash_Q \mathbf{0} \neq \mathbf{j}$, i.e., that $\vdash_Q \mathbf{0} \neq \mathbf{n'}$. But this immediately follows from $Q2$. If $i = m'$, then $j = n'$ and $m < n$, for some n. By the induction hypothesis, $\vdash_Q \mathbf{m} \neq \mathbf{n}$, and hence by $Q1$, $\vdash_Q \mathbf{m'} \neq \mathbf{n'}$, i.e., $\vdash_Q \mathbf{i} \neq \mathbf{j}$.

Lemma 14.8

Let $A(x_1, x_2, x_3) =$ the formula

$$(x_1 = x_2 \,\&\, x_3 = \mathbf{1}) \vee (x_1 \neq x_2 \,\&\, x_3 = \mathbf{0}).$$

Then $A(x_1, x_2, x_3)$ represents $f_=$ in Q.

Proof. If $f_=(i,j) = 1$, then $i = j$. So $\vdash_Q \mathbf{i} = \mathbf{j} \,\&\, \mathbf{1} = \mathbf{1}$, so $\vdash_Q A(\mathbf{i}, \mathbf{j}, \mathbf{1})$, whence $\vdash_Q \forall x_3(A(\mathbf{i}, \mathbf{j}, x_3) \leftrightarrow x_3 = \mathbf{1})$, as $\forall x_3 (A(\mathbf{i}, \mathbf{j}, x_3) \rightarrow x_3 = \mathbf{1})$ is a logical consequence of $A(\mathbf{i}, \mathbf{j}, \mathbf{1})$ when $i = j$. If $f_=(i,j) = 0$, then $i \neq j$. By 14.7 $\vdash_Q \mathbf{i} \neq \mathbf{j}$. So $\vdash_Q \mathbf{i} \neq \mathbf{j} \,\&\, \mathbf{0} = \mathbf{0}$, whence $\vdash_Q \forall x_3(A(\mathbf{i}, \mathbf{j}, x_3) \leftrightarrow x_3 = \mathbf{0})$.

Thus $f_=$ is also representable in Q. We now show that any function obtained by composition from functions representable in Q is also representable in Q.

Suppose that $A(x_1, \ldots, x_m, x)$ represents f in Q, and that

$$B_1(\mathsf{x}, x_{n+1}), \ldots, B_m(\mathsf{x}, x_{n+1})$$

represent $g_1, \ldots, g_m$, respectively. Then if h is obtained from $f, g_1, \ldots, g_m$, by composition,

$$C(\mathsf{x}, x), = \exists y_1 \ldots \exists y_m(B_1(\mathsf{x}, y_1) \,\&\, \ldots \,\&\, B_m(\mathsf{x}, y_m) \,\&\, A(y_1, \ldots, y_m, x)),$$

represents h.

For if $g_1(\mathsf{p}) = i_1, \ldots, g_m(\mathsf{p}) = i_m$, and $f(i_1, \ldots, i_m) = j$, then $h(\mathsf{p}) = j$, and

$$\vdash_Q B_1(\mathbf{p}, \mathbf{i}_1), \tag{1}$$

$$\vdash_Q \forall x_{n+1}(B_1(\mathbf{p}, x_{n+1}) \rightarrow x_{n+1} = \mathbf{i}_1), \tag{2}$$

$$\vdots$$

$$\vdash_Q B_m(\mathbf{p}, \mathbf{i}_m), \tag{2m-1}$$

$$\vdash_Q \forall x_{n+1}(B_m(\mathbf{p}, x_{n+1}) \rightarrow x_{n+1} = \mathbf{i}_m), \tag{2m}$$

$$\vdash_Q A(\mathbf{i}_1, \ldots, \mathbf{i}_m, \mathbf{j}), \text{ and} \tag{2m+1}$$

$$\vdash_Q \forall x(A(\mathbf{i}_1, \ldots, \mathbf{i}_m, x) \rightarrow x = \mathbf{j}). \tag{2m+2}$$

(1), (3), ..., $(2m-1)$, and $(2m+1)$ clearly entail that $\vdash_Q C(\mathbf{p}, \mathbf{j})$. And (2), (4), ..., $(2m)$, and $(2m+2)$ entail that $\vdash_Q \forall x(C(\mathbf{p}, x) \to x = \mathbf{j})$. We may see this as follows: Assume we have $B_1(\mathbf{p}, y_1), \ldots, B_m(\mathbf{p}, y_m)$, and $A(y_1, \ldots, y_m, x)$. From (2), we have $y_1 = \mathbf{i}_1, \ldots,$ and from $(2m)$ we have $y_m = \mathbf{i}_m$. So we have $A(\mathbf{i}_1, \ldots, \mathbf{i}_m, x)$, whence from $(2m+2)$ we have $x = \mathbf{j}$. Thus

$$\vdash_Q \forall x(\exists y_1 \ldots \exists y_m(B_1(\mathbf{p}, y_1) \,\&\, \ldots \,\&\, B_m(\mathbf{p}, y_m) \,\&\, A(y_1, \ldots, y_m, x)) \to x = \mathbf{j}),$$

i.e. $\vdash_Q \forall x(C(\mathbf{p}, x) \to x = \mathbf{j})$, and therefore C represents h.

Lemma 14.9

For each i, $\vdash_Q \forall x\, x' + \mathbf{i} = x + \mathbf{i}'$.

Proof. Induction on i. If $i = 0$, $\forall x\, x' + \mathbf{0} = x + \mathbf{0}'$ follows from

$$\forall x\,(x' + \mathbf{0} = x' = (x + 0)' = x + \mathbf{0}'),$$

which follows from $Q4$ and $Q5$. If $i = m'$, then by the induction hypothesis $\vdash_Q \forall x\, x' + \mathbf{m} = x + \mathbf{m}'$, whence by $Q5$, $\vdash_Q \forall x\,(x' + \mathbf{m}' = (x' + \mathbf{m})' = (x + \mathbf{m}')' = x + \mathbf{m}'')$, and hence

$$\vdash_Q \forall x\, x' + \mathbf{i} = x + \mathbf{i}'.$$

We now define $x_1 < x_2$ to be the formula $\exists x_3\, x_3' + x_1 = x_2$.

Lemma 14.10

If $i < j$, then $\vdash_Q \mathbf{i} < \mathbf{j}$.

Proof. Suppose $i < j$. Then for some m, $m' + i = j$. By 14.3,

$$\vdash_Q \mathbf{m}' + \mathbf{i} = \mathbf{j}, \quad \text{and so} \quad \vdash_Q \exists x_3\, x_3' + \mathbf{i} = \mathbf{j}, \text{ i.e. } \vdash_Q \mathbf{i} < \mathbf{j}.$$

Lemma 14.11

For each i, $\vdash_Q \forall x(x < \mathbf{i} \to x = \mathbf{0} \vee \ldots \vee x = \mathbf{i} - \mathbf{1})$ (where, if $i = 0$, the consequent is an empty disjunction and hence is to be regarded as equivalent to $\mathbf{0} \neq \mathbf{0}$).

Proof. Induction on i. Basis step: $i = 0$. We must show $\vdash_Q \forall x - x < \mathbf{0}$. By $Q3$ we have $x = \mathbf{0} \vee \exists y\, x = y'$. Assume $x < \mathbf{0}$, i.e., $\exists w\, w' + x = \mathbf{0}$. If $x = \mathbf{0}$ holds, we have $w' = w' + \mathbf{0}$ (by $Q4$) $= w' + x = \mathbf{0}$, which is impossible by $Q2$. If $x = y'$ holds, we have $(w' + y)' = w' + y'$ (by $Q5$) $= w' + x = \mathbf{0}$ which is again impossible by $Q2$. Thus $\vdash_Q \forall x - x < \mathbf{0}$.

Induction step. We suppose $\vdash_Q \forall x\,(x < \mathbf{i} \to x = \mathbf{0} \vee \ldots \vee x = \mathbf{i} - \mathbf{1})$. We must show $\vdash_Q \forall x\,(x < \mathbf{i}' \to x = \mathbf{0} \vee x = \mathbf{0}' \vee \ldots \vee x = \mathbf{i})$. Assume we have

$x<\mathbf{i}'$, i.e., $\exists w\, w'+x = \mathbf{i}'$. By $Q3$ we have $\exists y\, x = y' \vee x = \mathbf{o}$. If $x = y'$ holds, then we have $\mathbf{i}' = w'+x = w'+y' = (w'+y)'$ (by $Q5$), whence by $Q1$ we have $\mathbf{i} = w'+y$, and therefore $y<\mathbf{i}$. By the induction hypothesis we have $y = \mathbf{o} \vee \ldots \vee y = \mathbf{i-1}$ (if $i = 0$, we have $\mathbf{o} \neq \mathbf{o}$), and therefore we have $x = \mathbf{o}' \vee \ldots \vee x = \mathbf{i}$ (if $i = 0$, we have $\mathbf{o} \neq \mathbf{o}$), and therefore we have $x = \mathbf{o} \vee x = \mathbf{o}' \vee \ldots \vee x = \mathbf{i}$, which we also have in case $x = \mathbf{o}$ holds.

Lemma 14.12

For each i, $\vdash_Q \forall x(\mathbf{i}<x \rightarrow x = \mathbf{i}' \vee \mathbf{i}'<x)$.

Proof. Assume $\mathbf{i} < x$, i.e. $\exists w\, w'+\mathbf{i} = x$. We have $w = \mathbf{o} \vee \exists y\, w = y'$ by $Q3$. From $w = \mathbf{o}$ and $w'+\mathbf{i} = x$, we have $\mathbf{o}'+\mathbf{i} = x$, whence by 14.3 we have $x = \mathbf{i}'$. From $w = y'$ and $w'+\mathbf{i} = x$, we have $y''+\mathbf{i} = x$, whence by 14.9 we have $y'+\mathbf{i}' = x$, and so we have $\mathbf{i}'<x$.

Lemma 14.13

For each i, $\vdash_Q \forall x(\mathbf{i}<x \vee x = \mathbf{i} \vee x<\mathbf{i})$.

Proof. Induction on i. Basis step: $i = 0$. Assume $x \neq \mathbf{o}$. By $Q3$ we then have $\exists y\, x = y'$, and so by $Q4$ we have $\exists y\, y'+\mathbf{o} = x$, i.e., $\mathbf{o}<x$. Induction step: we suppose $\vdash_Q \forall x(\mathbf{i}<x \vee x = \mathbf{i} \vee x<\mathbf{i})$. We must show $\vdash_Q \forall x(\mathbf{i}'<x \vee x = \mathbf{i}' \vee x<\mathbf{i}')$. By 14.12

$$\vdash_Q \forall x(\mathbf{i}<x \rightarrow x = \mathbf{i}' \vee \mathbf{i}'<x). \text{ By 14.10} \qquad \text{(I)}$$

$$\vdash_Q \forall x(x = \mathbf{i} \rightarrow x<\mathbf{i}'). \text{ And by 14.11 and 14.10} \qquad \text{(II)}$$

$$\vdash_Q \forall x(x<\mathbf{i} \rightarrow x<\mathbf{i}'). \qquad \text{(III)}$$

But from (I), (II), (III), and the induction hypothesis, it follows that

$$\vdash_Q \forall x(\mathbf{i}'<x \vee x = \mathbf{i}' \vee x<\mathbf{i}').$$

We can now show that the result g of applying minimization to any regular $(n+1)$-place function f that is representable in Q is also representable in Q.

Suppose that f is a regular $(n+1)$-place function, and that

$$A(\mathsf{x}, x_{n+1}, x_{n+2})$$

represents f in Q. Let $B(\mathsf{x}, x_{n+1})$ = the formula

$$(A(\mathsf{x}, x_{n+1}, \mathbf{o}) \,\&\, \forall w\,(w<x_{n+1} \rightarrow -A(\mathsf{x}, w, \mathbf{o}))).$$

Then B represents g in Q.

For suppose that $g(\mathsf{p}) = i$. Then $f(\mathsf{p}, i) = 0$, and for any $j < i, f(\mathsf{p}, j) \neq 0$. Since A represents f in Q, we have

$$\vdash_Q A(\mathbf{p}, \mathbf{i}, \mathbf{0}), \text{ and (if } i > 0), \qquad (i)$$
$$\vdash_Q -A(\mathbf{p}, \mathbf{0}, \mathbf{0}), \qquad (0)$$
$$\vdots$$
$$\vdash_Q -A(\mathbf{p}, \mathbf{i-1}, \mathbf{0}). \qquad (i-1)$$

(0), ..., $(i-1)$, and 14.11 entail that

$$\vdash_Q \forall w(w < \mathbf{i} \rightarrow -A(\mathbf{p}, w, \mathbf{0})), \qquad (i+1)$$

which, together with (i), entails that $\vdash_Q B(\mathbf{p}, \mathbf{i})$.

We must show that $\vdash_Q \forall x_{n+1}(B(\mathbf{p}, x_{n+1}) \rightarrow x_{n+1} = \mathbf{i})$. Assume $B(\mathbf{p}, x_{n+1})$, i.e., $A(\mathbf{p}, x_{n+1}, \mathbf{0}) \,\&\, \forall w(w < x_{n+1} \rightarrow -A(\mathbf{p}, w, \mathbf{0}))$. From (i) and

$$\forall w(w < x_{n+1} \rightarrow -A(\mathbf{p}, w, \mathbf{0})), \text{ we have } -\mathbf{i} < x_{n+1}.$$

From $A(\mathbf{p}, x_{n+1}, \mathbf{0})$ and $(i+1)$, we have $-x_{n+1} < \mathbf{i}$. Thus by 14.13 we have $x_{n+1} = \mathbf{i}$. So $\vdash_Q \forall x_{n+1}(B(\mathbf{p}, x_{n+1}) \rightarrow x_{n+1} = \mathbf{i})$.

Exercises

14.1 Verify the following assertion: all recursive functions are representable in the theory ('R') whose language is L and whose theorems are the consequences in L of the following infinitely many sentences:

$\mathbf{i} \neq \mathbf{j}$ for all i, j such that $i \neq j$;

$\mathbf{i} + \mathbf{j} = \mathbf{k}$ for all i, j, k such that $i + j = k$;

$\mathbf{i} \cdot \mathbf{j} = \mathbf{k}$ for all i, j, k such that $i \cdot j = k$;

$\forall x(x < \mathbf{i} \rightarrow x = \mathbf{0} \vee \ldots \vee x = \mathbf{i-1})$ for all i;

and $\forall x(x < \mathbf{i} \vee x = \mathbf{i} \vee \mathbf{i} < x)$, for all i.

14.2 Show that none of the following sentences are theorems of Q:

(*a*) $\forall x\, x \neq x'$,

(*b*) $\forall x \forall y \forall z\, x + (y + z) = (x + y) + z$,

(*c*) $\forall x \forall y\, x + y = y + x$,

(*d*) $\forall x\, \mathbf{0} + x = x$,

(*e*) $\forall x\, x < x'$,

(*f*) $\forall x \forall y - (x < y \,\&\, y < x)$,

(*g*) $\forall x \forall y \forall z\, x \cdot (y \cdot z) = (x \cdot y) \cdot z$,

(*h*) $\forall x \forall y\, x \cdot y = y \cdot x$,

(*i*) $\forall x\, \mathbf{0} \cdot x = x$,

(*j*) $\forall x \forall y \forall z\, x \cdot (y + z) = x \cdot y + x \cdot z$.

Hint: Let a and b be two objects that are not natural numbers, and consider the following successor, addition, and multiplication tables:

x	x'
i	i'
a	a
b	b

$+$	j	a	b
i	$i+j$	b	a
a	a	b	a
b	b	b	a

$\cdot$	0	$j \neq 0$	a	b
0	0	0	a	b
$i \neq 0$	0	$i \cdot j$	a	b
a	0	b	b	b
b	0	a	a	a

15
Undecidability, indefinability and incompleteness

We are now in a position to give a unified treatment of some of the central negative results of logic: Church's theorem on the undecidability of logic, Tarski's theorem on the indefinability of truth, and Gödel's first theorem on the incompleteness of systems of arithmetic. These theorems can all be seen as more or less direct consequences of the result of the last chapter, that all recursive functions are representable in *Q*, and a certain exceedingly ingenious lemma ('the diagonal lemma'), the idea of which is due to Gödel, and which we shall prove below. The first notion that we have to introduce is that of a *gödel numbering*.

A *gödel numbering* is an assignment of natural numbers (called 'gödel numbers') to expressions (in some set) that meets these conditions: (1) different gödel numbers are assigned to different expressions: (2) it is effectively calculable what the gödel number of any expression is; (3) it is effectively decidable whether a number is the gödel number of some expression in the set, and, if so, effectively calculable which expression it is the gödel number of.

Gödel numberings enable one to regard interpreted languages supposed to be 'about' the natural numbers – i.e. having the set of natural numbers as the domain of their intended interpretation – as also referring to the numbered expressions. The possibility then arises that certain sentences, ostensibly referring to certain numbers, could be seen as referring, via the gödel numbering, to certain expressions that are *identical* with those very sentences themselves. The state of affairs just described is no mere possibility; the proof of the diagonal lemma shows how it arises, and succeeding theorems show how it may be exploited.

We shall consider a particular set of expressions and a particular gödel numbering, to which we appropriate the words 'expression' and 'gödel number'. There is nothing special about our particular gödel numbering; the theorems and proofs that we are going to give with respect to the one we use could have been given with respect to any number of others. Our expressions are finite sequences of these (distinct) symbols.

We'll make the following 'conventions' about the identity of certain symbols: we stipulate that $x_0 = x$, $x_1 = y$, $f_0^0 = \mathbf{0}$, $f_0^1 = {}'$, $f_0^2 = +$, $f_1^2 = \cdot$, i.e. (by Church's thesis), that there is an effective procedure for deciding whether a given symbol may occur in some sentence in the language of the theory.

Here's the diagonal lemma:

Lemma 2

Let T be a theory in which diag is representable. Then for any formula $B(y)$ (of the language of T, containing just the variable y free), there is a sentence G such that

$$\vdash_T G \leftrightarrow B(\ulcorner G \urcorner).$$

Proof. Let $A(x, y)$ represent diag in T. Then for any n, k, if diag $(n) = k$, $\vdash_T \forall y\,(A(\mathbf{n}, y) \leftrightarrow y = \mathbf{k})$.

Let F be the expression $\exists y\,(A(x, y)\,\&\,B(y))$. F is a formula of the language of T that contains just the variable x free.

Let n be the gödel number of F.

Let G be the expression $\exists x\,(x = \mathbf{n}\,\&\,\exists y\,(A(x, y)\,\&\,B(y)))$. As $\mathbf{n} = \ulcorner F \urcorner$, G is the diagonalization of F and a sentence of the language of T. Since G is logically equivalent to $\exists y\,(A(\mathbf{n}, y)\,\&\,B(y))$, we have

$$\vdash_T G \leftrightarrow \exists y\,(A(\mathbf{n}, y)\,\&\,B(y)).$$

Let k be the gödel number of G. Then

$$\text{diag}(n) = k, \quad \text{and} \quad \mathbf{k} = \ulcorner G \urcorner.$$

So $\vdash_T \forall y(A(\mathbf{n}, y) \leftrightarrow y = \mathbf{k})$.

So $\vdash_T G \leftrightarrow \exists y\,(y = \mathbf{k}\,\&\,B(y))$.

So $\vdash_T G \leftrightarrow B(\mathbf{k})$, i.e., $\vdash_T G \leftrightarrow B(\ulcorner G \urcorner)$.

A theory is called *consistent* if there is no theorem of the theory whose negation is also a theorem. Equivalently, a theory is consistent iff there is some sentence in its language that is not a theorem, iff the theory is satisfiable.

A set θ of natural numbers is said to be *definable in* theory T if there is a formula $B(x)$ of the language of T such that for any number k, if $k \in \theta$, then $\vdash_T B(\mathbf{k})$, and if $k \notin \theta$, then $\vdash_T -B(\mathbf{k})$. The formula $B(x)$ is said to define θ in T. A two-place relation R of natural numbers is likewise definable in T if there is a formula $C(x, y)$ of the language of T such that for any numbers k, n, if kRn, then $\vdash_T C(\mathbf{k}, \mathbf{n})$, and if $k \not R n$, then $\vdash_T -C(\mathbf{k}, \mathbf{n})$, and $C(x, y)$ is then said to define R in T. (A perfectly analogous definition of definability can be given for three- and more-place relations on natural numbers; we won't need this more general notion, however.)

A theory T is called an *extension* of theory S if S is a subset of T, i.e., if any theorem of S is a theorem of T. If f is a function that is representable in S, and T is an extension of S, then f is representable in T, and indeed is represented in T by the same formula that represents it in S. Similarly, any formula that defines a set in some theory defines it in any extension of that theory.

Lemma 3

If T is a consistent extension of Q, then the set of gödel numbers of theorems of T is not definable in T.

Proof. Let T be an extension of Q. Then diag is representable in T; for as diag is a recursive function, and all recursive functions are representable in Q, diag is representable in Q, and hence is representable in any extension of Q.

Suppose now that $C(y)$ defines the set θ of gödel numbers of theorems of T. By the diagonal lemma, there is a sentence G such that

$$\vdash_T G \leftrightarrow -C(\ulcorner G \urcorner).$$

Let $k = \text{gn}(G)$. Then

$$\vdash_T G \leftrightarrow -C(\mathbf{k}). \qquad (*)$$

Then $\vdash_T G$. For if G is not a theorem of T, then $k \notin \theta$, and so, as $C(y)$ defines θ, $\vdash_T -C(\mathbf{k})$, whence by (*), $\vdash_T G$.

So $k \in \theta$. So $\vdash_T C(\mathbf{k})$, as $C(y)$ defines θ. So, by (*), $\vdash_T -G$, and T is therefore inconsistent.

A set of expressions is called *decidable* if the set of gödel numbers of its members is a recursive set. Thus a theory T is decidable iff the set θ of gödel numbers of its theorems is recursive, iff the characteristic function of θ is recursive.

If a theory is decidable, then an effective method exists for deciding whether any given sentence is a theorem of the theory. For to determine whether a sentence is a theorem, calculate its gödel number first and then calculate the value of the (recursive, hence calculable) characteristic function for the gödel number as argument. The sentence is a theorem iff the value is 1.

Conversely, if a theory is not decidable, then *unless Church's thesis is false*, no effective method exists for deciding whether a given sentence is a theorem of the theory. For if there were such a method, then the characteristic function of the set of gödel numbers of theorems would also be effectively calculable, and hence recursive, by Church's thesis.

Theorem 1

No consistent extension of Q is decidable.

Proof. Suppose T is a consistent extension of Q. Then by Lemma 3, the set θ of gödel numbers of theorems of T is not definable in T. Now if $A(x,y)$ represented the characteristic function f of θ in T, then $A(x, \mathbf{1})$ would define θ in T. (For then if $k \in \theta, f(k) = 1$, whence $\vdash_T A(\mathbf{k}, \mathbf{1})$; and if $k \notin \theta, f(k) = 0$, whence $\vdash_T \forall y\,(A(\mathbf{k}, y) \leftrightarrow y = \mathbf{0})$, whence, as $\vdash_Q \mathbf{0} \neq \mathbf{1}$, $\vdash_T -A(\mathbf{k}, \mathbf{1})$.) Thus the characteristic function of θ is not representable in T, and therefore, as T is an extension of Q, not representable in Q either, and hence not recursive. So T is not decidable.

Lemma 4

Q is not decidable.

Proof. Q is a consistent extension of Q.

We can now give another proof of the proposition that first-order logic has no decision procedure, a proof that is rather different from the one given in Chapter 10.

Let L be the theory in L, the language of arithmetic, whose theorems are just the valid sentences in L. All theorems of L are theorems of Q, of course, but as not all of (indeed, none of) the axioms of Q are valid, L is not an extension of Q, and we cannot therefore apply theorem 1. But because Q has only finitely many axioms, we can nonetheless prove that L is not decidable, and hence that there is no effective method for deciding whether or not a first-order sentence is valid.

Theorem 2 (*Church's undecidability theorem*)

L is not decidable.

Proof. Let C be a conjunction of the axioms of Q. Then a sentence A is a theorem of Q iff C implies A, iff $(C \to A)$ is valid, iff $(C \to A)$ is a theorem of L. (So, intuitively, a test for validity would yield a test for theoremhood in Q: to decide whether A is a theorem of Q, test $(C \rightarrow A)$ for validity.)

Let q be the gödel number of C. Let f be defined by:

$$f(n) = 1*(q*(399*(n*2))).$$

f is recursive. If n is the gödel number of A, then $f(n)$ is the gödel number of $(C \rightarrow A)$.

Let λ be the set of gödel numbers of theorems of L. If λ is recursive, then so is $\{n | f(n) \in \lambda\}$. But $\{n | f(n) \in \lambda\}$ is the set of gödel numbers of theorems of Q, which, by Lemma 4 is not recursive. Thus λ is not recursive and L is not decidable.

By *arithmetic* we shall understand that theory whose language is L and whose theorems are just the sentences of L that are *true* in the standard interpretation $\mathcal{N}$, in which the domain is the set of all natural numbers, and **o**, ′, +, and · are assigned zero, successor, addition, and multiplication, respectively.

Theorem 3

Arithmetic is not decidable.

Proof. Arithmetic is a consistent extension of Q, and by Theorem 1 no consistent extension of Q is decidable.

Thus, unless Church's thesis is false, there is no effective method for deciding whether an arbitrary sentence in the language of arithmetic is true or false in $\mathcal{N}$. This negative result is in contrast to Presburger's theorem, proved in Chapter 21, that an effective method exists for deciding whether an arbitrary sentence in the language of arithmetic *not containing* '·' is true or false (in $\mathcal{N}$).

Theorem 4 (*Tarski's indefinability theorem*)

The set of gödel numbers of sentences true in $\mathcal{N}$ is not definable in arithmetic.

Proof. Since the theorems of arithmetic are just the sentences true in $\mathcal{N}$, Theorem 4 follows from Lemma 3.

As any formula $B(x)$ will be true (in $\mathcal{N}$) of the number k if and only if $B(\mathbf{k})$ is a theorem of arithmetic, another way to put Theorem 4 is to say that there is no formula of the language of arithmetic (with one free variable) which is true of just those natural numbers that are gödel numbers of truths of arithmetic, or, more briefly, 'arithmetical truth is not arithmetically definable'.

Lemma 5

Any recursive set is definable in arithmetic.

Proof. Suppose θ is a recursive set. Then the characteristic function of θ is recursive, and hence representable in Q. As in the proof of Theorem 1, θ is then definable in Q, and hence definable in arithmetic, which is an extension of Q.

Lemma 5 shows that Theorem 4 is at least as strong a result as Theorem 3, as Theorem 3 says that the set of gödel numbers of truths of $\mathcal{N}$ is not recursive. Since the converse of Lemma 5 does not hold (cf. Exercise 3), Theorem 4 is actually stronger than Theorem 3.

A theory T is called *complete* if for every sentence A (in the language of T), either A or $-A$ is a theorem of T. A theory T is consistent and complete, then, iff for any sentence A, exactly one of A and $-A$ is a theorem. Arithmetic is a consistent, complete extension of Q.

A theory T is called *axiomatizable* if there is a decidable subset of T whose consequences (in the language of T) are just the theorems of T. If there is a finite, and hence decidable, subset with this property, the theory is said to be *finitely axiomatizable*. It is clear from the definition of axiomatizability that any decidable theory is axiomatizable; Q is an example of a (finitely) axiomatizable theory that is not decidable.

The version of Gödel's incompleteness theorem that we shall prove is the assertion that there is no complete, consistent, axiomatizable extension of Q. That there is none will follow from Theorem 1 and the proposition (Theorem 5) that any axiomatizable complete theory is decidable.

This last proposition should be confused neither with the statement that every complete decidable theory is axiomatizable, which is trivially true, nor with the statement that every decidable axiomatizable theory is complete, which is false (counterexample: the theory whose non-logical symbols are the sentence letters p and q, and whose theorems are the consequences in this language of p).

Theorem 5

Any axiomatizable complete theory is decidable.

Proof. Let T be any theory whatsoever. Since the set of symbols that may occur in sentences of the language of T is decidable (as we have assumed earlier), the set of sentences of the language of T is itself decidable, i.e. there exists a Turing machine which, when given a number as input, yields 1 as output iff the number is the gödel number of a sentence of the language of T, and yields 0 otherwise.

Now suppose that T is axiomatizable and complete. If T is inconsistent, then the theorems of T are just the sentences in the language of T, which, we have just noted, form a decidable set. We may therefore suppose that T is consistent also.

Since T is axiomatizable, there is a decidable set S of sentences whose consequences (in the language of T) are just the theorems of T. Let A be a sentence (in the language of T). We shall say that a sentence is *A-interesting* if it is a conditional of which the antecedent is a conjunction of members of S and the consequent is either A or $-A$. Then, A is a theorem of T iff there is a *valid* A-interesting sentence whose consequent is A itself. And, since T is consistent and complete, A is a theorem of T iff $-A$ is not a theorem of T, iff there is *no* valid A-interesting sentence whose consequent is $-A$. Since S is decidable, so is the set of A-interesting sentences (for each sentence A).

We shall show that T is decidable by showing that there exists a Turing machine M which, when given a sentence A of the language of T as input, yields 1 as output iff A is a theorem of T, and yields 0 as output otherwise.

In Chapter 12 we established the existence of a Turing machine M^* which, when given any sentence as input, terminated after a finite number of steps with the production of the words 'yes, valid' iff the given sentence was valid.

Our machine M works by first writing down the number 1 after the input sentence A and then going into a loop consisting of a sequence of *subroutines*. In the nth of these, M writes down those $k (\leqslant n)$ sentences with gödel numbers $\leqslant n$ that are A-interesting and then 'imitates' M^* k times, each time performing n steps in the operation of M^* when given as input (a new) one of the k A-interesting sentences that have been written down. If one or more of these k sentences is shown valid (by the production of the words 'yes, valid') after n such steps, M picks one of them and determines whether its consequent is A or $-A$ (as T is consistent, the case cannot arise in which one sentence has consequent A and another has $-A$), and then yields as output 1 or 0, accordingly.

But if not, M erases everything except A and the number after it, to which it adds 1, and then goes into the $n+1$st subroutine.

Since either A or $-A$ is a theorem of T, but not both, there is a valid A-interesting sentence C, with gödel number i, and M^*, when applied to C, will terminate with 'yes, valid' after some finite number of steps, say j. M, therefore, when applied to A, will go into at most $\max(i,j)$ subroutines before yielding 1 or 0 as output, and will yield 1 iff the consequent of C is A. So M yields 1 as output iff A is valid, and yields 0 otherwise. T is therefore decidable.

Theorem 6 (*Gödel's first incompleteness theorem*)

There is no consistent, complete, axiomatizable extension of Q.

Proof. Theorem 6 is an immediate consequence of Theorems 1 and 5.

Corollary

Arithmetic is not axiomatizable.

The import of Gödel's first incompleteness theorem is sometimes expressed in the words 'any sufficiently strong formal theory (or system) of arithmetic is incomplete (if it is consistent)'. A 'formal' theory may be taken to be one whose theorems are deducible via the usual rules of logic from an axiom system. Since an axiom system is here understood to be a set of sentences for which an effective procedure for determining membership exists, and since the usual rules of logic are sound and complete, that is, since all and only the logical consequences of a set of sentences can be deduced from the set by means of the rules, 'formal theory' can be considered synonymous with 'axiomatizable theory'. 'A formal theory of arithmetic' can therefore be taken to be an axiomatizable theory all of whose theorems are truths in some interpretation whose domain is the set of natural numbers and in which those of $\mathbf{0}$, $'$, $+$, $\cdot$, $<$, ^{2}, etc. that occur in the theorems have their familiar, standard meanings.

Theorem 6 thus represents a sharpening of the above statement of Gödel's theorem in that it indicates a sufficient condition for 'sufficient strength', *viz.*, *being an extension of* Q. Q, as we have seen, is a rather weak theory (cf. Exercise 14.2), and Theorem 6 is thus a correspondingly strong result. It follows from Theorem 6 that any consistent mathematical theory of which the theorems are just the consequences of some effectively specified set of axioms, and among which are the seven axioms of Q, is incomplete; hence for any interpretation of the language of the theory there will be truths in that interpretation which are not theorems of the theory. And perhaps the most significant consequence of Theorem 6 is what it says about the notions of *truth* (in the standard interpretation for the language of arithmetic) and *theoremhood*, or *provability* (in any particular formal theory): *that they are in no sense the same.*

Exercises

15.1 A formula $B(y)$ is called a truth-predicate for T if for any sentence G of the language of T, $\vdash_T G \leftrightarrow B(\ulcorner G \urcorner)$. Show that if T is a consistent theory in which diag is representable, then there is no truth-predicate for T.

15.2 Show that all functions representable in Q are recursive.

15.3 A set S of natural numbers is called *recursively enumerable* (*r.e.*) if there is a (two-place) recursive relation R such that $S = \{x \mid \exists y\, Rxy\}$. Show that for any set S, S is recursive iff both S and $\bar{S}$ are r.e. (Kleene's theorem). Are all r.e. sets definable in arithmetic? (Yes. Why?) Give some examples of r.e. sets and some examples of non-r.e. sets.

15.4 (*Craig*) Show that a theory T is axiomatizable if T is r.e., i.e. if the set of gödel numbers of members of T is r.e.

15.5 Let $B_1(y)$ and $B_2(y)$ be two formulas of the language of T with y as sole free variable. Show how to construct sentences A_1 and A_2 such that $\vdash_T A_1 \leftrightarrow B_1(\ulcorner A_2 \urcorner)$ and $\vdash_T A_2 \leftrightarrow B_2(\ulcorner A \urcorner_1)$.

Solution to 15.2 (*Using Church's thesis*)

15.2 Suppose $A(\mathsf{x}, y)$ represents f in Q. Since Q is consistent and $\mathbf{m} \neq \mathbf{n}$ is a theorem of Q whenever $m \neq n$, $\vdash_Q \forall y\,(A(\mathbf{p}, y) \leftrightarrow y = \mathbf{m})$ iff $f(\mathsf{p}) = m$. In order to calculate $f(\mathsf{p})$, then, one may use a 'search procedure' similar to the one used in the proof of Theorem 5 to determine for which m the conditional whose antecedent is some fixed conjunction of the axioms of Q and whose consequent is $\forall y\,(A(\mathbf{p}, y) \leftrightarrow y = \mathbf{m})$ is valid. That m – it will be unique – is $f(\mathsf{p})$.

Solution to 15.4 (very tricky)

Suppose that R is a recursive relation and

$$\{x \mid x \text{ is the gödel number of a member of } T\} = \{x \mid \exists y\, Rxy\}.$$

For any sentence A and natural number y, let A^y be the conjunction $(A\&...(A\&A)...)$ of $y+2$ occurrences of A. Thus, e.g.

$$A^2 = (A\&(A\&(A\&A)))$$

and $A^0 = (A\&A)$. Let $U = \{A^y | R\text{gn}(A)y\}$. If $A \in T$, then for some y, $R\text{gn}(A)y$ and $A^y \in U$; and if $A^y \in U$, then $A \in T$. Since A and A^y are equivalent, T and U imply the same sentences, and the set of sentences in the language of T that follow from U is thus T itself. To show that T is axiomatizable, then, we need only show that U is decidable. But U *is* decidable: to decide whether an arbitrary sentence B is in U, we may apply the following effective procedure. Determine whether B is the conjunction $(A\&...(A\&A)...)$ of z occurrences of some sentence A, for some $z \geqslant 2$. If not, $B \notin U$. But if so, find A and z, and let $x = \text{gn}(A)$ and $y = z-2$. Determine whether Rxy. (R is recursive.) If so, $B \in U$; if not, $B \notin U$.

16

Provability predicates and the unprovability of consistency

We learned in the last chapter that no consistent extension of Q was decidable, and that any complete axiomatizable theory was decidable; we concluded that no consistent, axiomatizable extension of Q was complete. In the present chapter we are going to discuss a certain theory, *Elementary Peano Arithmetic*, or Z, as we shall call it (following Hilbert–Bernays). Z's language is L, the language of arithmetic. The '*non-logical' axioms of Z are* $Q1$ through $Q7$, the seven axioms of Q, together with all *induction axioms*: sentences obtained from formulas of L of the form:

$$([A(\mathbf{o}) \,\&\, \forall x\,(A(x) \rightarrow A(x'))] \rightarrow \forall x\, A(x))$$

by the prefixing of universal quantifiers. The theorems of Z are the logical consequences in L of the axioms of Q and the induction axioms. ($Q3$ actually follows from an induction axiom. For let

$$A(x) = (x \neq \mathbf{o} \rightarrow \exists y\, x = y').$$

Then $A(\mathbf{o})$ and $\forall x\,(A(x) \rightarrow A(x'))$ are logical truths; and so

$$\forall x\, A(x), = Q3, = \forall x\,(x \neq \mathbf{o} \rightarrow \exists y\, x = y'),$$

is a theorem of Z, and thus its inclusion as a non-logical axiom of Z was not necessary.) Like the axioms of Q, the induction axioms are all true in the standard interpretation $\mathcal{N}$, and hence Z is consistent; Z is even more evidently an axiomatizable extension of Q. Z is thus incomplete.

But though incomplete, because of the presence of the induction axioms, Z is a much more powerful theory than Q: all of the sentences mentioned in Exercise 14.2, for example, are theorems of Z but not of Q. And if Z is supplied with a notion of *proof* (such as the one given below), then the proofs of a great many mathematical theorems, including much (notably) of the theory of numbers, can be 'reproduced' or 'carried out' in Z. Unlike Q, Z is a theory in which large portions of actual mathematics can be adequately formalized. And, as Z is an extension of Q, all recursive functions (sets or relations) are representable (definable) in Z.

In the present chapter we are going to investigate two closely related questions about Z and other theories: whether the consistency of Z is provable in Z, and whether a certain sentence, which may be taken to assert its own provability (in Z) is in fact true (and thus provable) or false (and thus unprovable). Unlike the question whether a sentence expressing its own unprovability (in Z) is provable or not – any such sentence must be both true and unprovable if everything provable is true – there seems to be no easy way to decide this second question *a priori*.

We shall now introduce, informally, the notion of a *proof of* a sentence in Z. We shall assume, but not prove, this fact: there is an effectively specifiable set of valid sentences, the 'logical' axioms of Z, such that a sentence A is a theorem of Z if and only if there is a finite sequence of sentences (of L) ending with A, each of which is either an axiom of Z (logical or non-) or the 'ponential' of two *earlier* sentences in the sequence. (D is the 'ponential' of C and $(C \to D)$.) Such a sequence shall be called a *proof in* Z *of* A.

We shall be a little bit more specific about the 'nature' of finite sequences of sentences: each such sequence is to be identified with the expression consisting of the same sentences in the same order, but separated by commas. As the comma has already been assigned the gödel number 29, every proof in Z acquires its own gödel number in consequence of this identification. The relation Proof, $= \{\langle m, n\rangle \mid m$ is the gödel number of a proof in Z of the sentence with gödel number $n\}$, is thus a recursive relation and is therefore definable in Z.

By 'straightforwardly transcribing' in L the definition of *proof in* Z *of* just given, making reference to gödel numbers instead of expressions, and utilizing, where necessary, the β-function of Chapter 14, we can construct a formula $\mathrm{Pr}(x, y)$ of L, which not only defines Proof in Z, but possesses certain other important properties as well. In order to describe these, we need a

Definition

$\mathrm{Prov}(y)$ is to be the formula $\exists x \, \mathrm{Pr}(x, y)$.

One important property that $\mathrm{Pr}(x, y)$ and $\mathrm{Prov}(y)$ have is that for any sentences A and C (of L),

(i) if $\vdash_Z A$, then $\vdash_Z \mathrm{Prov}(\ulcorner A \urcorner)$;
(ii) $\vdash_Z \mathrm{Prov}(\ulcorner A \to C \urcorner) \to [\mathrm{Prov}(\ulcorner A \urcorner) \to \mathrm{Prov}(\ulcorner C \urcorner)]$; and
(iii) $\vdash_Z \mathrm{Prov}(\ulcorner A \urcorner) \to \mathrm{Prov}(\ulcorner \mathrm{Prov}(\ulcorner A \urcorner) \urcorner)$.

(Later we shall express this property of $\mathrm{Prov}(y)$ by saying that $\mathrm{Prov}(y)$ is a *provability predicate* for Z. The consequent of the sentence mentioned in (iii) is the result of substituting $\mathbf{k}$ for y in $\mathrm{Prov}(y)$, k being the gödel number of Prov (⌜A⌝), which is itself the result of substituting the numeral for the gödel number of A for y in Prov (y).) It is also the case that for any sentence A,

(iv) if $\vdash_Z$ Prov (⌜A⌝), then $\vdash_Z A$.

That Prov(y) satisfies (i) follows directly from the fact that Pr (x, y) defines Proof in Z. For suppose that $\vdash_Z A$. Then there is a proof of A in Z. Let the gödel number of the proof be m. Then $\vdash_Z$ Pr (**m**, ⌜A⌝); and so $\vdash_Z \exists x$ Pr (x, ⌜A⌝), i.e., $\vdash_Z$ Prov (⌜A⌝).

A proof of C can be obtained from proofs of A and $(A \rightarrow C)$ by writing down the proof of A, then a comma, then the proof of $(A \rightarrow C)$, then another comma, and then C; this argument can be formalized in Z. Its formalization would show that Prov (y) satisfies (ii).

Showing that Prov (y) satisfies (iii) is much harder, and we shall only say that it involves showing that the argument that Prov (y) satisfies (i) (with '$\vdash_Z$' understood as meaning 'provable in Z') can be formalized in Z (for any given A). And showing *that* involves showing, in Z, that for any m, n, if m Proof n, then Pr (**m**, **n**) is provable in Z.†

As for (iv), suppose that $\vdash_Z$ Prov (⌜A⌝), i.e. $\vdash_Z \exists x$ Pr (x, ⌜A⌝). Then $\exists x$ Pr (x, ⌜A⌝) is true in $\mathcal{N}$, and so for some m, m is the gödel number of a proof in Z of A. So A is provable in Z.

A further important, if non-mathematical, property of Pr(x, y) is that it can plausibly be regarded as *meaning* or *saying* that the number represented by x is the gödel number of a proof in Z of the sentence whose gödel number is represented by y, or more concisely but somewhat inaccurately, as meaning that x is a proof of y. Not every formula that defines Proof has this property: let Dr (x, y) be the formula

$$(\mathrm{Pr}(x,y) \,\&\, y \neq \ulcorner \mathbf{0} = \mathbf{1} \urcorner).$$

Then, as Z is consistent, Dr (x, y) also defines Proof, but cannot be said to mean simply that x is a proof of y; it may be regarded as meaning that x is a proof of y, but y is not the sentence $\mathbf{0} = \mathbf{1}$. In brief and roughly, it is because it is the 'straightforward transcription' of the definition of *proof in Z of* that Pr (x, y) can be considered to mean what it does.

Prov (y) can similarly be taken as meaning that (the number represented by) y is (the gödel number of) a provable sentence. In what follows we shall concern ourselves with the questions whether a sentence that asserts its own provability in Z is provable (in Z) or not, and whether the

† For a complete account the reader may refer to Hilbert–Bernays, *Grundlagen der Mathematik*, and M. Löb 'Solution of a Problem of Leon Henkin', *Journal of Symbolic Logic* **20** (1955).

sentence that says that Z is consistent can be proved in Z. We may and shall take these, and other, questions, whose formulations involve notions like *asserting its own provability* or *expressing the non-provability of* $\mathbf{0} = \mathbf{1}$, as questions about various sentences constructed in certain specific ways from Prov(y), and not from some other formula with some but not all of Prov (y)'s properties.

From now on, T will be an extension of Q, not necessarily consistent, and $B(y)$ will be a formula of the language of T with the sole free variable y. If A is a sentence of the language of T with gödel number n, then $B(\ulcorner A \urcorner)$ is the result of everywhere substituting $\mathbf{n}$ for all free occurrences of y in $B(y)$.

We shall call $B(y)$ a *provability predicate for* T if for all sentences A and C (of the language of T), $B(y)$ meets the following three conditions:

(i) if $\vdash_T A$, then $\vdash_T B(\ulcorner A \urcorner)$;

(ii) $\vdash_T B(\ulcorner A \rightarrow C \urcorner) \rightarrow [B(\ulcorner A \urcorner) \rightarrow B(\ulcorner C \urcorner)]$; and

(iii) $\vdash_T B(\ulcorner A \urcorner) \rightarrow B(\ulcorner B(\ulcorner A \urcorner) \urcorner)$.

Prov(y) is a provability predicate for Z, but provability predicates need not have very much to do with provability: any formula $S(y)$ that defines $\{n|n$ is the gödel number of a sentence of $L\}$ in Q is a provability predicate for Q. (Note that $\vdash_Q S(\ulcorner \mathbf{0} = \mathbf{1} \urcorner)$, but not $\vdash_Q \mathbf{0} = \mathbf{1}$: an analogue of (iv) (above) fails for Q, $S(y)$ and $A = \mathbf{0} = \mathbf{1}$.)

The thought that whatever is provable had better be true might make it surprising that a further condition was not included in the definition of provability predicate, namely,

$$\text{For every sentence } A, \vdash_T B(\ulcorner A \urcorner) \rightarrow A.$$

It will become clear, though, that no provability predicate meets this extra condition unless T is actually inconsistent!

The diagonal lemma of Chapter 15 provided us with a way, when given any formula $B(y)$, of finding a sentence A such that $\vdash_T A \leftrightarrow B(\ulcorner A \urcorner)$. If we take T to be Z and $B(y)$ to be $-\text{Prov}(y)$ or Prov(y), the diagonal lemma enables us to construct sentences G and H such that

$$\vdash_Z G \leftrightarrow -\text{Prov}(\ulcorner G \urcorner) \quad \text{and} \quad \vdash_Z H \leftrightarrow \text{Prov}(\ulcorner H \urcorner).$$

G and H can be taken as expressing their own unprovability and provability, respectively. In what follows we shall answer the questions whether G and H are true or false, i.e., whether or not $\vdash_Z G$ and $\vdash_Z H$. We shall also answer the question whether or not

$$\vdash_Z -\text{Prov}(\ulcorner \mathbf{0} = \mathbf{1} \urcorner); \qquad -\text{Prov}(\ulcorner \mathbf{0} = \mathbf{1} \urcorner)$$

can be held to express the consistency of Z. Let us consider G first.

If $B(y)$ satisfies condition (i) of the definition of *provability predicate* (whether or not it satisfies (ii) and (iii)), and T is consistent, then if $\vdash_T A \leftrightarrow -B(\ulcorner A \urcorner)$, then not: $\vdash_T A$. For if $\vdash_T A$, then by (i), $\vdash_T B(\ulcorner A \urcorner)$, whence $\vdash_T -A$, and T is inconsistent. Thus since Z is consistent, G is not a theorem of Z (and is thus true). It is of some interest to observe that if $\mathrm{Pr}^*(x, y)$ defines Proof in Z, $\mathrm{Prov}^*(y) = \exists x\, \mathrm{Pr}^*(x, y)$, and

$$\vdash_Z G^* \leftrightarrow -\mathrm{Prov}^*(\ulcorner G^* \urcorner),$$

then not: $\vdash_Z G^*$; thus that $\mathrm{Prov}(y)$ actually satisfies (ii) and (iii) is irrelevant to whether or not $\vdash_Z G$. It is not irrelevant to whether or not $\vdash_Z H$ and $\vdash_Z -\mathrm{Prov}(\ulcorner \mathbf{0} = \mathbf{1} \urcorner)$.

In order to answer these two questions we shall state and prove the principal result of this chapter, Löb's theorem. The proof of Löb's theorem that we shall give resembles a certain 'proof' of the existence of Santa Claus:

Let S be the sentence 'If S is true, Santa exists'.

Assume

(1) S is true.

Then by the logic of identity,

(2) 'If S is true, Santa exists' is true.

From (2) we obtain

(3) If S is true, Santa exists.

And from (1) and (3), we infer

(4) Santa exists.

Having deduced (4) from the assumption (1), we may assert outright

(5) If S is true, Santa exists.

From (5) we obtain

(6) 'If S is true, Santa exists' is true.

And by the logic of identity again we have

(7) S is true.

From (5) and (7) we conclude

(8) Santa exists.

Löb's theorem

If $B(y)$ is a provability predicate for T, then for any sentence A, if $\vdash_T B(\ulcorner A \urcorner) \rightarrow A$, then $\vdash_T A$.

Proof. Suppose that

$$\vdash_T B(\ulcorner A \urcorner) \rightarrow A. \tag{1}$$

Let $D(y)$ be the formula $(B(y) \rightarrow A)$. The diagonal lemma, applied to $D(y)$, gives us a sentence C such that $\vdash_T C \leftrightarrow D(\ulcorner C \urcorner)$, i.e.,

$$\vdash_T C \leftrightarrow (B(\ulcorner C \urcorner) \rightarrow A). \tag{2}$$

So $$\vdash_T C \rightarrow (B(\ulcorner C \urcorner) \rightarrow A). \tag{3}$$

So by virtue of (i) of the definition of provability predicate,

$$\vdash_T B(\ulcorner C \rightarrow (B(\ulcorner C \urcorner) \rightarrow A) \urcorner). \tag{4}$$

By virtue of (ii),

$$\vdash_T B(\ulcorner C \rightarrow (B(\ulcorner C \urcorner) \rightarrow A) \urcorner) \rightarrow [B(\ulcorner C \urcorner) \rightarrow B(\ulcorner B(\ulcorner C \urcorner) \rightarrow A \urcorner)]. \tag{5}$$

So from (4) and (5) it follows that

$$\vdash_T B(\ulcorner C \urcorner) \rightarrow B(\ulcorner B(\ulcorner C \urcorner) \rightarrow A \urcorner). \tag{6}$$

By virtue of (ii) again,

$$\vdash_T B(\ulcorner B(\ulcorner C \urcorner) \rightarrow A \urcorner) \rightarrow [B(\ulcorner B(\ulcorner C \urcorner) \urcorner) \rightarrow B(\ulcorner A \urcorner)]. \tag{7}$$

So from (6) and (7),

$$\vdash_T B(\ulcorner C \urcorner) \rightarrow [B(\ulcorner B(\ulcorner C \urcorner) \urcorner) \rightarrow B(\ulcorner A \urcorner)]. \tag{8}$$

By virtue of (iii),

$$\vdash_T B(\ulcorner C \urcorner) \rightarrow B(\ulcorner B(\ulcorner C \urcorner) \urcorner). \tag{9}$$

So from (8) and (9),

$$\vdash_T B(\ulcorner C \urcorner) \rightarrow B(\ulcorner A \urcorner). \tag{10}$$

From (1) and (10),

$$\vdash_T B(\ulcorner C \urcorner) \rightarrow A. \tag{11}$$

From (2) and (11),

$$\vdash_T C. \tag{12}$$

By virtue of (i) again,

$$\vdash_T B(\ulcorner C \urcorner). \tag{13}$$

And so finally, from (11) and (13) we have that $\vdash_T A$.

The converse of Löb's theorem is of course trivial. It follows that a necessary and sufficient condition for a sentence A to be a theorem of Z is that $(\text{Prov}(\ulcorner A \urcorner) \rightarrow A)$ be a theorem of Z.

Corollary 1

Suppose that $B(y)$ is a provability predicate for T. Then if $\vdash_T A \leftrightarrow B(\ulcorner A \urcorner)$, then $\vdash_T A$.

Corollary 1 shows us that H, which 'says of itself that it is provable in Z', is provable in Z, and thus true.

Corollary 2. (*Gödel's second incompleteness theorem*)

Suppose that $B(y)$ is a provability predicate for T. Then if T is consistent, not: $\vdash_T -B(\ulcorner \mathbf{0} = \mathbf{1} \urcorner)$.

Proof. Suppose $\vdash_T -B(\ulcorner \mathbf{0} = \mathbf{1} \urcorner)$. Then $\vdash_T B(\ulcorner \mathbf{0} = \mathbf{1} \urcorner) \rightarrow \mathbf{0} = \mathbf{1}$. By Löb's theorem, then, $\vdash_T \mathbf{0} = \mathbf{1}$, and as T extends Q, T is inconsistent.

Corollary 2 shows that $-\text{Prov}(\ulcorner \mathbf{0} = \mathbf{1} \urcorner)$, which 'expresses the consistency of Z', is *not* provable in Z. ('If Z is consistent', one might be tempted to add. But Z is consistent.)

The connection between Löb's theorem, with

$$T = Z \quad \text{and} \quad B(y) = \text{Prov}(y),$$

and the provability of consistency in Z and its extensions may be stated as follows (this way of viewing the matter was suggested to us by Saul Kripke): Let C be a sentence of L, and let $Z+C$ be the theory whose theorems are the consequences in L of $Z \cup \{C\}$. Then the statement that $\vdash_Z \text{Prov}(\ulcorner A \urcorner) \rightarrow A$ is true if and only if $-\text{Prov}(\ulcorner A \urcorner)$ is a theorem of $Z + -A$; the statement that $\vdash_Z A$ is true if and only if $Z + -A$ is inconsistent. Let us call the sentence $-\text{Prov}(\ulcorner A \urcorner)$ *the consistency of* $Z + -A$. Löb's theorem then amounts to the assertion that if the consistency of $Z + -A$ is a theorem of $Z + -A$, then $Z + -A$ is inconsistent.

A formula $\text{Tr}(x)$ is called a *truth-predicate for* T if for every sentence A of the language of T, $\vdash_T A \leftrightarrow \text{Tr}(\ulcorner A \urcorner)$.

Corollary 3

If T is consistent, then T has no truth-predicate.

Proof. Suppose that Tr (x) is a truth-predicate for T. Then Tr (x) is a provability predicate for T. (Why?) Moreover, since Tr(x) is a truth-predicate, for every sentence A, $\vdash_T \mathrm{Tr}(\ulcorner A \urcorner) \rightarrow A$. It follows from Löb's theorem that every sentence A of the language of T is a theorem of T and thus that T is inconsistent.

Of course, Corollary 3 immediately follows from the diagonal lemma, when applied to the negation of any truth-predicate for T.

We conclude by showing that a formula may satisfy (i) – even with 'if' strengthened to 'iff' – without satisfying (ii) and (iii) and so counting as a provability predicate. For suppose that T is consistent and $B(y)$ is some formula such that for any sentence A, $\vdash_T A$ if and only if $\vdash_T B(\ulcorner A \urcorner)$. Let $D(y)$ be the formula $(B(y) \,\&\, y \neq \ulcorner \mathbf{0} = \mathbf{1} \urcorner)$. Then

$$\vdash_T -(B(\ulcorner \mathbf{0} = \mathbf{1} \urcorner) \,\&\, \ulcorner \mathbf{0} = \mathbf{1} \urcorner \neq \ulcorner \mathbf{0} = \mathbf{1} \urcorner),$$

i.e., $\vdash_T -D(\ulcorner \mathbf{0} = \mathbf{1} \urcorner)$. It follows from Corollary 2 that $D(y)$ is not a provability predicate for T. However, if $\vdash_T A$, then, since T is consistent, A is not the same sentence as $\mathbf{0} = \mathbf{1}$, and so $\ulcorner A \urcorner \neq \ulcorner \mathbf{0} = \mathbf{1} \urcorner$, and therefore (as T extends Q), $\vdash_T \ulcorner A \urcorner \neq \ulcorner \mathbf{0} = \mathbf{1} \urcorner$; but since $\vdash_T A$, $\vdash_T B(\ulcorner A \urcorner)$, so $\vdash_T (B(\ulcorner A \urcorner) \,\&\, \ulcorner A \urcorner \neq \ulcorner \mathbf{0} = \mathbf{1} \urcorner)$, i.e. $\vdash_T D(\ulcorner A \urcorner)$. Conversely, if $\vdash_T D(\ulcorner A \urcorner)$, then $\vdash_T B(\ulcorner A \urcorner)$, whence $\vdash_T A$. So we have shown that although $D(y)$ is not a provability predicate, for any sentence A, $\vdash_T A$ if and only if $\vdash_T D(\ulcorner A \urcorner)$.

Exercises

16.1 Suppose that $B(y)$ is a provability predicate for T and that $D(y)$ is the formula $(B(y) \,\&\, y \neq \ulcorner \mathbf{0} = \mathbf{1} \urcorner)$. Show that $D(y)$ meets condition (iii) but not condition (ii) of the definition of *provability predicate* unless T is inconsistent.

16.2 Suppose that $B(y)$ is a provability predicate for T. Use the existence of a sentence G such that $\vdash_T G \leftrightarrow -B(\ulcorner G \urcorner)$ to construct an 'alternative' proof that if T is consistent, then not: $\vdash_T -B(\ulcorner \mathbf{0} = \mathbf{1} \urcorner)$. Suggestion: show $\vdash_T B(\ulcorner B(\ulcorner G \urcorner) \urcorner) \rightarrow B(\ulcorner -G \urcorner)$, $\vdash_T B(\ulcorner G \urcorner) \rightarrow B(\ulcorner -G \urcorner)$, and

$$\vdash_T B(\ulcorner G \urcorner) \rightarrow [B(\ulcorner -G \urcorner) \rightarrow B(\ulcorner \mathbf{0} = \mathbf{1} \urcorner)].$$

Conclude that if $\vdash_T -B(\ulcorner \mathbf{0} = \mathbf{1} \urcorner)$, then $\vdash_T -B(\ulcorner G \urcorner)$, $\vdash_T G$, $\vdash_T B(\ulcorner G \urcorner)$.

16.3 Suppose that T is consistent and that for every sentence A of the language of T, $\vdash_T A$ if and only if $\vdash_T B(\ulcorner A \urcorner)$. Let $D(y)$ be the formula $(B(y) \,\&\, y \neq y)$. Then the diagonal lemma supplies a C such that

$$\vdash_T C \leftrightarrow D(\ulcorner C \urcorner).$$

Let $E(y)$ be the formula $(B(y) \,\&\, y \neq \ulcorner C \urcorner)$. Show that for every sentence A, $\vdash_T A$ if and only if $\vdash_T E(\ulcorner A \urcorner)$, $\vdash_T C \leftrightarrow E(\ulcorner C \urcorner)$, but not: $\vdash_T C$. Explain why there is no conflict with Corollary 1. What happens if $D^*(y)$ is the formula $(B(y) \vee y = y)$, C^* is such that $\vdash_T C^* \leftrightarrow D^*(\ulcorner C^* \urcorner)$, and $E^*(y)$ is the formula $(B(y) \vee y = \ulcorner C^* \urcorner)$?

16.4 Explain why any truth predicate for T is a provability predicate for T. Explain why T is inconsistent if some provability predicate $B(y)$ is such that $\vdash_T B(\ulcorner A \urcorner) \rightarrow A$, for every sentence A.

16.5 Suppose that $B(y)$ is a provability predicate for T and that $\vdash_T B(\ulcorner A \urcorner) \rightarrow C$ and $\vdash_T B(\ulcorner C \urcorner) \rightarrow A$. Show that $\vdash_T A$ and $\vdash_T C$.

16.6 Show that if $B(y)$ is a provability predicate for T, then

$$\vdash_T B(\ulcorner (B(\ulcorner A \urcorner) \rightarrow A) \urcorner) \rightarrow B(\ulcorner A \urcorner).$$

(*Hint*: show $\vdash_T B(\ulcorner L \urcorner) \rightarrow L$, where $L = B(\ulcorner (B(\ulcorner A \urcorner) \rightarrow A) \urcorner) \rightarrow B(\ulcorner A \urcorner)$.)

17
Non-standard models of arithmetic

We are now going to take up the topic of models of arithmetic. We know that there is at least one model of arithmetic, $\mathscr{N}$, the standard interpretation for the language L of arithmetic. Of course, that $\mathscr{N}$ is a model of arithmetic is true by definition: arithmetic is just the set of sentences of L that are true in $\mathscr{N}$. We shall enquire whether there is any sense at all in which $\mathscr{N}$ is the *unique* model of arithmetic.

Of course, no satisfiable set of sentences has exactly one model, literally speaking: given any model, one can construct another that is *isomorphic* but not identical, to it by 'replacing' some element of the domain by another object that is nowhere in the domain.

The definition of isomorphism of interpretations is this: An interpretation $\mathscr{I}$ *is isomorphic to* an interpretation $\mathscr{J}$ if

(1) $\mathscr{I}$ and $\mathscr{J}$ are interpretations of the same languages;

(2) $\mathscr{I}$ and $\mathscr{J}$ assign the same sentence letters the same truth-values; and

(3) there is a one–one function h with domain the domain of $\mathscr{I}$ and range the domain of $\mathscr{J}$ such that

(*a*) if $\mathscr{I}$ assigns a name the designation d, then $\mathscr{J}$ assigns it $h(d)$;

(*b*) if $\mathscr{I}$ assigns a function symbol the n-place function f, then $\mathscr{J}$ assigns it the function g such that for any $d_1, \ldots, d_n, d$ in the domain of $\mathscr{J}$, $f(d_1, \ldots, d_n) = d$ iff $g(h(d_1), \ldots, h(d_n)) = h(d)$;

(*c*) if $\mathscr{I}$ assigns a predicate letter the n-place characteristic function ϕ, then $\mathscr{J}$ assigns it the characteristic function ψ such that for any $d_1, \ldots, d_n$ in the domain of $\mathscr{I}$, $\phi(d_1, \ldots, d_n) = \psi(h(d_1), \ldots, h(d_n))$.

We leave it to the reader to verify that the relation *is isomorphic to* is an equivalence relation, and that the same sentences are true in isomorphic interpretations.

If T is a consistent theory of which any two models are *isomorphic*, then T comes as close to the ideal of having exactly one model as any theory could reasonably be expected to. Such theories are said to 'characterize' their models 'up to isomorphism' and to have 'essentially' one model. A definition, then: T is a *categorical* theory if any two models of T (that are interpretations of T's language) are isomorphic. So we might wonder whether arithmetic is categorical.

Every inconsistent theory is categorical (!), but there are consistent categorical theories as well: let T be the set of consequences (in the language containing just P) of $\exists y\, \forall x\, x = y \,\&\, \forall x\, Px$. Then T is categorical, for the domain of any model $\mathscr{I}$ of T will contain exactly one member, which $\mathscr{I}$ specifies that P is to be true of; and any two such models are isomorphic.

However, all models of T are finite (i.e. have finite domains). And this is no accident. For according to the 'upward' and 'downward' Skolem–Löwenheim theorems, any theory with an infinite model, has a model whose domain is enumerably infinite and also one whose domain is non-enumerably infinite. Since there is never any one–one function whose domain is an enumerably infinite set and whose range is a non-enumerably infinite set, the only consistent categorical theories are those each of whose models has a (fixed) finite number of elements in its domain.

Categoricity is thus not a particularly useful notion, and arithmetic, having an infinite model, is not categorical. A more interesting notion is that of *aleph-null-categoricity*: a theory T in some language is aleph-null-categorical if any two interpretations of that language which are models of T and which have domains that are enumerably infinite ('of cardinal number aleph-null') are isomorphic.

Suppose that T is a consistent theory with no finite models. If T is not complete, then T is not aleph-null-categorical, for if neither A (in T's language) nor $-A$ follows from T, then there is a model $\mathscr{I}$ of $T \cup \{A\}$, which by the Skolem–Löwenheim theorem may be assumed to have an enumerably infinite domain, and there is also a model $\mathscr{J}$ of $T \cup \{-A\}$ with enumerably infinite domain. Thus there are two non-isomorphic models $\mathscr{I}$ and $\mathscr{J}$ of T, both with enumerably infinite domains, and so T is not aleph-null-categorical.

$\vdash_Q \mathbf{m} \neq \mathbf{n}$ whenever $m \neq n$. Thus any model of Q, and therefore of any of its extensions, is infinite. And no consistent axiomatizable extension of Q is complete. So no consistent axiomatizable extension of Q is aleph-null-categorical. In particular, Z is not aleph-null-categorical.

But consistent aleph-null-categorical theories do exist: a simple example is the set of consequences of $\forall x\, Px$ (in the language containing just P). A more interesting example is given in Exercise 17.1. And, although arithmetic is a consistent extension of Q, because it is not axiomatizable, we might hope that, unlike Z, arithmetic would be aleph-null-categorical. Alas, like the hopes that arithmetic might be decidable or axiomatizable, this hope must also be dashed.

Theorem

Arithmetic is not aleph-null-categorical.

Proof. We shall show that there is an interpretation $\mathcal{I}$ of L that is a model of arithmetic, has an enumerably infinite domain, and is not isomorphic to $\mathcal{N}$.

Let a be a name different from $\mathbf{0}$. Let $A_0, A_1, A_2, \ldots$ be an enumeration of all sentences of L that are true in $\mathcal{N}$. Consider the set S of sentences $\{A_0, a \neq \mathbf{0}, A_1, a \neq \mathbf{1}, A_2, a \neq \mathbf{2}, \ldots\}$.

For any arbitrary finite subset S_0 of S there is a number n such that $a \neq \mathbf{n}$ is not in S_0, and hence the interpretation that is just like $\mathcal{N}$ except that it also assigns a the designation n is a model of S_0.

Thus every finite subset of S has a model, and so, by the compactness theorem, S has a model, and hence by the Skolem–Löwenheim theorem, S has a model $\mathcal{J}$ with an enumerable domain.

For any m, n if $m \neq n$, then $\mathbf{m} \neq \mathbf{n}$ is in S. As $\mathcal{J}$ is a model of S, $\mathcal{J}$ is a model of arithmetic whose domain is enumerably infinite.

Let e be the designation in $\mathcal{J}$ of a. As all of $a \neq \mathbf{0}$, $a \neq \mathbf{1}$, $a \neq \mathbf{2}, \ldots$ are true in $\mathcal{J}$, for no natural number n is e identical with the denotation in $\mathcal{J}$ of $\mathbf{n}$.

Let $\mathcal{I}$ be the interpretation of L that has the same domain as $\mathcal{J}$, that assigns $\mathbf{0}$, $'$, $+$, and $\cdot$ whatever $\mathcal{J}$ assigns them (but does not assign a any designation at all). $\mathcal{I}$ is an interpretation of L in which the same sentences of L are true as are true in $\mathcal{J}$. Since $\mathcal{J}$ is a model of arithmetic whose domain is enumerably infinite, so is $\mathcal{I}$.

But $\mathcal{I}$ is not isomorphic to $\mathcal{N}$. For the element e of the domain of $\mathcal{I}$ is not the denotation in $\mathcal{J}$, and hence not in $\mathcal{I}$, of $\mathbf{n}$, for any natural number n. But in $\mathcal{N}$, and consequently in any interpretation isomorphic to it, every domain element is denoted by $\mathbf{n}$, for some natural number n.

Interpretations in which the same sentences are true as are true in $\mathcal{N}$, but which are not isomorphic to $\mathcal{N}$, are called *non-standard models of arithmetic*. So we have just proved the existence of non-standard models of arithmetic with enumerably infinite domains.

What do non-standard models of arithmetic look like? Suppose $\mathcal{I}$ is a non-standard model of arithmetic. We'll call the objects in the domain of $\mathcal{I}$ NUMBERS. $\mathcal{I}$ assigns some NUMBER z to $\mathbf{0}$ and some functions s, $\oplus$, and $\odot$ to $'$, $+$, and $\cdot$. Let $x<y$ be the formula $\exists w\, w'+x = y$. We'll say that one NUMBER c is LESS THAN another d if $\mathcal{I}^{a\,b}_{c\,d}(a<b) = 1$.

First of all, no NUMBER is LESS THAN itself. For no (natural) number is less than itself; so $\forall x -x < x$ is true in $\mathcal{N}$, and hence true in $\mathcal{I}$; and so no NUMBER is LESS THAN itself. This argument illustrates our main technique for obtaining information about the 'appearance' of $\mathcal{I}$: observe that the natural numbers have a certain property, conclude that a certain sentence of L is true in $\mathcal{N}$, infer that it must also be true in $\mathcal{I}$ (since the same sentences are true in $\mathcal{N}$ and $\mathcal{I}$), and decipher the sentence 'over' $\mathcal{I}$. In this way we can conclude that exactly one of any two NUMBERS is LESS THAN the other, and that if one NUMBER is LESS THAN another, which is LESS THAN a third, then the first is LESS THAN the third. LESS THAN is thus a linear ordering of the NUMBERS, just as less than is a linear ordering of the numbers. Any NUMBERS c is LESS THAN sc.

0 is the least number; so z is the LEAST NUMBER. $0'$ is the next-to-least number; so sz is the next-to-LEAST NUMBER, etc. So there is an initial segment, z, sz, ssz, $sssz$, ... of the relation LESS THAN that is isomorphic to the series of (natural) numbers.

We call z, sz, ssz, $sssz$, ... the *standard* NUMBERS. Any others are *non-standard*. The standard NUMBERS are precisely those that can be obtained from z by applying the operation s a *finite* number of times. Any standard NUMBER is LESS THAN any non-standard NUMBER (for 3, e.g. is less than any number except for 0, 1, and 2.)

Any number other than 0 is the successor of some unique number; so any NUMBER other than z is s of some unique NUMBER. So we define r, a function from NUMBERS to NUMBERS, by $rz = z$ and $rsc = c$. Then if $c \neq z, src = c = rsc$.

If c is standard, then rc and sc are standard too, and if c is non-standard, then rc and sc are also non-standard. Moreover if c is non-standard, then rc is LESS THAN c.

We'll now define an equivalence relation, eq, on the NUMBERS. If c and d are NUMBERS we'll say that c eq d if for some *standard* (!) NUMBER e, either $c \oplus e = d$ or $d \oplus e = c$. Intuitively speaking, c eq d if c and d are a *finite* distance away from each other, i.e., in case one can get from c to d by applying r or s a finite number of times. Every standard NUMBER bears eq to all and only the standard NUMBERS.

We'll call the equivalence class under eq of any NUMBER c, c's *block*. c's block is then $\{\ldots, rrrc, rrc, rc, c, sc, ssc, sssc, \ldots\}$. c's block is infinite in both directions iff c is non-standard.

Suppose that c is LESS THAN d, and that c and d are in different blocks. Then, since sc is LESS THAN or equal to d, and c and sc are in the same block, *sc* is LESS THAN *d*. Similarly, *c* is LESS THAN *rd*. It follows that if there is even one member of a block *C* that is LESS THAN some member of a different block *D*, then every member of *C* is LESS THAN every member of *D*. If this is the case we shall say that *block C* is LESS THAN *block D*. A *block* is *non-standard* iff it contains some non-standard NUMBER. The standard block is the LEAST block.

There is no LEAST non-standard block, however. For suppose that *d* is a non-standard NUMBER. Then there is a *c* LESS THAN *d* such that either $c \oplus c = d$ or $c \oplus c \oplus sz = d$. (For any natural number $j \neq 0$, there is an $i < j$ such that either $i+i = j$ or $i+i+1 = j$.) Let's suppose $c \oplus c = d$. (The other case is similar.) If *c* is standard, so is $c \oplus c = d$. So *c* is non-standard. And *c* is not in the same block as *d*: for if $c \oplus e = d$ for some standard *e*, then $c \oplus e = c \oplus c$, whence $c = e$, contradicting the fact that *c* is non-standard. (The laws of addition hold in $\mathscr{N}$ and $\mathscr{I}$.) So *c*'s block is LESS THAN *d*'s. Similarly there is no GREATEST block.

Finally, if one block *C* is LESS THAN another *E*, then there is a third block *D*, which *C* is LESS THAN, and which is LESS THAN *E*. For suppose that *c* is in *C*, *e* is in *E*, and *c* is LESS THAN *e*. Then there is a *d* such that *c* is LESS THAN *d*, *d* is LESS THAN *e*, and either $c \oplus e = d \oplus d$ or $c \oplus e \oplus sz = d \oplus d$. (*Averages*, to within a margin of error of $\frac{1}{2}$, always exist in $\mathscr{N}$; *d* is the AVERAGE in $\mathscr{I}$ of *c* and *e*.) Suppose $c \oplus e = d \oplus d$. (The argument is similar in the other case.) If *d* is in *C*, then $d = c \oplus f$, for some standard *f*, and so

$$c \oplus e = c \oplus f \oplus c \oplus f,$$

and so $e = c \oplus f \oplus f$ (laws of addition), from which it follows, as $f \oplus f$ is standard, that *e* is in *C*. So *d* is not in *C*, and, similarly, not in *E* either. We may thus take *D* to be the block of *d*.

To sum up: the elements of the domain of any non-standard model $\mathscr{I}$ of arithmetic are going to be linearly ordered by LESS THAN. This ordering will have an initial segment that is isomorphic to the series of natural numbers, followed by a sequence of segments, each of which is isomorphic to the series of all integers (negative, zero, and positive). There is neither an earliest nor a latest member of this sequence, and between any two segments in it there lies a third. And if the domain of $\mathscr{I}$ is enumerable, there will be enumerably many segments, and the order of the non-initial ones will be isomorphic to that of the rational numbers. (See Exercise 17.1.)

Exercises

17.1 An interesting example of an aleph-null-categorical theory is the set of consequences (in the language containing '$<$' alone) of these six sentences:

(1) $\forall x - x < x$;
(2) $\forall x \forall y \forall z (x < y \,\&\, y < z \rightarrow x < z)$;
(3) $\forall x \forall y (x < y \vee x = y \vee y < x)$;
(4) $\forall x \exists y\, x < y$;
(5) $\forall x \exists y\, y < x$;
(6) $\forall x \forall y (x < y \rightarrow \exists z (x < z \,\&\, z < y))$.

Show that this theory is aleph-null-categorical. (*Hint:* suppose that $\mathscr{I}$ and $\mathscr{J}$ are two of its enumerably infinite models, that $a_0, a_1, \ldots$ ($b_0, b_1, \ldots$) is an enumeration of the domain of $\mathscr{I}$ ($\mathscr{J}$), and that $\mathscr{I}$ ($\mathscr{J}$) assigns '$<$' (the characteristic function of) the relation R (S). Define an isomorphism between the domains inductively, switching back and forth between the *a*s and the *b*s in such a way as not to overlook any: pair a_0 and b_0 off; suppose that a_n has not yet been paired off with any b. a_n will bear R to some (maybe none) of the already paired-off a's and some (maybe none) of the already paired-off *a*s will bear R to a_n. Pair a_n off with the earliest untaken b (in the enumeration $b_0, b_1 \ldots$) which bears S to those *b*s that are paired off with the *a*s to which a_n bears R and to which S is borne by those *b*s that are paired off with the *a*s which bear R to a_n. Then pair off the earliest hitherto neglected b with a suitable a. Continue in this manner.) Conclude that if the domain of $\mathscr{I}$ is the set of reals, that of $\mathscr{J}$ the set of rationals, and both $\mathscr{I}$ and $\mathscr{J}$ specify that '$<$' is to be true of c, d iff c is less than d (and specify nothing else), then $\mathscr{J}$ is an elementarily equivalent subinterpretation of $\mathscr{I}$.

17.2 Show that there is no formula $S(x)$ of L such that for any non-standard model $\mathscr{I}$ of arithmetic and any d, $\mathscr{I}^a_d(S(a)) = 1$ iff d is a non-standard NUMBER in the domain of $\mathscr{I}$.

18
Second-order logic

In Chapter 15 we learned that the set of sentences of L that are true in the standard model $\mathcal{N}$ was not the set of consequences in L of some effectively decidable set of sentences, and *a fortiori*, not the set of consequences in L of some single sentence (or finite set of sentences). In Chapter 17 we saw that there were at least two non-isomorphic models, indeed two non-isomorphic enumerable models, of the set of sentences of L that are true in $\mathcal{N}$.

We are now going to see that neither of these results is any longer the case if we broaden the notion of a *sentence* of a language in a certain way, and correspondingly extend our account of the conditions under which a sentence is true in an interpretation. We are going to introduce *second-order* sentences. The study of the conditions under which first- and second-order sentences are true is called *second-order logic*.

Second-order logic† differs dramatically from *first-order* (or *elementary*) logic, which we have been studying and will continue to study. Six of the more striking differences are these:

(1) There is no effective positive test for validity of second-order sentences.

(2) There is an enumerable, unsatisfiable set of sentences, every finite subset of which is satisfiable; one of the sentences in the set is a second-order sentence. (Thus the compactness theorem fails for second-order logic.)

(3) There is a single second-order sentence whose models are just those interpretations with non-enumerable domains. (The Skolem–Löwenheim theorem fails.)

(4) There is a single second-order sentence whose (first- and second-order) consequences in L are precisely the sentences true in $\mathcal{N}$.

(5) There is a single second-order sentence of which an interpretation of L is a model if and only if it is isomorphic to $\mathcal{N}$.

† Standard, or 'real', second-order logic, that is. In non-standard second-order logic, which we do not consider at all, the definition of interpretation is changed: separate domains are introduced for function and predicate variables. Outside of this chapter, we discuss second-order sentences and formulas only in the first half of Chapter 19 and in one paragraph in Chapter 23.

(6) There is a second-order formula which is true in $\mathcal{N}$ of precisely the gödel numbers of the first-order sentences true in $\mathcal{N}$.

We are going to demonstrate the truth of (1) through (5) in this chapter; we'll save (6) for the next. We prove (3) first, then (5). (1), (2) and (4) are quite direct consequences of (5).

At the outset we stress that in second-order logic we change neither the definition of *language* nor the definition of *interpretation*. A language is still an enumerable set of non-logical symbols, and an interpretation is still exactly the same sort of thing it has been all along (cf. Chapter 9). Changes are made in the notions of *formula* and *sentence*: more expressions are allowed to count as formulas or sentences. And the description, given in Chapter 9, of the conditions under which a sentence is true in one of its interpretations is also extended to cover the new sorts of sentences.

What is a second-order sentence? Let's refer to what we have been calling 'variables' as *individual variables*. We now introduce some new kinds of variable: function variables, sentence variables, and predicate variables. Just as we have one-, two-, three-, and more-place function symbols and predicate letters, we also have one-, two-, three-, and more-place function and predicate variables. We suppose that no symbol of any sort is also a symbol of any other sort. We extend the definition of formula by allowing function, sentence, or predicate variables to occur in those positions in formulas where previously only function symbols, sentence letters, or predicate letters (respectively!) could occur, and also allowing the new kinds of variable to occur after '∃' and '∀', in quantifiers. 'Free occurrence' and 'bound occurrence' are defined for the new kinds of variable exactly as they were defined for individual variables. Sentences, as always, are formulas in which no variables (individual, function, sentence or predicate) occur free. A second-order formula, then, is a formula that contains at least one occurrence of a function, sentence, or predicate variable, and a second-order sentence is a second-order formula that is a sentence. A formula or sentence of a language, whether first- or second-order, is, as before, one whose non-logical symbols all belong to the language. We should remark that sentence variables are not of particular importance and are included mainly for the sake of symmetry (but see Chapter 23).

Thus, in first-order logic we could identify a particular function as the identity function: $\forall x f(x) = x$. But in second-order logic, we can assert the existence of the identity function: $\exists u \forall x\, u(x) = x$. Similarly, where in first-order logic we could assert that two particular individuals share a certain property: $Pa \,\&\, Pb$, in second-order logic we can assert that every two individuals share some property or other: $\forall x \forall y \exists X (Xx \,\&\, Xy)$.

Finally, in first-order logic we can assert that if two particular individuals are identical, then they must either both have or both lack a particular property: $a = b \rightarrow (Pa \leftrightarrow Pb)$.

But in second-order logic, we can *define* identity via the Leibnizian principle of the identity of indiscernibles: $a = b \leftrightarrow \forall X (Xa \leftrightarrow Xb)$.

(In the examples u is used as a one-place function variable and X as a one-place predicate variable.) Each of the three second-order sentences which we have exhibited above is valid: true in each of its interpretations.

When is a second-order sentence S true in an interpretation $\mathscr{I}$? We answer this question by adding six more cases to the definition of $\mathscr{I}(S)$ of Chapter 9. Throughout F will be a formula which may contain free occurrences of the function variable u (the sentence variable p, the predicate variable X), and $f(P, R)$ will be a function symbol (a sentence letter, a predicate letter) that does not occur in F. f (R) is supposed to have the same number of places as u (X). $\mathscr{I}^f_{\mathfrak{f}}(\mathscr{I}^P_i, \mathscr{I}^R_\phi)$ will be an interpretation which differs from $\mathscr{I}$ (if at all) only in assigning the function $\mathfrak{f}$ to f (the truth-value $i (= 0, 1)$ to P, the characteristic function ϕ to R). And $F_u f (F_p P, F_X R)$ will be the sentence obtained from F by substituting $f(P, R)$ for all free occurrences of $u (p, X)$ in F.

Case 10. $\mathscr{I}(\forall uF) = 1$ if $\mathscr{I}^f_{\mathfrak{f}}(F_u f) = 1$ for every function $\mathfrak{f}$ which has the appropriate number of arguments and is defined everywhere on the domain of $\mathscr{I}$ and takes all its values in the domain of $\mathscr{I}$.

$\mathscr{I}(\forall uF) = 0$ if $\mathscr{I}^f_{\mathfrak{f}}(F_u f) = 0$ for even one such $\mathfrak{f}$.

Case 11. $\mathscr{I}(\exists uF) = 1$ if $\mathscr{I}^f_{\mathfrak{f}}(F_u f) = 1$ for even one $\mathfrak{f}$ as in case 10.

$\mathscr{I}(\exists uF) = 0$ if $\mathscr{I}^f_{\mathfrak{f}}(F_u f) = 0$ for every $\mathfrak{f}$ as in case 10.

Case 12. $\mathscr{I}(\forall pF) = 1$ if $\mathscr{I}^P_0(F_p P) = \mathscr{I}^P_1(F_p P) = 1$.

$\mathscr{I}(\forall pF) = 0$ if either $\mathscr{I}^P_0(F_p P) = 0$ or $\mathscr{I}^P_1(F_p P) = 0$ or both.

Case 13. $\mathscr{I}(\exists pF) = 1$ if either $\mathscr{I}^P_0(F_p P) = 1$ or $\mathscr{I}^P_1(F_p P) = 1$ or both.

$\mathscr{I}(\exists pF) = 0$ if $\mathscr{I}^P_0(F_p P) = \mathscr{I}^P_1(F_p P) = 0$.

Case 14. $\mathscr{I}(\forall XF) = 1$ if $\mathscr{I}^R_\phi(F_X R) = 1$ for every characteristic function ϕ which has the appropriate number of arguments and is defined everywhere on the domain of $\mathscr{I}$.

$\mathscr{I}(\forall XF) = 0$ if $\mathscr{I}^R_\phi(F_X R) = 0$ for even one such ϕ.

Case 15. $\mathscr{I}(\exists XF) = 1$ if $\mathscr{I}^R_\phi(F_X R) = 1$ for even one ϕ as in case 14.
$\mathscr{I}(\exists XF) = 0$ if $\mathscr{I}^R_\phi(F_X R) = 0$ for every ϕ as in case 14.

The definitions of validity, satisfiability, and implication are also unchanged for second-order sentences. Any sentence, first- or second-order, is valid iff true in all its interpretations, and satisfiable iff true in at least one of them. A set Δ of sentences implies a sentences S iff there is no interpretation in which all of the sentences in Δ are true but S is false.

Example 18.1. A definition of identity

As Whitehead and Russell point out (*Principia Mathematica*, Vol. 1, p. 57), the Leibnizian definition of identity which was given above can be simplified, for the sentence

$$a = b \leftrightarrow \forall X(Xa \rightarrow Xb) \qquad (1)$$

is valid. We *don't* need a biconditional on the right!

Proof. The left-to-right direction is trivial: If $\mathscr{I}(a = b) = 1$, then $\mathscr{I}(a) = \mathscr{I}(b)$, and then, for any (one-place) ϕ, $\mathscr{I}^R_\phi(Ra \rightarrow Rb) = 1$, and therefore $\mathscr{I}(\forall X(Xa \rightarrow Xb)) = 1$. For the right-to-left direction: If $\mathscr{I}(\forall X(Xa \rightarrow Xb)) = 1$, then for the particular ϕ which assigns the value 1 to $\mathscr{I}(a)$ and only to $\mathscr{I}(a)$, $\mathscr{I}^R_\phi(Ra \rightarrow Rb) = 1$. Since $\mathscr{I}^R_\phi(Ra) = 1$, $\mathscr{I}^R_\phi(Rb) = 1$. So $\mathscr{I}(b) = \mathscr{I}(a)$, and therefore $\mathscr{I}(a = b) = 1$. Thus, $\mathscr{I}$ assigns 1 to the left-hand side of (1) iff it assigns 1 to the right-hand side, and the proof is complete. (*Intuitively*: (1) is valid because among the properties of *a* is the property of *being identical with a*; then if *b* is to have *all* of *a*'s properties, it must have that one in particular.)

Example 18.2. An 'axiom' of enumerability

Let Ax En be the sentence

$$\exists z \exists u \forall X(Xz \,\&\, \forall x(Xx \rightarrow Xu(x)) \rightarrow \forall x Xx).$$

Ax En is a sentence that is true in an interpretation iff the domain of the interpretation is enumerable.

Proof. Let $\mathscr{I}$ be an interpretation and D the domain of $\mathscr{I}$.

Suppose $\mathscr{I}(\text{Ax En}) = 1$. Then for some a in D and some function f with domain D and range a subset of D

$$\mathscr{I}^{\mathbf{0}\,\mathbf{s}}_{a\,f}(\forall X(X\mathbf{o} \,\&\, \forall x(Xx \rightarrow X\mathbf{s}(x)) \rightarrow \forall x Xx)) = 1.$$

Let $A = \{a, f(a), f(f(a)), f(f(f(a))), \ldots\}$. A is an enumerable subset of D that contains a and also contains $f(b)$ whenever it contains b. Let ϕ be the characteristic function of A. Then

$$\mathscr{I}^{\mathbf{0}\,\mathbf{s}\,\mathbf{N}}_{a\,f\,\phi}(\mathbf{No}\,\&\,\forall x\,(\mathbf{N}x \rightarrow \mathbf{Ns}(x)) \rightarrow \forall x\,\mathbf{N}x) = 1.$$

Since a is in A, $\mathscr{I}^{\mathbf{0}\,\mathbf{s}\,\mathbf{N}}_{a\,f\,\phi}(\mathbf{No}) = 1$. Since A contains $f(b)$ whenever it contains b (for any b in D),

$$\mathscr{I}^{\mathbf{0}\,\mathbf{s}\,\mathbf{N}}_{a\,f\,\phi}(\forall x\,(\mathbf{N}x \rightarrow \mathbf{Ns}(x))) = 1.$$

Therefore $\mathscr{I}^{\mathbf{0}\,\mathbf{s}\,\mathbf{N}}_{a\,f\,\phi}(\forall x\,\mathbf{N}x) = 1$. Therefore, for any b in D, b is in A. So D has the same members as A and hence is enumerable.

Conversely, suppose that D is enumerable. D is nonempty, since D is the domain of $\mathscr{I}$. Let $a_0, a_1, \ldots$ be an enumeration without repetition of all members of D. (Possibly for some n, $a_0, a_1, \ldots = a_0, a_1, \ldots, a_n$.) Set $a = a_0$. Define f, a function with domain D and range a subset of D, as follows: If $a_0, a_1, \ldots$ is infinite, set $f(x) = y$ iff for some i, $x = a_i$ and $y = a_{i+1}$. But if for some n, $a_0, a_1, \ldots = a_0, a_1, \ldots, a_n$, set $f(x) = y$ iff either for some $i < n$, $x = a_i$ and $y = a_{i+1}$ or $x = a_n$ and $y = a_n$. In either case, if A is a subset of D that contains a and also contains $f(b)$ whenever it contains b, then every member of D is in A.

It follows that for every subset A of D with characteristic function ϕ,

$$\mathscr{I}^{\mathbf{0}\,\mathbf{s}\,\mathbf{N}}_{a\,f\,\phi}(\mathbf{No}\,\&\,\forall x\,(\mathbf{N}x \rightarrow \mathbf{Ns}(x)) \rightarrow \forall x\,\mathbf{N}x) = 1,$$

and thus that

$$\mathscr{I}^{\mathbf{0}\,\mathbf{s}}_{a\,f}(\forall X\,(X\mathbf{o}\,\&\,\forall x\,(Xx \rightarrow X\mathbf{s}(x)) \rightarrow \forall x\,Xx)) = 1,$$

and thus that

$$\mathscr{I}(\exists z\,\exists u\,\forall X(Xz\,\&\,\forall x(Xx \rightarrow Xu(x)) \rightarrow \forall x\,Xx)) = 1,$$

i.e. that $\mathscr{I}(\text{Ax En}) = 1$.

So Ax En is true in an interpretation iff its domain is enumerable.

We can now see that the Skolem–Löwenheim theorem fails for second-order logic: $-$Ax En is true in every interpretation whose domain is non-enumerably infinite (and such interpretations exist, so $-$Ax En *is* satisfiable!), but in no interpretation whose domain is enumerable. Thus assertion (3) holds.

We come now to the proof of assertion (5). Let Ind be the sentence

$$\forall X([X\mathbf{o}\,\&\,\forall x(Xx \rightarrow Xx')] \rightarrow \forall x\,Xx).$$

Ind is a second-order sentence of L which, when interpreted over $\mathscr{N}$, formalizes the principle of mathematical induction. Thus Ind is true in

$\mathcal{N}$. All of the enumerably many induction axioms of Chapter 16 are logical consequences of the one second-order sentence Ind. We noted in Chapter 16 that $Q3$, the third axiom of Q, follows from an induction axiom; $Q3$ therefore follows from Ind. We let PA be the conjunction of the other six axioms of Q and Ind. PA is true in $\mathcal{N}$.

Theorem

$\mathcal{N}$ is isomorphic to any interpretation of L which is a model of PA.

Proof. Suppose that $\mathcal{I}$ is an interpretation of L that is a model of PA. Let D be the domain of $\mathcal{I}$, and let e, s, p, and t be what $\mathcal{I}$ assigns to $\mathbf{0}$, $'$, $+$, and $\cdot$, respectively.

Because $Q1$, $Q2$, Ind, $Q4$, $Q5$, $Q6$, and $Q7$ are all true in $\mathcal{I}$, it follows that for any a, b in D and any subset A of D,

(i) if $s(a) = s(b)$, then $a = b$;

(ii) $e \neq s(a)$;

(iii) if both e is in A and $s(c)$ is in A whenever c is in A (for all c in D), then $A = D$;

(iv) $p(a, e) = a$;

(v) $p(a, s(b)) = s(p(a, b))$;

(vi) $t(a, e) = e$; and

(vii) $t(a, s(b)) = p(t(a, b), a)$.

Define h inductively by: $h(0) = e$; $h(n') = s(h(n))$.

We must show that h is a one–one function with domain the set of all natural numbers and range D, with the further properties that

$$h(m+n) = p(h(m), h(n)) \quad \text{and} \quad h(m \cdot n) = t(h(m), h(n)),$$

for all natural numbers m, n. This is straightforward if tedious. Here goes.

As e is an element of D and s is a function with domain D and range a subset of D, h is a function whose domain is the set of all natural numbers and whose range is a subset of D.

h is one–one. For otherwise let m be the least natural number such that for some $k > m$, $h(m) = h(k)$, and let n be $> m$ and such that $h(m) = h(n)$. As $m < n$, $n = j'$ for some j. $h(n) = h(j') = s(h(j))$, and if $m = 0$, $h(m) = h(0) = e$. But by (ii), $e \neq s(h(j))$. So $m \neq 0$. So $m = i'$ for some i, and then $i' < j'$, and so $i < j$, whence, as $i < m$, by the leastness of m, $h(i) \neq h(j)$, and so, by (i), $s(h(i)) \neq s(h(j))$; so

$$h(m) = h(i') = s(h(i)) \neq s(h(j)) = h(j') = h(n).$$

Contradiction.

Moreover e is in the range of h, and if c is in the range, then $c = h(n)$ for some n, whence $h(n') = s(c)$, and so $s(c)$ is in the range. It follows from (iii) that the range of $h = D$.

For any m, n, $h(m+n) = p(h(m), h(n))$. For $h(m+0) = h(m)$, which, by (iv), $= p(h(m), e) = p(h(m), h(0))$; and

$$h(m+n') = h((m+n)') = s(h(m+n)),$$

which, by the inductive hypothesis, $= s(p(h(m), h(n)))$, which by (v), $= p(h(m), s(h(n))) = p(h(m), h(n'))$.

And for any m, n, $h(m \cdot n) = t(h(m), h(n))$. For $h(m \cdot 0) = h(0) = e$, which, by (vi), $= t(h(m), e) = t(h(m), h(0))$; and $h(m \cdot n') = h(m \cdot n + m)$, which, by the foregoing, $= p(h(m \cdot n), h(m))$, which, by the inductive hypothesis, $= p(t(h(m), h(n)), h(m))$, which, by (vii),

$$= t(h(m), s(h(n))) = t(h(m), h(n')).$$

Corollary 1

Suppose that A is a (first- or second-order) sentence of L. Then $PA \vdash A$ iff A is true in $\mathcal{N}$.

Proof. Suppose $PA \vdash A$. Then A is true in all models of PA. But $\mathcal{N}$ is a model of PA. So A is true in $\mathcal{N}$. Conversely, suppose A is true in $\mathcal{N}$. Then if $\mathcal{I}$ is an interpretation of L that is a model of PA, then, by the theorem, $\mathcal{N}$ is isomorphic to $\mathcal{I}$, and so A is true in $\mathcal{I}$. So $PA \vdash A$.

Thus assertion (4) holds.

The significance of Corollary 1 should not be overestimated, for although a sentence (of L) is true in $\mathcal{N}$ if and only if it is a consequence of the single sentence PA, we cannot therefore conclude that arithmetic (the set of first-order truths in $\mathcal{N}$) is really decidable after all. To be able to conclude any such thing, we should have to know that there was a positive effective test for second-order validity or consequence, and Corollary 1 and the undecidability of arithmetic combine to provide a proof that there is no such positive test.

Corollary 2

There is no effective positive test for validity of second-order sentences.

Proof. Let A be a (first-order) sentence of L. Then $(PA \rightarrow A)$ is valid iff $PA \vdash A$, iff (by Corollary 1) A is true in $\mathcal{N}$, iff $-A$ is not true in $\mathcal{N}$, iff (by Corollary 1 again) $PA \nvdash -A$, iff $(PA \rightarrow -A)$ is not valid. If an effective positive test for second-order validity existed, then a decision procedure for truth in $\mathcal{N}$ would also exist: to test A for truth, apply the positive test for validity to both $(PA \rightarrow A)$ and $(PA \rightarrow -A)$. As exactly one of these is valid, whichever one of $(PA \rightarrow A)$ and $(PA \rightarrow -A)$ was valid would be so identified by the test. Then A would be true in $\mathcal{N}$ iff the test identified $(PA \rightarrow A)$ as valid. Since there is no decision procedure for truth in $\mathcal{N}$, there is no effective positive test for second-order validity either.

This result is sometimes formulated, rather misleadingly as follows: 'Second-order logic is incomplete.' A less misleading formulation along the same lines would go: 'no sound formalization of second-order logic is complete'. (It's not the logic that's incomplete; it's our attempted formalizations that are incomplete.)

Thus assertion (1) holds. It remains to show that (2) does too, that is, that the compactness theorem fails for second-order logic.

Corollary 3

Let $U = \{PA,\ a \neq \mathbf{0},\ a \neq \mathbf{1},\ a \neq \mathbf{2}, \ldots\}$. Then U is unsatisfiable, but every finite subset of U is satisfiable.

Proof. If U_0 is a finite subset of U, then, for some n, the sentence $a \neq \mathbf{n}$ is not in U_0, and thus the interpretation that is just like $\mathcal{N}$ except for assigning n to the name a is a model of U_0. (Cf. the theorem of Chapter 17.)

Suppose that $\mathcal{J}$ is a model of U. Let $\mathcal{I}$ be the interpretation of L that has the same domain as $\mathcal{J}$ and assigns $\mathbf{0}$, $'$, $+$, and $\cdot$ whatever $\mathcal{J}$ assigns them (but assigns nothing to a).

By Corollary 1, all truths in $\mathcal{N}$ are consequences of PA, and therefore hold in $\mathcal{J}$. Because all of $a \neq \mathbf{0}$, $a \neq \mathbf{1}$, $a \neq \mathbf{2}, \ldots$ are also true in $\mathcal{J}$, $\mathcal{I}$ cannot be isomorphic to $\mathcal{N}$, as the proof of the theorem of Chapter 17 shows. But then, as $\mathcal{J}(PA) = 1$, $\mathcal{I}(PA) = 1$, and therefore $\mathcal{I}$ must be isomorphic to $\mathcal{N}$. Contradiction. Therefore U has no model.

Second-order logic is, then, as Corollaries 2 and 3 and Example 18.2 tell us, 'unformalizable', incompact, and 'un-Skolem–Löwenheim-ish'. And more. Let us suppose that gödel numbers are assigned to second-order sentences in some reasonable way. Then the set of gödel numbers of valid second-order sentences is not (even) definable in arithmetic.

For suppose that $V(y)$ defined this set. Let f be some recursive function such that if n is the gödel number of A, then $f(n)$ is the gödel number of $(PA \rightarrow A)$. Let $B(x, y)$ represent f in arithmetic, and let $F(x)$ define the (recursive) set of gödel numbers of sentences of L in arithmetic. Then

$$F(x) \,\&\, \exists y\,(B(x,y) \,\&\, V(y))$$

defines the set of gödel numbers of theorems of arithmetic in arithmetic, contradicting Tarski's indefinability theorem. Much stronger results along these lines are known. See Exercise 18.4.

Exercises

18.1 Does it follow from the fact that $\exists x\, Fx \,\&\, \exists x - Fx$ is satisfiable that $\exists X[\exists x\, Xx \,\&\, \exists x - Xx]$ is valid?

18.2 Let us write '$aR_{*}b$' as a definitional abbreviation of

$$\text{`}\forall X\,([Xa \,\&\, \forall x\, \forall y\,(Xx \,\&\, xRy \rightarrow Xy)] \rightarrow Xb)\text{'}.$$

Show that (*a*), (*b*), and (*c*) are valid:

(*a*) $aR_{*}a$,

(*b*) $aRb \rightarrow aR_{*}b$,

(*c*) $aR_{*}b \,\&\, bR_{*}c \rightarrow aR_{*}c$.

Suppose that aRb iff a is a child of b. Under what conditions do we have $aR_{*}b$?

18.3 A set A is said to be Dedekind infinite if there is a one–one function whose domain is A and whose range is a proper subset of A. Let Ax D Inf be the sentence

$$\exists z\, \exists u\,(\forall x\, z \neq u(x) \,\&\, \forall x\, \forall y\,(x \neq y \rightarrow u(x) \neq u(y))).$$

Show that Ax D Inf is a sentence that is true in an interpretation iff the domain is Dedekind infinite. For those who know some set theory: if a set is Dedekind infinite, it is infinite; the axiom of (dependent) choice implies the converse, that if a set is infinite, it is Dedekind infinite. Show (without using any choice principles) that A is Dedekind infinite iff A has an enumerably infinite subset, that A is enumerably infinite iff A is enumerable and Dedekind infinite, and that A is finite iff A is enumerable but not Dedekind infinite. Devise a second-order sentence true in just the interpretations with finite domain. Do the same for those with enumerably infinite domain.

18.4 If $B(x)$ is a first- or second-order formula, then $B(x)$ is said to *define* a set θ *in second-order arithmetic* if for any natural number k,

$k \in \theta$ iff $B(\mathbf{k})$ is true in $\mathcal{N}$, and θ is then said to be definable in second-order arithmetic. Suppose our gödel numbering extended so that second-order sentences are assigned gödel numbers. Show that neither the set of gödel numbers of true second-order sentences of L nor the set of gödel numbers of valid second-order sentences is definable in second-order arithmetic.

18.5 (*Henkin*) Which of the eight combinations $\{Q1, Q2, \text{Ind}\}$, $\{Q1, Q2, -\text{Ind}\}, \ldots, \{-Q1, -Q2, -\text{Ind}\}$ are satisfiable? $Q1$ and $Q2$ are defined on p. 158; Ind, on p. 201.

18.6 Show that $\forall x \forall y \forall z (xRy \,\&\, yRz \rightarrow xRz)$ follows from

$$\forall X \forall z [\forall y (yRz \rightarrow [\forall x (xRy \rightarrow Xx) \rightarrow Xy]) \rightarrow \forall y (yRz \rightarrow Xy)].$$

19
On defining arithmetical truth

From now on we shall use 'V' to refer to the set of gödel numbers of first-order sentences of L that are true in $\mathcal{N}$. Tarski's indefinability theorem asserts that V is not definable in arithmetic. We want to show that this negative result is poised, so to speak, between two positive results: on the one hand, that each of certain 'approximations' V_n to V *is* definable in arithmetic, and, on the other, that V itself is 'definable in second-order arithmetic' (this notion is explained below).

V_n is the set of gödel numbers of first-order sentences of L that are true in $\mathcal{N}$ and contain *at most n* occurrences of any of the *logical operators* '$-$', '&', 'v', '$\rightarrow$', '$\leftrightarrow$', '$\exists$', and '$\forall$'. For any n, then, V_n is a proper subset of V; if $m < n$, $V_m \subsetneq V_n$; and a number k is in V iff for some n, k is in V_n. Our first theorem asserts that *for any n, V_n is definable in arithmetic.*

Recall that a (first-order) formula $B(x)$ of L is said to define a set θ (of natural numbers) in arithmetic if for any k, $k \in \theta$ iff $\mathcal{N}(B(\mathbf{k})) = 1$. And if there is such a $B(x)$, θ is said to be definable in arithmetic. Similarly, a (first-order) formula $C(x, y)$ of L is said to define a relation R (of natural numbers) in arithmetic if for any k, n, k bears R to n iff

$$\mathcal{N}(C(\mathbf{k}, \mathbf{n})) = 1,$$

and R is then said to be definable in arithmetic.

The true assertion that for any n, the set V_n is definable in arithmetic must be distinguished from the false assertion that the two-place relation R which k bears to n iff k is in V_n (for all k, n) is definable in arithmetic. R is not definable in arithmetic, for if $C(x, y)$ defined R, then $\exists y\, C(x, y)$ would define V, contradicting Tarski's indefinability theorem. So although for each n, there is a formula of L (containing one free variable) that defines the set V_n, there is no formula of L (containing two free variables) that defines the two-place relation R just described.

In this chapter and the next, we shall assume that '&', '$\rightarrow$', '$\leftrightarrow$', '$\forall$', and '$\neq$' are *defined* (non-primitive) symbols and that only '$-$', 'v', '$\exists$', and '$=$' properly occur in formulas and sentences of L. Theorems 1 and 2 hold good without this assumption, but it is then (inessentially) more complicated to prove them.

Theorem 1

For any n, the set V_n is definable in arithmetic.

The proof of both of Theorems 1 and 2 is given at the end of the chapter.

A set θ is said to be *definable in second-order arithmetic* if there is a formula $B(x)$ of L, *which may be* (but need not be) *a second-order formula*, such that for any natural number k, $k \in \theta$ iff $\mathcal{N}(B(\mathbf{k})) = 1$. $B(x)$ then *defines* θ in second-order arithmetic. All sets definable in arithmetic are definable in second-order arithmetic, but not conversely, for V, as we shall see, is definable in second-order arithmetic.

In this and the next chapter we shall be discussing sets of sets of natural numbers. To keep the levels straight, we shall always use the word 'class' to mean 'set of sets (of natural numbers)'; sets (of natural numbers) will always be so called.

The language $L+G$ has the same non-logical symbols as L (*viz.*, $\mathbf{0}$, $'$, $\mathbf{+}$, and $\cdot$), plus one other, the one-place predicate letter G. If A is a set of natural numbers, $\mathcal{N}_A^G$ is the interpretation of $L+G$ which is just like $\mathcal{N}$ except that it assigns G (the characteristic function of) the set A.

If $L+b$ is the language with the same non-logical symbols as L plus the name b, then a *set* A will be definable in arithmetic iff for some (first-order) sentence F of $L+b$, $A = \{n \mid \mathcal{N}_n^b(F) = 1\}$.

Analogously, a *class* C is said to be definable in arithmetic if there is a first-order sentence F of $L+G$ such that $C = \{A \mid \mathcal{N}_A^G(F) = 1\}$.

So, for example, the class of all sets containing 2, the class of all sets containing only odd numbers, and the class of all sets whatsoever are all definable in arithmetic, for the first $= \{A \mid \mathcal{N}_A^G(G\mathbf{2}) = 1\}$, the second $= \{A \mid \mathcal{N}_A^G(\forall x(Gx \to \exists y(y+y)' = x)) = 1\}$, and the third

$$= \{A \mid \mathcal{N}_A^G(\forall x\, x = x) = 1\}.$$

If the set θ is defined in arithmetic by the formula $B(x)$, then the class C whose sole member is θ is also definable in arithmetic, for C is then just $\{A \mid \mathcal{N}_A^G(\forall x(B(x) \leftrightarrow Gx)) = 1\}$. However, there are certain one-member classes that are definable in arithmetic, whose sole member is *not* definable in arithmetic. As our second theorem shows, the class $\{V\}$ whose sole member is V is one of these. An obvious question to ask is whether the class of sets that are definable in arithmetic is itself definable in arithmetic. We shall devote the next chapter to showing that the answer to this question is 'no'.

It is an immediate consequence of Theorem 2, which asserts that $\{V\}$

is definable in arithmetic, that V is definable in second-order arithmetic. To see this assume that $\{V\} = \{A \mid \mathcal{N}_A^G(F) = 1\}$, where F is some first-order sentence of $L+G$. Let $F(X)$ be the result of replacing each occurrence in F of G by an occurrence of X. Then $F(G) = F$. Then if

$$B(x) = \forall X(F(X) \rightarrow Xx),$$

then $B(x)$ defines V in second-order arithmetic.

The reason: $k \in V$ iff

for every set A, if $A \in \{V\}$, then $k \in A$, iff
for every set A, if $\mathcal{N}_A^G(F) = 1$, then $\mathcal{N}_A^G(G\mathbf{k}) = 1$, iff
for every set A, if $\mathcal{N}_A^G(F(G)) = 1$, then $\mathcal{N}_A^G(G\mathbf{k}) = 1$, iff
for every set A, $\mathcal{N}_A^G(F(G) \rightarrow G\mathbf{k}) = 1$, iff
$\mathcal{N}(\forall X(F(X) \rightarrow X\mathbf{k})) = 1$, iff
$\mathcal{N}(B(\mathbf{k})) = 1$.

The formula $\exists X(F(X) \,\&\, Xx)$ also defines V in second-order arithmetic, as is shown by the proof obtained from the one just given by replacing 'every' by 'some', 'if... then—' by '... and—', '$\rightarrow$' by '&', and '$\forall X$' by '$\exists X$'.

Theorem 2

$\{V\}$ is definable in arithmetic.

Proof of Theorems 1 and 2

We shall need the following facts about recursiveness (which are evident from Church's thesis):

(1) The set S of gödel numbers of sentences of L is recursive.

(2) For each n, the set S_n of gödel numbers of sentences of L containing at most n occurrences of logical operators is recursive.

(3) There exist recursive functions ν, δ, η, and σ such that if B and C are sentences of L, F is a formula of L, v is a variable, and the gödel numbers of B, C, F, and v are i, j, p, and q, respectively, then $\nu(i)$ is the gödel number of $-B$, $\delta(i,j)$ is the gödel number $(B \vee C)$, $\eta(p,q)$ is the gödel number of $\exists vF$, and $\sigma(p,q,m)$ is the gödel number of the result of substituting an occurrence of the numeral $\mathbf{m}$ for each free occurrence of v in F. (These functions are to have the value 0, which is not the gödel number of anything, for arguments not of the indicated sort. E.g. if q is not the gödel number of a variable, then $\sigma(p,q,m) = 0$.)

Let $S(x)$, $S^0(x)$, $S^1(x)$, $S^2(x)$, ... be formulas that define S, S_0, S_1, S_2, ... in arithmetic. And let Nu(x,y), De(x,y,z), Et(x,y,z), and Si(x,y,z,w) be formulas that represent ν, δ, η, and σ in arithmetic.

It is also evident from Church's thesis that V_0 is a recursive set, for an atomic sentence of L, i.e., a sentence of L in which there are no occurrences of logical operators, is of the form $s = t$, where s and t are terms of L. In order to decide whether a given number belongs to V_0, we have only to decide whether the number is the gödel number of a sentence $s = t$, where s has the same denotation in $\mathcal{N}$ as t. As it is effectively calculable what the denotation in $\mathcal{N}$ of any term of L is, it is effectively decidable whether any given number belongs to V_0, and hence V_0 is recursive, and thus definable in arithmetic. Let $V^0(x)$ be the formula that defines V_0 in arithmetic.

Now a sentence that contains $n+1$ occurrences of logical operators is either the negation of a sentence that contains n such occurrences, or the disjunction of two sentences, each of which contains at most n such occurrences, or the existential quantification of some formula that contains n such occurrences. Moreover, if F is a formula that contains n occurrences of operators, then any sentence obtained from F by substituting occurrences of a numeral $\mathbf{m}$ for free occurrences of variables in F also contains n occurrences of operators.

As every element of the domain of $\mathcal{N}$ is denoted in $\mathcal{N}$ by some numeral $\mathbf{m}$, a sentence $\exists vF$ is true in $\mathcal{N}$ iff for some number m, the result of substituting an occurrence of $\mathbf{m}$ for each free occurrence of v in F – if there are none, the result is F itself – is true in $\mathcal{N}$.

In terms of V_n, we can therefore characterize V_{n+1} as the set of those natural numbers k such that k is in S_{n+1} and either k is in V_n, or for some i, $\nu(i) = k$ and i is not in V_n, or for some i, j, $\delta(i,j) = k$ and either i is in V_n or j is in V_n, or for some p, q, $\eta(p,q) = k$ and for some m, $\sigma(p,q,m)$ is in V_n.

So if $V^n(x)$ defines V_n, V_{n+1} is defined by

$$S^{n+1}(x) \,\&\, \{V^n(x) \vee \exists y\,[\mathrm{Nu}(y,x) \,\&\, -V^n(y)] \vee \exists y \exists z\,[\mathrm{De}(y,z,x) \,\&\, (V^n(y) \vee V^n(z))] \vee \exists y_1 \exists z_1\,[\mathrm{Et}(y_1,z_1,x) \,\&\, \exists w \exists u(\mathrm{Si}(y_1,z_1,w,u) \,\&\, V^n(u))]\}.$$

It follows by induction that for each n, V_n is definable in arithmetic. Theorem 1 is thus proved.

The set of sentences true in $\mathcal{N}$ can be characterized as the unique set Γ meeting these conditions:

(1) Γ contains only sentences of L;

(2) Γ contains an atomic sentence B iff B's gödel number is in V_0;

(3) For any sentence B, $-B$ is in Γ iff B is not in Γ.

(4) For any sentences B, C, $(B \vee C)$ is in Γ iff at least one of B and C is in Γ; and

(5) For any formula F and variable v, if $\exists vF$ is a sentence, then $\exists vF$ is in Γ iff for some number m, the result of substituting an occurrence of $\mathbf{m}$ for each free occurrence of v in F is in Γ.

V is thus the unique set A such that

for any i, if i is in A, i is in S;
for any i, if i is in S_0, then i is in A iff i is in V_0;
for any i, if i is in S, then $\nu(i)$ is in A iff i is not in A;
for any i, j, if i and j are in S, then $\delta(i, j)$ is in A iff either i or j is in A; and
for any p, q, if $\eta(p, q)$ is in S, then $\eta(p, q)$ is in A iff for some m, $\sigma(p, q, m)$ is in A.

So if $F =$

$$\forall x\,(Gx \to S(x))\,\&$$
$$\forall x\,(S^0(x) \to (Gx \leftrightarrow V^0(x)))\,\&$$
$$\forall x\,\forall y(S(x)\,\&\,\mathrm{Nu}\,(x, y) \to (Gy \leftrightarrow -Gx))\,\&$$
$$\forall x\,\forall y\,\forall z\,(S(x)\,\&\,S(y)\,\&\,\mathrm{De}\,(x, y, z) \to (Gz \leftrightarrow (Gx \vee Gy)))\,\&$$
$$\forall x\,\forall y\,\forall z\,(\mathrm{Et}\,(x, y, z)\,\&\,S(z) \to (Gz \leftrightarrow \exists w\,\exists u\,(\mathrm{Si}\,(x, y, w, u)\,\&\,Gu))),$$

which is a sentence of $L+G$, then V is the unique set A such that

$$\mathcal{N}_A^G(F) = 1,$$

and Theorem 2 is proved.

Exercise

Show that for each n, the set of gödel numbers of true prenex sentences of L that contain at most n quantifiers is definable in arithmetic. Show it with the qualification 'prenex' omitted. (*Hint*: show that any formula containing at most n occurrences of '$-$', '&', 'v', '$\to$', '$\leftrightarrow$', '$\exists$', and '$\forall$' is logically equivalent to a prenex formula containing at most 4^n quantifiers.)

20
Definability in arithmetic and forcing

Having seen that $\{V\}$ is definable in arithmetic even though V is not, we now want to show that the class of sets that are definable in arithmetic is not definable in arithmetic. The proof of this theorem, which is due to Addison, uses methods that were invented by Cohen and applied by him in his celebrated demonstration that the continuum hypothesis† does not follow from the usual axioms of set theory. The proof of Addison's theorem provides an excellent introduction to these methods, which are basic to present-day research in set theory. The first notion we need is that of a *condition*.

A *condition* is a finite, consistent set of sentences of either of the forms $G\mathbf{m}$ or $-G\mathbf{m}$. $\varnothing$, the null set, is a condition. Other examples of conditions are $\{G\mathbf{17}\}$, $\{G\mathbf{2}, -G\mathbf{17}\}$, and $\{-G\mathbf{0}, -G\mathbf{1}, \ldots, -G\mathbf{999\,999}, G\mathbf{1\,000\,000}\}$. '$p$', '$q$', and '$r$' will be used as variables for conditions. We suppose conditions to have been assigned gödel numbers in some reasonable way.

We shall say that a condition q *extends* (or is an *extension* of) a condition p if p is a subset of q. Every condition, then, extends itself and extends $\varnothing$.

Forcing is a relation between certain conditions and certain sentences of $L+G$. We shall write '$p \Vdash S$' to mean 'the condition p forces the sentence S'. The relation: $p \Vdash S$ is inductively defined by the following five stipulations:

(1) If S is an atomic sentence of L, then $p \Vdash S$ iff S is true in $\mathcal{N}$.

(2) If t is a term of L and m is the denotation of t in $\mathcal{N}$, then if S is the sentence Gt, then $p \Vdash S$ iff $G\mathbf{m}$ is in p.

(3) If S is a disjunction $(B \vee C)$, then $p \Vdash S$ iff either $p \Vdash B$ or $p \Vdash C$.

(4) If S is an existential quantification $\exists x\, B(x)$, then $p \Vdash S$ iff for some n, $p \Vdash B(\mathbf{n})$.

(5) If S is a negation $-B$, then $p \Vdash S$ iff for every q that extends p, it is not the case that $q \Vdash B$.

† The continuum hypothesis, first conjectured by Cantor, is the assertion that every set of real numbers either is enumerable or has the same cardinal number as the set of all reals (the continuum). Two sets have the same cardinal number if there is a one-to-one correspondence between them. That the continuum hypothesis is consistent with set theory was established by Gödel before Cohen's result, by methods of an entirely different sort.

Clause 5 bears repeating: a condition forces the negation of a sentence iff no extension of the condition forces the sentence. It follows that no condition forces some sentence and its negation and also that either a condition forces the negation of a sentence or some extension forces the sentence. We shall shortly show that if a condition forces some sentence, then any extension of it also forces that sentence.

It follows from clauses 2 and 5 that $p \Vdash -G\mathbf{m}$ iff $-G\mathbf{m}$ is in p. For if $-G\mathbf{m}$ is not in p, then $p \cup \{G\mathbf{m}\}$ is an extension of p that forces $G\mathbf{m}$. So if $p \Vdash -G\mathbf{m}$, i.e. no extension of p forces $G\mathbf{m}$, then $-G\mathbf{m}$ is in p. And if $-G\mathbf{m}$ is in p, then $G\mathbf{m}$ is in no extension of p, and hence no extension of p forces $G\mathbf{m}$ and so $p \Vdash -G\mathbf{m}$.

So $\{G\mathbf{3}\}$ does not force either $G\mathbf{11}$ or $-G\mathbf{11}$, and therefore does not force $(G\mathbf{11} \vee -G\mathbf{11})$. Thus a condition may imply a sentence without forcing it. We shall see that the converse is also possible, that, for example, $\varnothing$ forces $--\exists x Gx$ (even though it does not force $\exists x\, Gx$).

Lemma 1

If $p \Vdash S$ and q extends p, then $q \Vdash S$.

Proof. Suppose that $p \Vdash S$ and q extends p. If S is an atomic sentence, either of L or of the form Gt, then, trivially, q also forces S. If $S = (B \vee C)$, then either $p \Vdash B$ or $p \Vdash C$, whence by the inductive hypothesis, either $q \Vdash B$ or $q \Vdash C$, and therefore $q \Vdash (B \vee C)$. If $S = \exists x\, B(x)$, then for some n, $p \Vdash B(\mathbf{n})$, whence by the i.h., $q \Vdash B(\mathbf{n})$, and therefore $q \Vdash \exists x\, B(x)$. Lastly, if $S = -B$, then no extension of p forces B, and therefore, as any extension of q is an extension of p, no extension of q forces B, whence $q \Vdash -B$.

It follows directly from Lemma 1 that if $p \Vdash B$, then $p \Vdash --B$, for any extension of p will then force B, and thus no extension of p will force $-B$. In like manner, if $p \Vdash -B$ and $p \Vdash -C$, then $p \Vdash -(B \vee C)$, for every extension of p will force both $-B$ and $-C$, and thus no extension of p will force $(B \vee C)$.

So if $p \Vdash B$ and $p \Vdash -C$, then $p \Vdash -(-(-B \vee -C) \vee -(B \vee C))$. For then $p \Vdash (-B \vee -C)$, and so $p \Vdash --(-B \vee -C)$. And also $p \Vdash (B \vee C)$, so $p \Vdash --(B \vee C)$. So $p \Vdash -(-(-B \vee -C) \vee -(B \vee C))$. Similarly if $p \Vdash -B$ and $p \Vdash C$, $p \Vdash -(-(-B \vee -C) \vee -(B \vee C))$. These two facts will be used in the proof of lemma 5, as will the truth-functional equivalence of $-(-(-B \vee -C) \vee -(B \vee C))$ with $-(B \leftrightarrow C)$.

Lemma 2

If S is a sentence of L, then for every p, $p \Vdash S$ iff $\mathcal{N}(S) = 1$.

Proof. If S is an atomic sentence of L, the lemma is true in virtue of clause 1 of the definition of 'forces'. If $S = (B \vee C)$, then $p \Vdash S$ iff either $p \Vdash B$ or $p \Vdash C$, iff by the inductive hypothesis, $\mathcal{N}(B) = 1$ or $\mathcal{N}(C) = 1$, iff $\mathcal{N}(S) = 1$. If $S = \exists x B(x)$, then $p \Vdash S$ iff for some n, $p \Vdash B(\mathbf{n})$, iff, by the i.h., for some n, $\mathcal{N}(B(\mathbf{n})) = 1$, iff $\mathcal{N}(S) = 1$. Lastly, if $S = -B$, then $p \Vdash S$ iff no extension of p forces B, iff, by the i.h., it is not the case that $\mathcal{N}(B) = 1$, iff $\mathcal{N}(S) = 1$.

Forcing is a curious relation: since $\varnothing$ does not contain any sentence $G\mathbf{n}$, for no n does $\varnothing$ force $G\mathbf{n}$, and therefore $\varnothing$ does not force $\exists x\, Gx$. But $\varnothing$ does force $--\exists x\, Gx$! For suppose that some p forced $-\exists x\, Gx$. Then no extension of p forces $\exists x\, Gx$. Let m be the least natural number such that $-G\mathbf{m}$ is not in p. Such a number will exist, as p is a finite set. Let $q = p \cup \{G\mathbf{m}\}$. q is a condition. q extends p. $q \Vdash G\mathbf{m}$. So $q \Vdash \exists x Gx$. Contradiction. Thus no p forces $-\exists x\, Gx$. So no extension of $\varnothing$ forces $-\exists x\, Gx$. So $\varnothing$ forces $--\exists x\, Gx$.

Some more definitions

We shall say that a condition p is *A-correct* if for any m, if $G\mathbf{m}$ is in p, then m is in A, and if $-G\mathbf{m}$ is in p, then m is not in A. p is A-correct, then, iff $\mathcal{N}_A^G$ is a model of p.

We shall say that a set A *Forces* a sentence S iff some A-correct condition forces S. Note that A cannot Force both S and $-S$, as the union of the A-correct conditions forcing S and $-S$ would be a condition that forced both.

A set A is called *generic* if for every sentence S of $L+G$, either A Forces S or A Forces $-S$.

The first fact about generic sets that we have to prove is that they exist! Indeed,

Lemma 3

For any p, there is a generic set A such that p is A-correct.

Proof. Let $S_0, S_1, S_2, \ldots$ be an enumeration of all sentences of $L+G$, and let $p_0, p_1, \ldots$ be an enumeration of all conditions. We inductively define a sequence $q_0, q_1, \ldots$ of conditions by: $q_0 = p$; $q_{i+1} = q_i$ if $q_i \Vdash -S_i$, and q_{i+1} = the first extension of q_i (in the enumeration $p_0, p_1, \ldots$) that forces S_i if not. (If q_i does not force $-S_i$, then some extension of q_i forces S_i.) If $i \leqslant j$, then q_j extends q_i. Let $A = \{m | \text{for some } i, G\mathbf{m} \text{ is in } q_i\}$. We claim that p is A-correct and that A is generic. Since for each i, either $q_{i+1} \Vdash S_i$ or $q_{i+1} \Vdash -S_i$, to show this we have only to show that for each i, q_i is A-correct, and hence have only to show that if $-G\mathbf{m}$ is in q_i, then m is not in A.

Well, if $-G\mathbf{m}$ is in q_i and m is in A, then for some j, $G\mathbf{m}$ is in q_j. Let $k = \max(i,j)$. Then both $G\mathbf{m}$ and $-G\mathbf{m}$ are in q_k, which is impossible, as q_k is consistent. So for each i, q_i is A-correct, and A is generic.

The next fact about genericity can be put: if A is generic, being true in $\mathcal{N}_A^G$ = being Forced by A.

Lemma 4

If S is a sentence of $L+G$ and A is a generic set, then A Forces S iff $\mathcal{N}_A^G(S) = 1$.

Proof. The proof is again an induction on the complexity of S. There are five cases to consider, corresponding to the five clauses of the definition of forcing.

Case 1. S is an atomic sentence of L. Then A Forces S iff some A-correct p forces S, iff $\mathcal{N}(S) = 1$, iff $\mathcal{N}_A^G(S) = 1$.

Case 2. $S = Gt$. Let m be the denotation of t in $\mathcal{N}$. Then A Forces S iff some A-correct p forces Gt, iff $G\mathbf{m}$ is in some A-correct p, iff $\{G\mathbf{m}\}$ is A-correct, iff m is in A, iff $\mathcal{N}_A^G(Gt) = 1$, iff $\mathcal{N}_A^G(S) = 1$.

Case 3. $S = (B \vee C)$. Then A Forces S iff some A-correct p forces $(B \vee C)$, iff some A-correct p either forces B or forces C, iff either some A-correct p forces B or some A-correct p forces C, iff either A Forces B or A Forces C, iff, by the inductive hypothesis, either $\mathcal{N}_A^G(B) = 1$ or $\mathcal{N}_A^G(C) = 1$, iff $\mathcal{N}_A^G(B \vee C) = 1$, iff $\mathcal{N}_A^G(S) = 1$.

Case 4. $S = \exists x\, B(x)$. Then A Forces S iff some A-correct p forces $\exists x\, B(x)$, iff for some A-correct p, there is an n such that p forces $B(\mathbf{n})$, iff for some n, there is an A-correct p such that p forces $B(\mathbf{n})$, iff for some n, A Forces $B(\mathbf{n})$, iff, by the i.h., for some n, $\mathcal{N}_A^G(B(\mathbf{n})) = 1$, iff $\mathcal{N}_A^G(\exists x\, B(x)) = 1$, iff $\mathcal{N}_A^G(S) = 1$.

Case 5. $S = -B$. No set Forces both B and $-B$. As A is generic, A Forces at least one of B and $-B$. So A Forces S iff A Forces $-B$, iff A does not

Force B, iff, by the i.h., it is not the case that $\mathscr{N}_A^G(B) = 1$, iff $\mathscr{N}_A^G(-B) = 1$, iff $\mathscr{N}_A^G(S) = 1$.

The last fact about generic sets to be proved is that none of them is definable in arithmetic.

Lemma 5

No generic set is definable in arithmetic.

Proof. Suppose otherwise. Then there are a generic set A and a formula $B(x)$ of L, such that for every n, n is in A iff $\mathscr{N}(B(\mathbf{n})) = 1$. So

$$\mathscr{N}_A^G(\forall x\,(Gx \leftrightarrow B(x))) = 1.$$

So $$\mathscr{N}_A^G(-\exists x\, F(x)) = 1,$$

where $$F(x) = -(-(-Gx \vee -B(x)) \vee -(Gx \vee B(x))).$$

By lemma 4, then, A Forces $-\exists x\, F(x)$, and so for some A-correct p, $p \Vdash -\exists x\, F(x)$. So for every q that extends p and every n, q does not force $F(\mathbf{n})$. Let k = the least natural number such that neither $G\mathbf{k}$ nor $-G\mathbf{k}$ is in p. Let $r = p \cup \{G\mathbf{k}\}$ if $B(\mathbf{k})$ is false in $\mathscr{N}$, and $= p \cup \{-G\mathbf{k}\}$ if $B(\mathbf{k})$ is true in $\mathscr{N}$. r is a condition. r extends p. If $B(\mathbf{k})$ is false in $\mathscr{N}$, then $r \Vdash G\mathbf{k}$ and, by Lemma 2, $r \Vdash -B(\mathbf{k})$, and so by the remarks between Lemmas 1 and 2, $r \Vdash F(\mathbf{k})$. Similarly, if $B(\mathbf{k})$ is true in $\mathscr{N}$, then $r \Vdash -G\mathbf{k}$ and $r \Vdash B(\mathbf{k})$, and so $r \Vdash F(\mathbf{k})$. Contradiction.

A set A is called *n-generic* if for every sentence S of $L+G$ that contains at most n occurrences of logical operators, either A Forces S or A Forces $-S$.

Lemma 6

For any n, there is an n-generic set A that is definable in arithmetic.

Proof. Let $S_0, S_1, S_2, \ldots$ be an enumeration of all sentences of $L+G$ that contain at most n occurrences of logical operators, and let $p_0, p_1, \ldots$ be an enumeration of all conditions. We may suppose that there are recursive functions f and g such that for any i, $f(i)$ and $g(i)$ are the gödel numbers of S_i and p_i, respectively. As in the proof of Lemma 3, we define a sequence $q_0, q_1, \ldots$ of conditions by: $q_0 = \varnothing$; $q_{i+1} = q_i$ if $q_i \Vdash -S_i$, and q_{i+1} = the first extension of q_i (in the enumeration $p_0, p_1, \ldots$) that forces S_i if not. And we set $A = \{m \mid \text{for some } i,\ G\mathbf{m} \text{ is in } q_i\}$. As in the proof of Lemma 3, A is n-generic. We must show that A is definable in arithmetic.

Now although the relation† $\{\langle i,j\rangle \mid i$ is the gödel number of a condition that forces the sentence with gödel number $j\}$ is not definable in arithmetic, for each n, the relation F_n, $=\{\langle i,j\rangle \mid i$ is the gödel number of a condition that forces the sentence with gödel number j and that sentence contains at most n occurrences of logical operators$\}$, *is* definable in arithmetic. The proof of this fact is quite similar to that of the fact that for each n, V_n is definable in arithmetic, and we omit it here.

We have assumed that the conditions were assigned gödel numbers in some 'reasonable' way. This means, at least, that the relations R, $= \{\langle i,j\rangle \mid i$ is the gödel number of a condition that extends the condition with gödel number $j\}$, and S, $= \{\langle k, m\rangle \mid k$ is the gödel number of a condition that contains $G\mathbf{m}\}$, are both recursive, and thus definable in arithmetic. By using R, F_{n+1}, f, g, and the β-function (or some other device for 'encoding' finite sequences of numbers as single numbers) to 'mimic' the inductive definition of the sequence $q_0, q_1, \ldots$, we can construct a formula $C(x,y)$ that defines the relation $\{\langle i, k\rangle \mid q_i$ has gödel number $k\}$ in arithmetic. But then, if $D(y,z)$ defines S, $\exists x \exists y\,(C(x,y)\ \&\ D(y,z))$ defines A in arithmetic.

Lemma 7

If S is a sentence of $L+G$ that contains at most n occurrences of logical operators, and A is an n-generic set, then A Forces S iff $\mathcal{N}_A^G(S) = 1$.

Proof. The proof of Lemma 4 proves Lemma 7.

At last we can prove Addison's theorem.

Addison's theorem

The class of sets definable in arithmetic is not definable in arithmetic.

Proof. Suppose otherwise. Then there is a sentence S of $L+G$ such that for any set A, $\mathcal{N}_A^G(S) = 1$ iff A is definable in arithmetic. Let n be the number of occurrences of logical operators in S. By Lemma 6, there exists an n-generic set A^* that is definable in arithmetic. So $\mathcal{N}_{A^*}^G(S) = 1$. So by Lemma 7, A^* Forces S. So for some A^*-correct p, $p \Vdash S$. By Lemma 3, there exists a generic set A_0 such that p is A_0-correct. As $p \Vdash S$, A_0 Forces S. So by Lemma 4, $\mathcal{N}_{A_0}^G(S) = 1$, and so A_0 is definable in arithmetic, which contradicts Lemma 5, according to which no generic set is definable in arithmetic.

† $\{\langle i,j\rangle \mid \ldots i \ldots j \ldots\}$ is the relation which, for all i, j, i bears to j iff $\ldots i \ldots j \ldots$

Exercises

20.1 Show that if $p \Vdash ---B$, $p \Vdash -B$.

20.2 Give an example of a sentence S such that the set of all even numbers Forces neither S nor $-S$.

20.3 Show that $\{\langle i,j\rangle \mid i$ is the gödel number of a condition that forces the sentence of $L+G$ with gödel number $j\}$ is not definable in arithmetic.

20.4 Where would our proof of Addison's theorem have broken down if we have chosen '&' and '∀' as primitive instead of 'v' and '∃', and made the obvious analogous stipulations in the definition of forcing?

20.5 Show that all subsets of a generic set that are definable in arithmetic are finite.

20.6 Show that if A is generic, then $\{A\}$ is not definable in arithmetic.

20.7 Show that $\{A \mid A$ is generic$\}$ is not definable in arithmetic.

20.8 Show that every generic set contains infinitely many prime numbers.

20.9 Show that the class of all sets of natural numbers and the class of all generic sets have the same cardinality.

20.10 Show that no generic set has a density. *Densities* are real numbers r, where $0 \leqslant r \leqslant 1$. A set A has density r if r is the limit as n goes to infinity of the ratio: $\dfrac{\text{number of members of } A < n}{n}$.

21

The decidability of arithmetic with addition, but not multiplication

Arithmetic is not decidable; but *arithmetic without multiplication* is decidable, as Presburger showed. *Arithmetic* (with addition, but) *without multiplication* is the theory whose theorems are the theorems of arithmetic that do not contain the multiplication sign '·', the '·'-free sentences of L true in $\mathcal{N}$. *Arithmetic* (with multiplication but) *without addition* is the theory whose theorems are the theorems of arithmetic that contain neither '**+**' nor '′'. Like arithmetic without multiplication, arithmetic without addition is a decidable theory, as Skolem showed. (If we discard '**+**' but not '′', we have an undecidable theory that is essentially no different from arithmetic, for addition is definable from successor and multiplication: $i+j=k$ iff $(a \cdot c)' \cdot (b \cdot c)' = ((a \cdot b)' \cdot (c \cdot c))')$, where $a = i'$, $b = j'$, and $c = k''$. We'll give a proof of Presburger's result.

In order to show how to decide whether or not a sentence containing just '**0**', '′', and '**+**' is true in $\mathcal{N}$, we shall consider an interpretation $\mathcal{J}$, whose domain is the set of *all* integers, positive, negative, and zero. $\mathcal{J}$ assigns '**0**' zero and '**1**' one. $\mathcal{J}$ assigns '**+**' the addition function (defined now on all integers). $\mathcal{J}$ assigns '**−**' the ordinary subtraction function. $\mathcal{J}$ specifies that '**<**' is to be true of i, j iff i is less than j. $\mathcal{J}$ also makes a specification about infinitely many *one-place* predicate letters $D_2, D_3, D_4, \ldots$: $\mathcal{J}$ specifies that D_m $(m \geqslant 2)$ is to be true of i iff i is divisible (without remainder, i.e., evenly) by m.

We shall show how to decide whether or not a sentence containing just '**0**', '**1**', '**+**', '**−**', '**<**', and any of the D_ms is true in $\mathcal{J}$. Once we have shown this, we will also have shown how to decide whether or not a sentence S of L not containing '·' is true in $\mathcal{N}$, for the result of replacing every occurrence of '′' by one of '**+1**' and *relativizing* all quantifiers $\exists v$ and $\forall v$ in S to the formula $(\mathbf{0} = v \vee \mathbf{0} < v)$ is true in $\mathcal{J}$ iff S is true in $\mathcal{N}$. (To relativize $\exists v$ ($\forall v$) to a formula F in a sentence S is to rewrite all contexts in S of the form: $\exists v \ldots$ ($\forall v \ldots$) as:

$$\exists v (F \& \ldots) \quad (\forall v (F \rightarrow \ldots)).)$$

Let us adopt some terminology. In this chapter a *term* shall be an 'open term': '**0**' or '**1**' or a variable or an expression obtained by filling the blanks of '+' and '−' by (shorter) open terms; (open) terms are thus the expressions obtainable from terms (in the usual sense) by substitution of (zero or more) variables for names. A *formula* (*sentence*) shall be a formula (sentence) whose only non-logical symbols are '**0**', '**1**', '+', '−', '<', and D_2, D_3, etc. 'Denotation' shall mean 'denotation in $\mathscr{I}$'; 'true', 'true in $\mathscr{I}$'. We shall say that two terms r and s (two formulas F and G) are *coextensive* if the sentence $\forall v_1, ..., \forall v_n r = s$ (the sentence

$$\forall v_1 \ldots \forall v_n(F \leftrightarrow G))$$

is true, $v_1, ..., v_n$ being all of the variables occurring (occurring free) in either r or s (either F or G).

If t is a term containing no variables, then, given t, we can effectively calculate the denotation of t. Given any *atomic* sentence, then, we can likewise effectively calculate the truth-value of the sentence, and therefore we can do the same thing for any *quantifier-free* sentence whatsoever. *We shall show how to decide whether or not a sentence S is true by showing how, given S, we can effectively find a quantifier-free sentence T which is coextensive with S (i.e., has the same truth-value as S)*: once T is found, its truth-value, which is also S's truth-value, can be effectively determined.

The method we shall use for finding our T from S is called *elimination of quantifiers* and consists in showing how to associate with each quantifier-free formula F (possibly containing x as well as some other variables free), a quantifier-free formula G, which is coextensive with $\exists xF$ and in which only such variables occur free as occur free in $\exists xF$. If for each quantifier-free F such a quantifier-free G can be found, then with each sentence S a quantifier-free sentence T, coextensive with S, can also be found: first put S into prenex normal form, and then replace each universal quantifier $\forall v$ in the prefix (the string of quantifiers) by $-\exists v-$. Then work from the inside of the sentence outward, successively replacing existential quantifications of quantifier-free formulas by coextensive quantifier-free formulas (in which no new variables occur free) until a sentence containing no bound variables, i.e., a quantifier-free sentence is obtained.

So let F be a quantifier-free formula. We obtain G, coextensive with $\exists xF$, and containing free occurrences of only such variables as occur free in $\exists xF$, by performing, in order, the following 30 operations, which replace formulas by coextensive ones with no new free variables:

(1) Put F into disjunctive normal form, i.e. rewrite F as a (logically equivalent) disjunction of conjunctions of atomic formulas contained in F and negations of atomic formulas contained in F.

Atomic formulas and their negations have the following six forms:

$$\begin{array}{lll} r = s & r \neq s & (r \text{ and } s \text{ are terms.}) \\ r < s & r \nless s & \\ D_m s & -D_m s & \end{array}$$

(2) Replace each occurrence of $r = s$ by an occurrence of

$$(r < (s+1) \,\&\, s < (r+1)).$$

(3) Replace each occurrence of $r \neq s$ by an occurrence of

$$(r < s \vee s < r).$$

(4) Use the distributive laws to put the result back into disjunctive normal form.

(5) Replace each occurrence of $r \nless s$ by an occurrence of $s < (r+1)$.

(6) Replace each occurrence of $-D_m s$ by an occurrence of

$$(D_m(s+1) \vee D_m(s+2) \vee \ldots \vee D_m(s+(m-1))).$$

(Here we have written '2' instead of '(1 + 1)' and '$(m-1)$' instead of

$$\text{`}\underbrace{(1+(\ldots+1)\ldots)}_{m-1 \text{ `1's}}\text{'.)}$$

The result is coextensive with the original, as any integer m ($\neq 0$) divides exactly one of $a, a+1, \ldots, a+(m-1)$ (for any integer a).

(7) Put the result back into disjunctive normal form.

At this point we have a formula that is a disjunction of conjunctions of atomic formulas of the forms $r < s$ and $D_m s$.

(8) Replace each occurrence of $r < s$ by an occurrence of $0 < (s-r)$.

We write '$-s$', etc. as short for '$(0-s)$', etc.

We shall say that a term is in *normal form* if it has one of these five forms:

$$\underbrace{(x+(\ldots+x)\ldots)}_{k \text{ `}x\text{'s}} \qquad (\text{`}kx\text{', for short})$$

$$\underbrace{-(x+(\ldots+x)\ldots)}_{k \text{ `}x\text{'s}} \qquad (\text{`}-kx\text{', for short})$$

$$\underbrace{((x+(\ldots+x)\ldots)+t)}_{k \text{ `}x\text{'s}} \qquad (\text{`}kx+t\text{', for short})$$

$$\underbrace{(-(x+(\ldots+x)\ldots)+t)}_{k\ \text{`}x\text{'s}} \quad (\text{`}-kx+t\text{', for short})$$

$$t,$$

where x does not occur in t at all.

Every term is coextensive with one in normal form.

(9) Replace all non-normal terms in the formula with coextensive normal terms.

(10) Replace each occurrence of $0 < -kx$ ($0 < kx + t$, $0 < -kx + t$) by an occurrence of $kx < 0$ ($-t < kx$, $kx < t$, respectively).

At this point all inequalities (formulas whose predicate letter is '$<$') that contain the variable x either have the form $t < kx$ or the form $kx < t$, where t is a term in which x does not occur and k is positive. We shall call inequalities of the first form, *lower* inequalities, and those of the second form, *upper* inequalities.

We now proceed to replace our formula by one in which each disjunct contains at most one lower inequality.

(11) Rearrange the order of conjuncts in each disjunct so that all lower inequalities occur on the left.

Observe that if '$t_1 < k_1 x \,\&\, t_2 k_1 \leqslant t_1 k_2$' holds, then so does

$$\text{`}t_1 k_2 < k_1 k_2 x\text{'}, \quad \text{and so does} \quad \text{`}t_2 k_1 < k_1 k_2 x\text{'},$$

and therefore so does '$t_2 < k_2 x$'.

$$(\text{`}t_2 k_1\text{' denotes the term } \underbrace{(t_2+(\ldots+t_2)\ldots)}_{k_1\ t_2\text{s}}.)$$

(12) If a conjunction $t_1 < k_1 x \;\&\; t_2 < k_2 x$ occurs in some disjunct, replace each occurrence of it in that disjunct by an occurrence of

$$(t_1 < k_1 x \,\&\, t_2 k_1 < t_1 k_2) \vee (t_1 < k_1 x \,\&\, t_2 k_1 = t_1 k_2) \vee (t_2 < k_2 x \,\&\, t_1 k_2 < t_2 k_1).$$

(13) Remove all occurrences of '$=$', as in 2.

(14) Put the result back into disjunctive normal form.

The maximum number of lower inequalities in each disjunct has now been reduced by one or reduced to zero or one. (Note that $t_2 k_1 < t_1 k_2$ is not a lower inequality since it does not contain x.)

(15) Repeat 11–14 until there is at most one lower inequality in each disjunct.

(16) Reduce the maximum number of upper inequalities in each disjunct to at most one in a similar way.

Recall that m divides n iff m divides $-n$.

(17) Replace each occurrence of

$$D_m kx\,(D_m - kx,\, D_m(kx+t),\, D_m(-kx+t))$$

by an occurrence of $D_m(kx-0)$ ($D_m(kx-0)$, $D_m(kx--t)$, $D_m(kx-t)$, respectively).

Observe that m divides $y-z$ iff either m divides both y and z or m divides both $y-1$ and $z-1$ or ... or m divides both $y-(m-1)$ and $z-(m-1)$.

(18) Replace each occurrence of $D_m(kx-t)$ by an occurrence of

$$([D_m(kx-0) \,\&\, D_m(t-0)] \vee [D_m(kx-1) \,\&\, D_m(t-1)] \vee \ldots$$
$$\vee\, [D_m(kx-(m-1)) \,\&\, D_m(t-(m-1))]).$$

(19) Put the result back into disjunctive normal form.

At this point all formulas of the form $D_m s$ in which x occurs are of the form $D_m(kx-i)$ where i and k denote non-negative integers. We shall call such formulas *congruences*. We now need to consider the question: when does m divide $kx-i$?

Let $A_{m,k,i}$ be the set of those integers y such that $0 \leqslant y < m$ and m divides $ky-i$. Given m, k, and i, we can effectively determine which (if any) of $0, 1, \ldots, m-1$ belong to $A_{m,k,i}$.

Lemma 1

m divides $kx-i$ iff for some y in $A_{m,k,i}$, m divides $x-y$.

Proof. Suppose m divides $kx-i$. If we divide x by m, we obtain integers y and w such that $x = wm+y$ and $0 \leqslant y < m$. So $wm = x-y$. So m divides $x-y$. So m divides $-k(x-y) = ky-kx$. So m divides

$$(ky-kx)+(kx-i) = ky-i.$$

So y is in $A_{m,k,i}$. Conversely, if y is in $A_{m,k,i}$ and m divides $x-y$, then m divides $k(x-y) = kx-ky$, and hence divides

$$(kx-ky)+(ky-i) = kx-i,$$

since m divides $ky-i$.

The following step is therefore justified:

(20) For each congruence $D_m(kx-i)$ occurring in the formula, determine which of $0, 1, \ldots, m-1$ belong to $A_{m,k,i}$. If none of $0, 1, \ldots, m-1$

belong to $A_{m,k,i}$, replace each occurrence of $D_m(kx-i)$ by an occurrence of $0 < 0$. Otherwise, replace each occurrence of $D_m(kx-i)$ by an occurrence of $(D_m(x-i_1) \vee \ldots \vee D_m(x-i_j))$, where $\{i_1, \ldots, i_j\} = A_{m,k,i}$.

(21) Put the result back into disjunctive normal form.

At this point all congruences in which x occurs are of the form $D_m(x-i)$.

We now wish to replace our formula with one (*) in which all congruences are still of the form $D_m(x-i)$, and of which it is true that if two congruences $D_{m_1}(x-i_1)$ and $D_{m_2}(x-i_2)$ occur in some one of its disjuncts, then m_1 and m_2 are relatively prime, i.e., no integer > 1 divides both m_1 and m_2.

(22) Replace each occurrence of $D_m(x-i)$ by an occurrence of

$$(D_{m_1}(x-i) \,\&\, \ldots \,\&\, D_{m_k}(x-i)),$$

where for some $p_1, \ldots, p_k, e_1, \ldots, e_k$, we have

$$m = m_1 \cdot \ldots \cdot m_k, \quad m_1 = p_1^{e_1}, \quad \ldots, \quad m_k = p_k^{e_k}, \quad p_1 < \ldots < p_k,$$

and $\qquad p_1, \ldots, p_k$

are prime. (Every positive integer is a product of a unique set of powers of primes.)

Observe now that if a divides b, then a divides $x-y$ and b divides $x-z$ iff a divides $z-y$ and b divides $x-z$.

(23) If for some prime p, and some e_1, e_2, with $e_1 \leqslant e_2$, there occur within some one disjunct the (two) congruences $D_{m_1}(x-i_1)$ and $D_{m_2}(x-i_2)$, where $m_1 = p^{e_1}$ and $m_2 = p^{e_2}$, replace each occurrence of $D_{m_1}(x-i_1)$ in that disjunct by an occurrence of $D_{m_1}(i_2-i_1)$, and delete all repetitions of conjuncts.

(24) Repeat operation 23 sufficiently often until no disjunct contains two congruences $D_{m_1}(x-i_1)$ and $D_{m_2}(x-i_2)$, where for some prime p and some $e_1, e_2, m_1 = p^{e_1}$, and $m_2 = p^{e_2}$.

At this point we have obtained a formula with property (*). We now wish to reduce the number of congruences that occur in each disjunct to zero or one.

(25) Rewrite all congruences to the left of any other conjuncts in each disjunct of the formula.

Each disjunct of the formula now has the form:

$$D_{m_1}(x-i_1) \,\&\, \ldots \,\&\, D_{m_k}(x-i_k) \,\&\, \text{—} \ldots \text{—}.$$

We now need a result from the theory of numbers called *the Chinese remainder theorem*. We write '$\text{rm}(i, m_j)$' to mean 'the remainder on dividing i by m_j'. (rm (i, m_j) is 'defined', i.e., there is such a number as rm (i, m_j), iff $m_j > 0$.)

The Chinese remainder theorem

Suppose that $i_1, \ldots, i_k$ are any natural numbers, that any two of $m_1, \ldots, m_k$ are relatively prime, and that for every $1 \leqslant j \leqslant k$, $i_j < m_j$. Let m be the product of $m_1, \ldots, m_k$. Then for some $i < m$, rm $(i, m_j) = i_j$ (for all $1 \leqslant j \leqslant k$).

Proof. (Enderton) Let $E(i) = \langle \text{rm } (i, m_1), \ldots, \text{rm } (i, m_k)\rangle$. There are n possible remainders (*viz.*, $0, 1, \ldots, n-1$) obtainable by dividing various integers by n, and so there are at most $m_1 \cdot \ldots \cdot m_k = m$ different possible values E can have. We must show that for some $i < m$, $E(i) = \langle i_1, \ldots, i_k\rangle$. And we shall have shown this if we can show that if $i < a < m$, then $E(i) \neq E(a)$, for then E will take each of its possible values, including $\langle i_1, \ldots, i_k\rangle$, at exactly one of the natural numbers $i < m$.

Suppose then that $i < a < m$ and $E(i) = E(a)$. Then $0 < a - i < m$ and for all $1 \leqslant j \leqslant k$, rm $(i, m_j) =$ rm (a, m_j). So each m_j divides $a - i$. But since any two of $m_1, \ldots, m_k$ are relatively prime, m must also divide $a - i$. But this is impossible, as $0 < a - i < m$. This proves the theorem.

We return to the proof of Presburger's theorem.

For every j $(1 \leqslant j \leqslant k)$, let $i_j^* = \text{rm}(i_j, m_j)$, the remainder on dividing i_j by m_j. Then $i_j^* < m_j$, and m_j divides $x - i_j$ iff m_j divides $x - i_j^*$. We now apply the Chinese remainder theorem to obtain an i such that

$$\text{rm}(i, m_j) = i_j^*,$$

and hence such that m_j divides $i - i_j^*$, for every j $(1 \leqslant j \leqslant k)$.

Let $m = m_1, \ldots m_k$. Then, as any two of $m_1, \ldots, m_k$ are relatively prime, m divides $x - i$ iff m_1 divides $x - i$ and ... and m_k divides $x - i$. But m_j divides $x - i$ iff m_j divides $(x-i)+(i-i_j^*) = x - i_j^*$, iff m_j divides $x - i_j$. So m divides $x - i$ iff m_1 divides $x - i_1$ and ... and m_k divides $x - i_k$.

(26) Replace each occurrence of $D_{m_1}(x-i_1) \,\&\, \ldots \,\&\, D_{m_k}(x-i_k)$ with an occurrence of an appropriate congruence $D_m(x-i)$.

At this point we have obtained a formula $(F_1 \vee \ldots \vee F_j)$ each of whose disjuncts contains at most one lower inequality, one upper inequality, and one congruence of the form $D_m(x-i)$.

(27) Rewrite each disjunct so that all conjuncts containing x occur on the left.

(28) Rewrite $\exists x(F_1 \vee \ldots \vee F_j)$ as $(\exists x F_1 \vee \ldots \vee \exists x F_j)$.

(29) Within each disjunct $\exists x F_k$, confine the quantifier to those three or fewer conjuncts in which x occurs; if there are none, delete the quantifier.

Thus, to complete the description of the procedure for replacing F by G, we need only show how to find a quantifier-free formula (**) that is coextensive with any given existential quantification (with respect to x) of a nonempty conjunction containing at most one lower inequality $s < jx$ (here $j \geqslant 1$ and x does not occur in s), at most one upper inequality $kx < t$, and at most one congruence of the form $D_m(x-i)$ and that contains no new free variables, for then operation 30 will give us our G:

(30) Replace occurrences of existential quantifications of the form described by occurrences of appropriate coextensive quantifier-free formulas.

To this end, let us note that

$$\exists x(D_m(x-i)\,\&\, s < jx \,\&\, kx < t) \tag{A}$$

is coextensive with

$$\exists x\,(D_{jkm}(jkx-jki)\,\&\, ks < jkx \,\&\, jkx < jt). \tag{B}$$

(Here, of course, 'jkx' denotes the term $\underbrace{(x+(\ldots+x)\ldots)}_{jk\text{ `}x\text{'s}}$.)

(B), in turn, is coextensive with

$$\exists x\,(D_{jkm}(x-jki)\,\&\, ks < x \,\&\, x < jt), \tag{C}$$

for if jkm divides $x-jki$, then jk divides $x-jki$, so jk divides x, and so for some x^*, $x = jkx^*$. (C) is of the form

$$\exists x\,(D_m(x-i)\,\&\, s < x \,\&\, x < t). \tag{D}$$

Any such formula is coextensive with the formula

$$((s+1) < t\,\&\, D_m((s+1)-i)) \vee \ldots \vee ((s+m) < t\,\&\, D_m((s+m)-i)),$$

which is quantifier-free and has the same free variables as (D); for given any two integers y and z, there will be an integer x, strictly between y and z, which is greater by i than some multiple of m, if and only if one of $y+1, \ldots, y+m$ is itself below z and greater by i than some multiple of m.

As for simpler formulas than (A), $\exists x(s < jx \,\&\, kx < t)$ is coextensive with $\exists x\,(ks < jkx \,\&\, jkx < jt)$, which is coextensive with

$$\exists x\,(D_{jk}(x-0)\,\&\, ks < x \,\&\, x < jt),$$

which is a formula of form (D). Since $jx \geqslant x$ if x is positive, and since there are arbitrarily large positive integers that are greater by i than some multiple of m, $\exists x(D_m(x-i)\,\&\, s < jx)$ is coextensive with the true quantifier-free formula '$0 < 1$'. Similarly,

$$\exists x(D_m(x-i)\,\&\, kx < t),\ \ \exists x\, D_m(x-i),\ \ \exists x\, s < jx,$$

and $$\exists x\, kx < t$$

are all coextensive with '$0 < 1$'.

As we have now shown how to find a quantifier-free formula with property (**) in all cases, we are done.

22
Dyadic logic is undecidable: 'eliminating' names and function symbols

The truth-table method is a well-known effective procedure for deciding whether or not an arbitrary sentence of the propositional calculus is valid. At the end of Chapter 12 we described an effective method for deciding whether or not an arbitrary *quantifier-free* sentence was valid, having shown in Chapter 10 that there can be no such method for arbitrary sentences of first-order logic. An examination of the proof in Chapter 10 gives us some more information: *there is no effective method for deciding the validity of an arbitrary sentence that may contain a name, a one-place function symbol, and (any number of) two-place predicate letters.* For the Q_is and S_js found in the sentences in $\Delta \cup \{H\}$ constructed in the proof were two-place predicate letters, and the only other non-logical symbols in those sentences were the name **o**, the one-place function symbol $'$, and the two-place predicate letter $<$.

In the present chapter we are going to improve this result by showing that there is no effective method for deciding the validity of an arbitrary *pure dyadic* sentence. A *pure* (or '"="-free') sentence is one not containing the equals-sign; a dyadic sentence is one whose non-logical symbols are all two-place predicate letters. So we shall show that 'pure dyadic logic is undecidable'. (In Chapter 25 we shall strengthen this result still further by showing that there is no effective procedure for deciding the validity of an arbitrary pure dyadic sentence *that contains only a single two-place predicate letter.*)

We are going to establish the undecidability of pure dyadic logic by showing that, in a certain weak sense, the equals-sign '=', names, and function symbols are *dispensable*. We cannot hope to show that any sentence containing '=', names, or function symbols is equivalent to one not containing them, for that is not true. Consider the sentence '$\exists x \forall y\, x = y$'. This sentence is true in exactly those interpretations whose domains contain exactly one member. By Corollary 3 of Chapter 13, however, any sentence not containing '=', if true in an interpretation whose domain contains one member, is true in one whose domain contains more than one member. '$\exists x \forall y\, x = y$' cannot therefore be equivalent to any sentence that does not contain '='.

We can, however, show that there is an effective procedure which, when applied to a sentence F, yields a sentence not containing '=', names, or function symbols, which is valid if and only if F is valid. Thus we shall show how to find (effectively) a sentence without '=', names, or function symbols, which is valid iff a given sentence is valid, without knowing whether or not the given sentence is valid.

In what follows, we shall not give names special treatment, for we can simply suppose that names are a special sort of function symbol, namely, 0-*place* function symbols. Whatever we say about function symbols in this chapter thus goes for names too.

We'll discuss the 'elimination' of function symbols first. The first fact that we shall need is that any sentence is *equivalent* to one in which all function symbols occur immediately to the right of '=' (and therefore to one in which no function symbols occur in the blanks to the right of predicate letters other than '=' or the blanks after other function symbols or the blank to the left of '=').

The proof is quite simple: Let us call an expression that is either a term or obtained from a term by substitution of variables for names an *open term* (cf. Chapter 21). Suppose now that F is a sentence, v is a variable that does not occur in F, t is an open term that occurs in some atomic formula D in F, G is the result of substituting an occurrence of v for a single occurrence of t in D, and H is the result of substituting an occurrence of $\exists v(G \& v = t)$ for an occurrence of D in F. Then H is equivalent to F. So if F is a sentence containing $n+1$ occurrences of function symbols that are not immediately to the right of '=', F is equivalent to a sentence H containing only n such occurrences, and hence, by mathematical induction, to one containing no such occurrences.

So for the remainder of this chapter, we'll assume that all function symbols in sentences occur immediately to the right of '='.

We shall now show how to 'dispense with' function symbols. We'll suppose that F is a sentence containing an n-place function symbol f and show how, given F, to find a sentence G, which is valid iff F is, which does not contain f, and which contains only function symbols that may otherwise be contained in F. Repeating the procedure to be described a sufficient number of times will yield a sentence that is valid iff F is, and that contains no function symbols. Here goes:

Let R be an $(n+1)$-place predicate letter not occurring in F.

Let H be the sentence obtained from F by replacing every occurrence in F of each formula $s_{n+1} = f(s_1, \ldots, s_n)$ by an occurrence of the formula $Rs_1 \ldots s_n s_{n+1}$. (Here the s_is are names or variables.)

If $\mathscr{I}$ and $\mathscr{J}$ are two interpretations with the same domain that agree in what they assign to all names, sentence letters, predicate letters and function symbols other than f and R, and $\mathscr{J}$ specifies that R is to be true of $c_1, \ldots, c_n, c_{n+1}$ iff $c_{n+1} = \mathfrak{f}(c_1, \ldots, c_n)$, where $\mathfrak{f}$ is the function $\mathscr{I}$ assigns to f, then H is true in $\mathscr{J}$ iff F is true in $\mathscr{I}$.

Let C be the sentence $\forall x_1 \ldots \forall x_n \exists y \forall z (Rx_1 \ldots x_n z \leftrightarrow z = y)$. C is true in an interpretation $\mathscr{J}$ iff the set of $(n+1)$-tuples of which $\mathscr{J}$ specifies that R is to be true is a *function*.

Let $G = (C \rightarrow H)$. G clearly does not contain f, and contains no other function symbols than those that may be contained in F.

Lemma

G is valid iff F is valid.

Proof. If G is not valid, then in some interpretation $\mathscr{J}$, C is true and H false. The set of $(n+1)$-tuples of which $\mathscr{J}$ specifies that R is to be true is then a function $\mathfrak{f}$. Let $\mathscr{I}$ be the interpretation that is just like $\mathscr{J}$ except that f is assigned $\mathfrak{f}$. F is then false in $\mathscr{I}$, and so not valid. Conversely, if F is not valid, there is an interpretation $\mathscr{I}$ in which F is false. f is assigned some function $\mathfrak{f}$ by $\mathscr{I}$. If $\mathscr{J}$ is just like $\mathscr{I}$ except that $\mathscr{J}$ specifies that R is to be true of $c_1, \ldots, c_n, c_{n+1}$ iff $c_{n+1} = \mathfrak{f}(c_1, \ldots, c_n)$, then C is true in $\mathscr{J}$, but H is false in $\mathscr{J}$, and so G is false in $\mathscr{J}$, and therefore not valid.

We now turn to the matter of finding a sentence G, not containing '$=$', that is valid iff a given sentence F is valid. Having shown how to 'dispense with' function symbols, we shall suppose that the given sentence F contains no one- or more-place function symbols (and so F may contain names).

So: Let E be a two-place predicate letter that does not occur in F. Let C be the conjunction of all of the following sentences:

$$\forall x\, xEx,$$

$$\forall x \forall y (xEy \rightarrow yEx),$$

$$\forall x \forall y \forall z ((xEy \,\&\, yEz) \rightarrow xEz),$$

and for each (n-place) predicate letter R that occurs in F, the sentence

$$\forall x_1 \ldots \forall x_n \forall y_1 \ldots \forall y_n ((x_1 E y_1 \,\&\, \ldots \,\&\, x_n E y_n) \rightarrow (Rx_1 \ldots x_n \leftrightarrow Ry_1 \ldots y_n)).$$

C is true in all (but not only) interpretations $\mathscr{J}$ in which $\mathscr{J}$ specifies that E has the identity relation (on $\mathscr{J}$'s domain) for its extension, i.e. in which $\mathscr{J}$ specifies that E is to be true of c_1, c_2 iff c_1 is identical with c_2.

Let H be the result of replacing every occurrence of '$=$' in F by one of E.

If $\mathscr{I}$ and $\mathscr{J}$ are two interpretations that differ (if at all) only in that $\mathscr{J}$ specifies that E has the identity relation as its extension, then H is true in $\mathscr{J}$ iff F is true in $\mathscr{I}$.

Our desired sentence G is then the conditional $(C \rightarrow H)$. G clearly does not contain '$=$'.

If F is not valid, and therefore false in some interpretation $\mathscr{I}$, then G is not valid, for C will be true, and H false, in the interpretation $\mathscr{J}$ that is just like $\mathscr{I}$ except that $\mathscr{J}$ specifies that E has the identity relation as its extension. We thus need only prove the following lemma.

Lemma

If G is not valid, F is not valid.

Proof. Suppose that G is not valid. Then there is an interpretation $\mathscr{I}$ in which C is true, but H false. We may suppose that $\mathscr{I}$ specifies nothing about non-logical symbols not occurring in G. By the Skolem–Löwenheim theorem, we may further suppose that the domain D of $\mathscr{I}$ is enumerable. We shall construct an interpretation $\mathscr{J}$ in which H has the same truth-value (*viz*. false) that it has in $\mathscr{I}$, and in which E has the identity relation for its extension. F will therefore have the same truth-value in $\mathscr{J}$ that H has, and thus not be valid.

Let $\sim$ be the relation on D that holds between members d_1 and d_2 of D if $\mathscr{I}$ specifies that E is to be true of d_1, d_2. Since C is true in $\mathscr{I}$, $\forall x\, xEx$ is too, and therefore $\sim$ is a reflexive relation on D. Since $\forall x \forall y(xEy \rightarrow yEx)$ is true in $\mathscr{I}$, $\sim$ is a symmetric relation, and since

$$\forall x \forall y \forall z\, ((xEy \,\&\, yEz) \rightarrow xEz)$$

is true in $\mathscr{I}$, $\sim$ transitive. $\sim$ is therefore an equivalence relation on D.

Moreover, if $c_1 \sim d_1, \ldots,$ and $c_n \sim d_n$, then because

$$\forall x_1 \ldots \forall x_n \forall y_1 \ldots \forall y_n((x_1 E y_1 \,\&\, \ldots \,\&\, x_n E y_n) \rightarrow (Rx_1 \ldots x_n \leftrightarrow Ry_1 \ldots y_n))$$

is true in $\mathscr{I}$, $\mathscr{I}$ specifies that R is to be true of $c_1, \ldots, c_n$ iff $\mathscr{I}$ specifies that R is to be true of $d_1, \ldots, d_n$.

We can now define $\mathscr{J}$:

(1) The domain of $\mathscr{J}$ is to be the set of equivalence classes $[d]$ of members of the domain of $\mathscr{I}$ under the relation $\sim$. (Cf. the proof of Lemma III in Chapter 12.)

(2) $\mathscr{J}$ assigns a sentence letter the same truth-value as $\mathscr{I}$.

(3) $\mathscr{J}$ assigns a name the designation $[d]$ iff $\mathscr{I}$ assigns it d.

(4) $\mathscr{J}$ specifies that an n-place predicate letter R is to be true of $[d_1], \ldots, [d_n]$ iff $\mathscr{I}$ specifies that R is to be true of $d_1, \ldots, d_n$. (This clause of the definition is independent of the choice of members $d_1, \ldots, d_n$ of the equivalence classes $[d_1], \ldots, [d_n]$.)

The domain of $\mathscr{J}$ is enumerable, since if $d_0, d_1, \ldots$ is an enumeration of all members of the domain of $\mathscr{I}$, $[d_0], [d_1], \ldots$ is an enumeration of all members of the domain of $\mathscr{J}$.

It follows from clause (4) that $\mathscr{J}$ specifies that E is to be true of $[d_1]$, $[d_2]$ iff $\mathscr{I}$ specifies that E is to be true of d_1, d_2, hence iff $d_1 \sim d_2$, hence iff $[d_1] = [d_2]$. Thus the extension of E according to $\mathscr{J}$ is just the identity relation on the domain of $\mathscr{J}$.

We want to see that H has the same truth-value in $\mathscr{J}$ as in $\mathscr{I}$. We do this as follows: for each d in D choose a new name a_d (different names for different members of D). Let $\mathscr{I}_1$ be just like $\mathscr{I}$ except that for each a_d, $\mathscr{I}_1(a_d) = d$; let $\mathscr{J}_1$ be just like $\mathscr{J}$ except that for each a_d, $\mathscr{J}_1(a_d) = [d]$. Every atomic '='-free sentence has the same truth-value in $\mathscr{I}_1$ as in $\mathscr{J}_1$. Since every member of the domain of $\mathscr{I}_1$ ($\mathscr{J}_1$) is denoted, according to $\mathscr{I}_1$ ($\mathscr{J}_1$), by some name, it follows that *every* '='-free sentence has the same truth-value in $\mathscr{I}_1$ as in $\mathscr{J}_1$. (Cf. the proof of Corollary 3, Chapter 13.) Any sentence that does not contain any names a_d has the same truth-value in $\mathscr{I}$ as in $\mathscr{I}_1$ and the same truth-value in $\mathscr{J}$ as in $\mathscr{J}_1$. Since H contains neither '=' nor any names a_d, H is true in $\mathscr{I}$ iff true in $\mathscr{I}_1$, iff true in $\mathscr{J}_1$, iff true in $\mathscr{J}$. The lemma is proved.

The undecidability of pure dyadic logic is now immediate. If we apply the 'elimination' procedures described above to the conditional of Chapter 10 (whose antecedent was some conjunction of all the members of Δ, whose consequent was H, and which was valid iff the given machine halted on the given input), we replace $'$ by a two-place predicate letter, '=' by a two-place predicate letter, and **o** by a one-place predicate letter. Thus, but for one one-place predicate letter A, the result would be a pure dyadic sentence that was valid iff its original was. And it was easy enough to 'eliminate' A: pick a new two-place predicate letter B, and replace each occurrence of each formula Av (v a variable) by an occurrence of the formula Bvv. The result is again valid iff the original was.

Exercises

22.1 Show that there is an effective procedure for finding a pure dyadic sentence, valid iff any given sentence is valid. *Hint*: use the machine M^* of Chapter 12.

22.2 Show that there is an effective procedure for finding a pure dyadic prenex sentence, whose prefix (the initial string of quantifiers) consists of three universal quantifiers followed by a sequence of existential quantifiers, that is valid iff any given sentence is valid. *Hint*: in placing function symbols to the right of '$=$', consider $\forall v(v=t \rightarrow G)$. If R is a two-place predicate letter, a suitable 'identity axiom' is

$$\forall x \forall y \forall z\,(xEy \rightarrow [(Rxz \leftrightarrow Ryz)\,\&\,(Rzx \leftrightarrow Rzy)]).$$

See (9.20).

22.3 (Presupposes Chapter 18.) Let A be a sentence. Let A^* be an '$=$'-free sentence that is satisfiable iff A is satisfiable. Let A^{**} be the result of replacing each non-logical symbol in A^* by a new variable of the same logical type. Let A^{***} be a sentence obtained by prefixing existential quantifiers to A^{**}. Let A^{****} = the sentence

$$((\text{Ax En}\,\&\,\text{Ax D Inf}) \rightarrow A^{***}).$$

(Cf. Exercise 18.3.) Show that A^{****} is valid iff A is satisfiable. Conclude (again) that there is no effective positive test for second-order validity.

23
The Craig interpolation lemma

If the sentence A implies the sentence C, there is a sentence B which A implies, which implies C, and which contains only such names as are contained in both of A and C. The reason is clear: let $a_1, \ldots, a_m$ be the names contained in A but not in C. (The a_is are supposed to be distinct; if $m = 0$, we may take $B = A$.) Let $v_1, \ldots, v_m$ be m completely new variables, and let A^* be the result of everywhere substituting each v_i for a_i in A. Then, since $(A \rightarrow C)$ is valid, so is $\forall v_1, \ldots, \forall v_m(A^* \rightarrow C)$, and hence so is $(\exists v_1 \ldots \exists v_m A^* \rightarrow C)$. $\exists v_1 \ldots \exists v_m A^*$ is then a suitable B, for A implies, it, it implies C, and all names it contains are contained in both A and C.

It might occur to us to ask whether this fact about names and implication can be subsumed under one about names, function symbols, sentence letters, and predicate letters, and implication; that is, to ask whether if A implies C there is always a sentence B, which A implies, which implies C, and which contains only names, function symbols, sentence letters and predicate letters, which are contained in both A and C.

The answer to this question, as stated, is *no*. Let A = '$\exists x\, Fx \,\&\, \exists x - Fx$' and C = '$\exists x \exists y\, x \neq y$'. A implies C, but there is no sentence at all containing only names, etc. contained in both A and C, and therefore none that is implied by A and implies C.

C in this example contains '$=$'. Suppose, then, that we disregard '$=$'. And let us now ask: if the sentence A implies the sentence C, is there a sentence B which A implies, which implies C, and which contains only non-logical symbols that are contained in both A and C? ('$=$' is a *logical* symbol; cf. Chapter 9.)

We can easily find a *second-order* sentence implied by A, implying C, and containing only non-logical symbols which are contained in both A and C: as before, replace each name, function symbol, sentence letter, or predicate letter other than '$=$' occurring in A but not in C by a new individual variable, function variable, sentence variable, or predicate variable of the appropriate number of places, and existentially quantify the result. (Cf. Chapter 18.)

But suppose that we are interested exclusively in first-order sentences: if the first-order sentence A implies the first-order sentence C, is there always a first-order sentence B, which A implies, which implies C, and which contains only non-logical symbols that are contained in both A and C?

The Craig interpolation lemma is the assertion that the answer to this question is *yes*. (B is called an 'interpolation sentence', whence the name.)

Our proof of Craig's lemma will have four parts. In the first, we will assume that A and C are sentences of the propositional calculus. In the second, A and C will not contain quantifiers, function symbols or the equals-sign. In the third (the hard part), we lift the assumption that A and C do not contain quantifiers (the proof we give is due to Dreben and Putnam); and in the last part, we lift the assumption that A and C don't contain function symbols or '='. At the outset, we may assume that A is satisfiable and that C is not valid. For if A is unsatisfiable, '$-\forall x\, x = x$' will do as B; if C is valid, '$\forall x\, x = x$' will do.

Part I

Our hypothesis is that A is satisfiable, C is not valid, A implies C, and each of A and C is a truth-functional compound of sentence letters.

As A is satisfiable, A is true in some interpretation $\mathscr{M}$ which assigns truth-values only to the sentence letters in A; as C is not valid, C is false in some interpretation $\mathscr{L}$ which assigns truth-values only to the sentence letters in C. It follows that there must be at least one sentence letter P contained in both A and C, for otherwise there would be an interpretation $\mathscr{K}$ that assigns any sentence letter whatever truth-value either $\mathscr{M}$ or $\mathscr{L}$ assigns it, and then A would not imply C, as $\mathscr{K}(A) = 1$ and $\mathscr{K}(C) = 0$.

Let T be the sentence $(P \vee -P)$; F, the sentence $(P \,\&\, -P)$.

We may assume that there is at least one sentence letter Q contained in A but not C. (If not, we may take $B = A$.) Let D_1 be the result of everywhere replacing occurrences of Q in A by T; D_2, by F. Note that if $\mathscr{I}$ interprets A and differs from $\mathscr{J}$ (if at all) only in that $\mathscr{I}(Q) = 1$, then $\mathscr{I}(A) = \mathscr{J}(D_1)$; if the only difference is that $\mathscr{I}(Q) = 0$, then $\mathscr{I}(A) = \mathscr{J}(D_2)$. Let

$$D = (D_1 \vee D_2).$$

A implies D. For suppose $\mathscr{I}(A) = 1$. Then if $\mathscr{I}(Q) = 1$, $\mathscr{I}(A) = \mathscr{I}(D_1)$, whence $\mathscr{I}(D) = 1$. And if $\mathscr{I}(Q) = 0$, $\mathscr{I}(A) = \mathscr{I}(D_2)$, whence $\mathscr{I}(D) = 1$.

D_1 implies C. For suppose that $\mathscr{J}(D_1) = 1$ but $\mathscr{J}(C) = 0$. Let $\mathscr{I}$ differ from $\mathscr{J}$ (if at all) only in that $\mathscr{I}(Q) = 1$. Then $\mathscr{I}(A) = 1$ and, since C does not contain Q, $\mathscr{I}(C) = \mathscr{J}(C) = 0$, contradicting the assumption that A implies C. Similarly D_2 implies C. Therefore D, $= (D_1 \vee D_2)$, *implies C.*

A implies D, D implies C, and every sentence letter in D is in A. Moreover A contains one more sentence letter foreign to C than D does. So repeating the construction sufficiently often, using the D of one stage as the A of the next, eventually yields a sentence B, implied by A, implying C, and containing no sentence letters not contained in both A and C.

Part II

Our hypothesis is that A is satisfiable, C is not valid, A implies C, and each of A and C is a truth-functional compound of sentence letters and atomic sentences obtained by attaching a string of names to a predicate letter (not '=').

We can obtain a suitable B by first replacing each atomic sentence in A and C that is not a sentence letter by a completely new sentence letter (different ones for different atomic sentences, but the same sentence letter for each atomic sentence wherever it occurs in A or C), then applying Part I to get an interpolation sentence, and finally replacing any new sentence letters contained in it by their originals. Not only will all of the names, sentence letters, and predicate letters occurring in the B obtained in this way occur in both A and C, all atomic sentences occurring in the B will also so occur.

(There is one tiny subtlety. We must see that if A^* and C^* are the sentences obtained from A and C by replacing atomic sentences by sentence letters, then A^* implies C^*. But if $\mathscr{I}(A^*) = 1$ and $\mathscr{I}(C^*) = 0$, then, as *none of the predicate letters in A or C is* '=', there is an interpretation $\mathscr{J}$ of A and C in which each name denotes *itself*, which assigns sentence letters whatever $\mathscr{I}$ assigns them, and which specifies that any predicate letter R in A or C is to be true (false) of $a_1, \ldots, a_n$ iff $\mathscr{I}(P) = 1$ (0), whenever P is the sentence letter that replaces $Ra_1 \ldots a_n$; and we will then have that $\mathscr{J}(A) = 1$ and $\mathscr{J}(C) = 0$. And therefore if A^* doesn't imply C^*, A doesn't imply C, *contra* our hypothesis.)

Part III

Our hypothesis is that A is satisfiable, C is not valid, A implies C, and neither A nor C contains function symbols or '='.

We may assume that A is in prenex normal form. Let $\bar{C}$ be a prenex equivalent of the negation of C. (The usual prenexing operations introduce no new non-logical symbols.) As A implies C, $\{A, \bar{C}\}$ is unsatisfiable. Let $\mathscr{D}$ be a canonical derivation from $\{A, \bar{C}\}$. Some finite set of quantifier-free sentences in $\mathscr{D}$ is unsatisfiable. Let m be the least integer such that the result of deleting all sentences after the mth in $\mathscr{D}$ is a derivation whose quantifier-free sentences form an unsatisfiable set. And let us call this derivation, derivation I.

Since neither A nor $\bar{C}$ contains function symbols, the instantial term in every application of UI or EI in $\mathscr{D}$ and hence in derivation I, is a name. (Cf. clause 5 of the definition of 'canonical derivation'.)

Let $a_1, \ldots, a_n$ be the names appearing in derivation I that do not occur in either A or $\bar{C}$. We suppose that $a_1, \ldots, a_n$ first occur in derivation I in the order $a_1, \ldots, a_n$. So $i < j$ if a_i occurs in a sentence that is earlier than the conclusion of an application of EI in which a_j is the instantial term.

We now 'split' derivation I into two derivations, II and V, as follows: Assign A 2. Assign $\bar{C}$ 5. Suppose that S is a sentence in derivation I and that each sentence in I earlier than S has been assigned 2 or 5, but not both. Assign S whichever of 2 and 5 its premise (in I) has been assigned. Derivation II (V) is the sequence of sentences in derivation I that have been assigned 2 (5), arranged in the same order that they were in I. (In this part we shall be forming several new derivations out of old ones. We shall assume that the annotations of the new derivations agree with the annotations of the old ones as far as possible in specifying which earlier sentences in new derivations later ones are to be regarded as inferred from. Of course some lines will have to be renumbered, for some 'gaps' will have to be closed up.)

The conjunction of quantifier-free sentences in derivation II (V) is satisfiable, for otherwise $A(\bar{C})$ would be unsatisfiable, and we are assuming that this is not the case. Let A^* be the conjunction of the quantifier-free sentences in derivation II, and let C^* be the negation of the conjunction of the quantifier-free sentences in derivation V. A^* is satisfiable, C^* is not valid, and A^* implies C^*. It follows from the version of Craig's lemma that was proved in Part II that there is a sentence B^* such that A^* implies B^*, B^* implies C^*, and B^* contains only names, sentence letters, and predicate letters contained in both A^* and C^*.

Let $b_1, \ldots, b_k$ be those of the a_is (arranged in the order of the a_is) that occur in B^*. Let $v_1, \ldots, v_k$ be k distinct variables, and let B^v be the formula obtained from B^* by everywhere replacing b_i in B^* by v_i.

Let $B = Q_1 v_1 \ldots Q_k v_k B^v$, where Q_i is $\exists$ if b_i is the instantial name in an application of EI whose premise and conclusion are in derivation II; otherwise Q_i is $\forall$.

Let $\bar{B} = Q'_1 v_1 \ldots Q'_k v_k - B^v$, where Q'_i is $\forall$ if Q_i is $\exists$, and vice versa. $\bar{B}$ is thus a prenex equivalent of the negation of B.

We must now show that A implies B and B implies C. We'll show this by showing that $\{A, \bar{B}\}$ and $\{B, \bar{C}\}$ are both unsatisfiable. Since B contains only names, sentence letters, and predicate letters contained in both of A and C, we will then have proved Craig's lemma under the assumption that A and C don't contain function symbols or '='.

Let derivation III (derivation VI) be the sequence of $k+1$ sentences of which the first is $\bar{B}(B)$ and the $(i+1)$st – for any i between 1 and k – is the result of deleting $Q'_i v_i$ $(Q_i v_i)$ from the ith sentence and replacing each occurrence of v_i by one of b_i. The last sentence of derivation III (VI) will then be $-B^*$ (B^*).

If we can somehow show that derivations II and III can be 'collated' so as to produce a derivation from $\{A, \bar{B}\}$, derivation IV, that contains just the sentences in either derivation II or III, we will then have seen that $\{A, \bar{B}\}$ is unsatisfiable, for IV will be a derivation that contains $-B^*$ and all of the conjuncts of A^*, and therefore the set of its quantifier-free sentences will be unsatisfiable. If, similarly, we can also collate derivations V and VI to produce a derivation from $\{B, \bar{C}\}$, derivation VII, containing exactly the sentences in either V or VI, we will have shown that $\{B, \bar{C}\}$ is unsatisfiable, for VII will contain B^* and all conjuncts in the conjunction whose negation is C^*, and therefore the set of its quantifier-free sentences will be unsatisfiable.

Before describing the collating procedure we need a definition. Let's call a sentence in any of derivations II, III, V and VI a *red* sentence (in that derivation) if it is the conclusion of an application of EI (in that derivation).

We have to observe that no instantial name in any red sentence in derivation II (i.e., no instantial name in any application of EI whose conclusion is a red sentence in II – similarly, for III, V and VI) is identical with any instantial name in any red sentence in derivation III, and that the same holds for derivations V and VI.

As for II and III, if b_i is the instantial name in some red sentence in III, then v_i is existentially quantified in $\bar{B}$, hence universally quantified in B, and therefore b_i is not the instantial name of an application of EI whose premise and conclusion are in II, i.e. b_i is not the instantial name in any red sentence in II. As for V and VI, if b_i is the instantial name in some red sentence in VI, then v_i is existentially quantified in B, and hence b_i is the instantial name of an application of EI whose premise and conclusion are in II; were b_i also the instantial name in some application of EI in V, then b_i would be the instantial name in two different applications of EI in I, which is impossible.

Here's how to do the collating:

Suppose that we have completed n stages of the collating procedure (possibly $n = 0$), during which we have removed some (possibly no) sentences from derivations II and III and begun a new sequence of sentences which will eventually become derivation IV. In the $(n+1)$st stage we search through what remains of derivations II and III, looking for a red sentence in each.

If there are red sentences in the remainders of both II and III, we look at the earliest ones in the remainders and see which of the two has the earlier instantial name (in the order of the a_is). As we have just seen, one of the two names will always be earlier than the other. We then transfer all sentences above the red sentence with the earlier name, together with the red sentence itself, to the bottom of what we have so far constructed of derivation IV. We then go on to stage $n+2$.

If there is no red sentence in what remains of II, we add the remainder of III to the bottom of what we have so far constructed of IV, and then add on the remainder of II. If there is a red sentence in the remainder of II, but none in the remainder of III, we add the remainder of II on to the bottom of what we have so far constructed of IV, and then add on the remainder of III. In either case there is no stage $n+2$, as derivation IV is now finished.

A picture of the situation is given on page 240.

The sequence of sentences that we have been calling 'derivation IV' really is a derivation from $\{A, \bar{B}\}$ whose quantifier-free sentences form an unsatisfiable set. For the sentences in IV are just those in II or III, and two sentences that are both in II (III) are in the same order in IV as in II (III). Every sentence in IV is either A or $\bar{B}$ or follows from an earlier sentence by EI or UI. Lastly, totally new names are always used as the instantial names in applications of EI in IV. For suppose that a_j is the instantial name in the conclusion G of an application of EI in IV, and that G is a sentence in II (III). The only names in sentences above G in IV are names that are (a) either in A or $\bar{B}$, or (b) in sentences in II (III) above G, or (c) in sentences above or identical to some red sentence

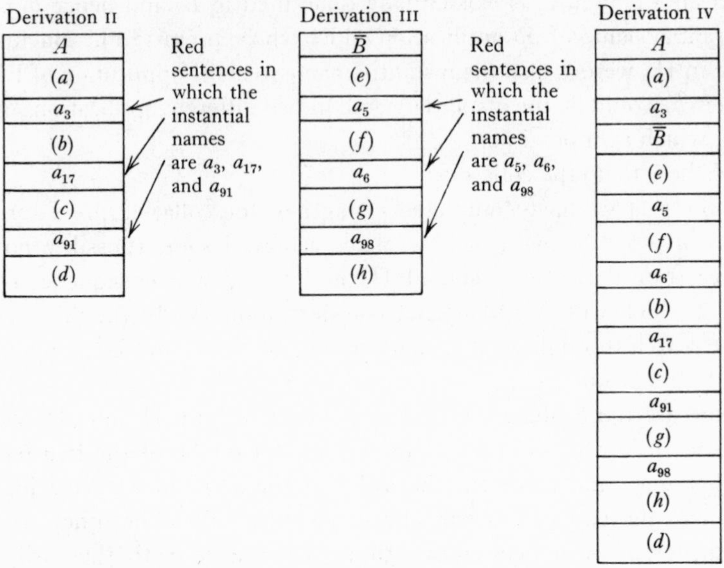

Figure 23-1

in III (II) in which the instantial name is earlier than a_j; and all of these are names different from a_j.

Derivation VII is constructed from derivations V and VI exactly as IV was constructed from II and III, and is thus a derivation from $\{B, \bar{C}\}$ whose quantifier-free sentences form an unsatisfiable set.

Part IV

Our hypothesis is that A is satisfiable, C is not valid, and A implies C.

We may safely assume that any function symbol that occurs in A or C occurs immediately to the right of '$=$'. (Cf. Chapter 22.) We now need a rather wordy lemma.

Lemma

Suppose that

$f_1, \ldots, f_m$ are m function symbols,
F is a sentence,
any function symbol in F is one of $f_1, \ldots, f_m$,
$R_1, \ldots, R_m$ are m predicate letters not occurring in F,

if f_i is an n-place function symbol, then R_i is an $(n+1)$-place predicate letter,

and E is a two-place predicate letter not occurring in F that is also different from all the R_is.

For any sentence G, define G^- to be the sentence obtained from G by first replacing every occurrence in G of a formula $s_{n+1} = f_i(s_1, \ldots, s_n)$ by one of the formula $R_i s_1 \ldots s_n s_{n+1}$, and then replacing every occurrence of '$=$' by one of E.

Then F is valid iff F^- follows from the conjunction S of all of these sentences:

$$\forall x\, xEx,$$

$$\forall x \forall y (xEy \rightarrow yEx),$$

$$\forall x \forall y \forall z ((xEy \,\&\, yEz) \rightarrow xEz),$$

and for each n, and each $(n+1)$-place R_i, the sentence

$$\forall x_1 \ldots \forall x_n \exists y \forall z (R_i x_1 \ldots x_n z \leftrightarrow yEz),$$

and for each n, and each n-place predicate letter R (other than E) occurring in F^-, the sentence

$$\forall x_1 \ldots \forall x_n \forall y_1 \ldots \forall y_n ((x_1 E y_1 \,\&\, \ldots \,\&\, x_n E y_n) \rightarrow (Rx_1 \ldots x_n \leftrightarrow Ry_1 \ldots y_n)).$$

Proof. If f is an n-place function symbol, then

$$\forall x_1 \ldots \forall x_n \exists y \forall z (z = f(x_1, \ldots, x_n) \leftrightarrow y = z)$$

is *valid*. So if $f_{j_1}, \ldots, f_{j_k}$ are those of $f_1, \ldots, f_m$ that do not occur in F, then F is equivalent to

$$\{[\forall x_1 \ldots \forall x_{n_1} \exists y \forall z (z = f_{j_1}(x_1, \ldots, x_{n_1}) \leftrightarrow y = z) \,\&\, \ldots$$
$$\&\, \forall x_1 \ldots \forall x_{n_k} \exists y \forall z (z = f_{j_k}(x_1, \ldots, x_{n_k}) \leftrightarrow y = z)] \rightarrow F\}.$$

This last sentence, call it (1), is a conditional whose antecedent is a conjunction of valid sentences and whose consequent is F. (If $k = 0$, (1) $= F$.) First applying the procedure for eliminating function symbols described in Chapter 22 from (1) (using R_i to replace f_i), and then applying the procedure for eliminating '$=$' (using E) yields a sentence, (2), that is equivalent to $(S \rightarrow F^-)$. Since by the main result of Chapter 22, (1) is valid iff (2) is valid, F is valid iff $(S \rightarrow F^-)$ is valid, and hence iff F^- follows from S. The lemma is proved.

Now consider the conditional $(A \to C)$. Call it F. F is valid. Let $f_1, \ldots, f_m$ be the m function signs occurring in F. Let $R_1, \ldots, R_m$ and E be suitable new predicate letters. And let F^- and S be formed as in the hypothesis of the lemma. Then, by the lemma, F^- follows from S.

F^- is a conditional. Its antecedent is A^-, and its consequent is C^-. So $S \to (A^- \to C^-)$ is valid.

We may assume that $S = S_1 \& \ldots \& S_i \& S_{i+1} \& \ldots \& S_j$, where $S_1, \ldots, S_i$ are the conjuncts of S *which contain only E and predicate letters contained in A^- but not in C^-.*

Then $(S_1 \& \ldots \& S_i \& A^-) \to (S_{i+1} \& \ldots \& S_j \to C^-)$ is valid.

$(S_1 \& \ldots \& S_i \& A^-)$ is satisfiable. For if it were unsatisfiable, then $-(A^-), = (-A)^-$, would follow from S, and hence (by the lemma) $-A$ would be valid, and A unsatisfiable, which it is not. Similarly,

$$(S_{i+1} \& \ldots \& S_j \to C^-) \text{ is not valid.}$$

The version of Craig's lemma proved in Part III can now be applied. It entails that there is a sentence $B^@$, implied by $(S_1 \& \ldots \& S_i \& A^-)$ and implying $(S_{i+1} \& \ldots \& S_j \to C^-)$, and containing only names, sentence letters, and predicate letters contained in both of these sentences. Since $S_1, \ldots, S_i$ contain no names or sentence letters at all, and, with the possible exception of E, only predicate letters contained in A^- but not in C^-, $B^@$ contains only names, sentence letters, and predicate letters contained in both A^- and C^- (again with the possible exception of E).

Let B be the sentence that comes from $B^@$ by replacing each occurrence of E in $B^@$ by one of '=', and each occurrence of each formula $R_i s_1 \ldots s_n s_{n+1}$ by one of the formula

$$s_{n+1} = f(s_1, \ldots, s_n). \text{ We then have that } B^- = B^@.$$

(Apart from ' = ') B contains only such predicate letters, names, and sentence letters as are contained in both A and C. If f_i is a function symbol in B, then R_i is in B^-, and so R_i is in both A^- and C^-, and therefore f_i is in both A and C. So B contains only such function symbols as are contained in both A and C as well.

We know that $(S_1 \& \ldots \& S_i \& A^-)$ implies B^- and B^- implies

$$(S_{i+1} \& \ldots \& S_j \to C^-).$$

It follows that both $(A^- \to B^-)$ and $(B^- \to C^-)$ follow from S. Since $(A^- \to B^-) = (A \to B)^-$ and $(B^- \to C^-) = (B \to C)^-$, we have that $(A \to B)$ and $(B \to C)$ are valid (the lemma again), and thus that A implies B and B implies C.

24
Two applications of Craig's lemma

In this chapter we are going to use Craig's interpolation lemma to demonstrate two results about theories, one about the conditions under which the union of two theories is satisfiable, the other about the conditions under which definitions are consequences of theories.

'Theory' will be understood in a very general way (cf. Chapter 9): A theory is just a set of sentences in some (first-order) language that contains every sentence of that language that is a logical consequence of the set.

Robinson's consistency theorem

We begin by showing that $T_1 \cup T_2$, the union of the theories T_1 and T_2, is satisfiable if and only if there is no sentence in T_1 whose negation is in T_2.

The 'only if' part is obvious: if there were a sentence in T_1 whose negation were in T_2, the union could not possibly be satisfiable; for there could be no interpretation in which both the sentence and its negation were true.

The 'if' part follows quickly from the compactness theorem and Craig's lemma: Suppose that the union of T_1 and T_2 is unsatisfiable. By the compactness theorem, there is a finite subset S_0 of the union which is unsatisfiable. If there are no members of S_0 that belong to T_1, then T_2 is unsatisfiable, and so '$\forall x\, x = x$' is a sentence in T_1 whose negation is in T_2; if no members of S_0 belong to T_2, T_1 is unsatisfiable, and so '$-\forall x\, x = x$' is a sentence in T_1 whose negation is in T_2. So we may suppose that S_0 contains some members of both T_1 and T_2. Let $F_1, \ldots, F_i$ be the members of S_0 that are in T_1; let $G_1, \ldots, G_j$ be the members of S_0 that are in T_2.

Let $A = (F_1 \& \ldots \& F_i)$; let $C = -(G_1 \& \ldots \& G_j)$. A implies C. By Craig's lemma, there is a sentence B implied by A, implying C, and containing only non-logical symbols contained in both A and C. B is therefore a sentence in the languages of both T_1 and T_2. Since A is in T_1 and implies B, B is in T_1. Since $(G_1 \& \ldots \& G_j)$ is in T_2, so is $-B$, as $(G_1 \& \ldots \& G_j)$ implies $-B$. So B is a sentence in T_1 whose negation is in T_2.

An extension T' of a theory T is called *conservative* if every sentence of the language of T that is a theorem of T' is a theorem of T. We shall now prove a theorem about conservative extensions:

Theorem

If $\mathscr{L}_i$ is the language of the theory T_i ($i = 0, 1, 2$), $\mathscr{L}_0 = \mathscr{L}_1 \cap \mathscr{L}_2$, T_3 is the set of sentences of $\mathscr{L}_1 \cup \mathscr{L}_2$ that are consequences of $T_1 \cup T_2$, and T_1 and T_2 are conservative extensions of T_0, then T_3 is also a conservative extension of T_0.

Proof. Suppose B is a sentence of $\mathscr{L}_0$ that is a theorem of T_3. We must show that B is a theorem of T_0. Let U_2 be the set of consequences in $\mathscr{L}_2$ of $T_2 \cup \{-B\}$. Since B is a theorem of T_3, $T_1 \cup T_2 \cup \{-B\}$ is unsatisfiable, and therefore $T_1 \cup U_2$ is unsatisfiable. Therefore there is a sentence D in T_1 whose negation $-D$ is in U_2. D is in $\mathscr{L}_1$; $-D$, in $\mathscr{L}_2$. Thus D and $-D$ are both in $\mathscr{L}_0$, and therefore so is $(-B \rightarrow -D)$. Since D is in T_1, which is a conservative extension of T_0, D is in T_0. And since $-D$ is in U_2, $(-B \rightarrow -D)$ is in T_2, which is also a conservative extension of T_0. Thus $(-B \rightarrow -D)$ is also in T_0, and therefore so is B, which follows from D and $(-B \rightarrow -D)$.

An immediate consequence is:

Theorem (*A. Robinson's consistency theorem*)

If $\mathscr{L}_i$ is the language of T_i ($i = 0, 1, 2$), $\mathscr{L}_0 = \mathscr{L}_1 \cap \mathscr{L}_2$, T_0 is a complete theory, and T_1 and T_2 are satisfiable extensions of T_0 (which is therefore also satisfiable), then $T_1 \cup T_2$ is satisfiable.†

Proof. A satisfiable extension of a *complete* theory is conservative, and a conservative extension of a satisfiable theory is satisfiable. Thus if T_0, T_1, T_2 satisfy the hypothesis of Robinson's consistency theorem, then T_3, defined as above, is a satisfiable extension of T_0, and therefore $T_1 \cup T_2$ is satisfiable.

Having shown Robinson's consistency theorem to follow from Craig's interpolation lemma, we conclude this section by showing how a 'double

† Neither the assumption that $\mathscr{L}_0 = \mathscr{L}_1 \cap \mathscr{L}_2$ nor the assumption that T_0 is complete can be dropped from the statement of Robinson's consistency theorem: Let T_0 (T_1, T_2) be the set of (propositional calculus) consequences in $\mathscr{L}_0$ ($\mathscr{L}_1$, $\mathscr{L}_2$) of $\{p\}$ ($\{p, q\}, \{p, -q\}$), where $\mathscr{L}_1 = \mathscr{L}_2 = \{p, q\}$. If $\mathscr{L}_0 = \{p\}$, then T_0 is complete but $\mathscr{L}_0 \neq \mathscr{L}_1 \cap \mathscr{L}_2$; on the other hand, if $\mathscr{L}_0 = \{p, q\}$, then $\mathscr{L}_0 = \mathscr{L}_1 \cap \mathscr{L}_2$ but T_0 is not complete.

compactness argument' yields Craig's lemma from Robinson's theorem. (William Craig and Abraham Robinson proved the lemma and the theorem independently of each other.)

Suppose that A implies C. Let $\mathscr{L}_1$ ($\mathscr{L}_2$) be the language consisting of the non-logical symbols occurring in A (C). Let $\mathscr{L}_0 = \mathscr{L}_1 \cap \mathscr{L}_2$. We want to show that there is a sentence B of $\mathscr{L}_0$ implied by A and implying C. Let Δ be the set of sentences of $\mathscr{L}_0$ that are implied by A. We first show that $\Delta \cup \{-C\}$ is unsatisfiable. Suppose that it is not and that $\mathscr{I}$ is a model of $\Delta \cup \{-C\}$. Let T_0 be the set of sentences of $\mathscr{L}_0$ that are true in $\mathscr{I}$. T_0 is a complete theory whose language is $\mathscr{L}_0$. Let T_1 (T_2) be the set of sentences of $\mathscr{L}_1$ ($\mathscr{L}_2$) that are consequences of $T_0 \cup \{A\}$ ($T_0 \cup \{-C\}$). T_2 is a satisfiable extension of T_0: $\mathscr{I}$ is a model of $T_0 \cup \{-C\}$, and hence of T_2. But $T_1 \cup T_2$ is not satisfiable: any model of $T_1 \cup T_2$ would be a model of $\{A, -C\}$, and since A implies C, there is no such model. Thus by Robinson's consistency theorem, T_1 is not a satisfiable extension of T_0, and therefore $T_0 \cup \{A\}$ is unsatisfiable. By the compactness theorem, there is a finite set of sentences in T_0 whose conjunction D, which is in $\mathscr{L}_0$, implies $-A$. Thus A implies $-D$, $-D$ is in $\mathscr{L}_0$, $-D$ is in Δ, and $-D$ is therefore true in $\mathscr{I}$. But this is a contradiction, as all of the conjuncts of D are in T_0 and are therefore true in $\mathscr{I}$. So $\Delta \cup \{-C\}$ is unsatisfiable, and, by the compactness theorem again, there is a finite set of members of Δ whose conjunction B implies C. B is in $\mathscr{L}_0$, since its conjuncts are, and, as A implies each of these, A implies B.

Beth's definability theorem

Beth's definability theorem is a theorem about the relation between two different explications, or ways of making precise, the notion of a *theory's giving a definition of one concept in terms of other concepts*. As one might expect, each of the explications discusses a relation that may or may not hold between a theory, a symbol in the language of that theory (which is supposed to 'represent' a certain concept), and other symbols in the language of the theory (which 'represent' other concepts), rather than directly discussing a relation that may or may not hold between a theory, a concept, and other concepts.

The supposition of Beth's theorem, then, is that α and $\beta_1, \ldots, \beta_n$ are non-logical symbols of the language of some theory T and that α is identical with none of $\beta_1, \ldots, \beta_n$.

The first explication incorporates the idea, due to Padoa, that a theory defines a concept in terms of others if 'any specification of the universe of discourse of the theory and of the meanings of the symbols representing the other concepts (that is compatible with the truth of all sentences in the theory) uniquely determines the meaning of the symbol representing that concept'. This sort of definition is called implicit definition (by a theory, of one concept in terms of others). This idea is explicated in the following definition. We say that α is implicitly definable from $\beta_1, \ldots, \beta_n$ in T if any two models of T with the same domain which agree in what they assign to $\beta_1, \ldots, \beta_n$ also agree in what they assign to α.

The second explication is somewhat more obvious than the first and embodies the idea that a theory defines a concept in terms of others when 'a definition of that concept in terms of the others is a consequence of that theory'. And this sort of definition is called explicit definition. We say that

α is explicitly definable from $\beta_1, \ldots, \beta_n$ in T if a definition of α from $\beta_1, \ldots, \beta_n$ is one of the sentences of T.

What's a definition of α from $\beta_1, \ldots, \beta_n$? It's a sentence

$$\forall x_1 \ldots \forall x_k (--\alpha-- \leftrightarrow \_\beta_1, \ldots, \beta_n\_), \text{ where}$$

all non-logical symbols occurring in $\_\beta_1, \ldots, \beta_n\_$ belong to $\{\beta_1, \ldots, \beta_n\}$;

all variables occurring free in $\_\beta_1, \ldots, \beta_n\_$ belong to $\{x_1, \ldots, x_k\}$, and $--\alpha--$ is

the formula $x_1 = \alpha$ if α is a name (in this case $k = 1$),

the sentence α if α is a sentence letter (in this case $k = 0$ and the definition is a biconditional),

the formula $\alpha x_1 \ldots x_k$ if α is a k-place predicate letter, and

the formula $x_k = \alpha(x_1, \ldots, x_{k-1})$ if α is a $(k-1)$-place function sign.

Beth's definability theorem states that α is explicitly definable from $\beta_1, \ldots, \beta_n$ in T if and only if α is implicitly definable from $\beta_1, \ldots, \beta_n$ in T; 'explicit definability = implicit definability'. Here's its proof.

We first form a new theory T' by replacing every non-logical symbol γ occurring in a sentence in T, *with the exception of* $\beta_1, \ldots, \beta_n$, by a completely new symbol γ' of the same logical type. That is, we replace names a by names a', 17-place function signs f by 17-place function signs f', 96-place predicate letters R by 96-place predicate letters R', etc. Since α is not one of the β_is, we replace it too by α'.

Suppose now that $\mathscr{I}$ and $\mathscr{J}$ are two models of T with the same domain which agree in what they assign to $\beta_1, \ldots, \beta_n$. Let $\mathscr{I} + \mathscr{J}$ be the interpretation whose domain is the common domain of $\mathscr{I}$ and $\mathscr{J}$, which assigns to any β_i whatever $\mathscr{I}$ and $\mathscr{J}$ both assign to it, assigns to any other non-logical symbol γ appearing in T whatever $\mathscr{I}$ assigns to it, and assigns to any non-logical symbol γ' appearing in T' whatever $\mathscr{J}$ assigns to γ. ($\mathscr{I}+\mathscr{J}$ assigns nothing to any other symbols.) Then $\mathscr{I}+\mathscr{J}$ is a model of $T \cup T'$.

Conversely, if $\mathscr{K}$ is a model of $T \cup T'$, $\mathscr{K}$ can clearly be 'decomposed' into two models of T, $\mathscr{I}$ and $\mathscr{J}$, which have the same domain and agree in what they assign to $\beta_1, \ldots, \beta_n$ as follows: the domain of $\mathscr{I}$ = the domain of $\mathscr{J}$ = the domain of $\mathscr{K}$; $\mathscr{I}$ and $\mathscr{J}$ both assign $\beta_1, \ldots, \beta_n$ whatever $\mathscr{K}$ assigns them; $\mathscr{I}$ assigns any non-logical symbol γ whatever $\mathscr{K}$ assigns γ; and $\mathscr{J}$ assigns γ whatever $\mathscr{K}$ assigns γ'.

We can now give a reformulation of the definition of implicit definability that has a somewhat more 'syntactic' character.

Lemma

α is implicitly definable from $\beta_1, \ldots, \beta_n$ in T iff

$$\forall x_1 \ldots \forall x_k (--\alpha-- \leftrightarrow --\alpha'--)$$

follows from $T \cup T'$ (where $--\alpha'--$ is the result of substituting α' for α in $--\alpha--$).

Proof. Left-to-right direction: suppose that α is implicitly definable from $\beta_1, \ldots, \beta_n$ in T. Suppose that $\mathscr{K}$ is a model of $T \cup T'$. Let $\mathscr{I}$ and $\mathscr{J}$ be the models of T into which $\mathscr{K}$ can be 'decomposed'. $\mathscr{I}$ and $\mathscr{J}$ have the same domain and agree in what they assign to the β_is. By the supposition, they therefore agree in what they assign to α. $\mathscr{K}$ therefore assigns the same thing to α as to α'. $\forall x_1 \ldots \forall x_k (--\alpha-- \leftrightarrow --\alpha'--)$ is therefore true in $\mathscr{K}$. Thus any model of $T \cup T'$ is a model of

$$\forall x_1 \ldots \forall x_k (--\alpha-- \leftrightarrow --\alpha'--),$$

which therefore follows from $T \cup T'$.

Right-to-left direction: suppose that $\forall x_1 \ldots \forall x_k (--\alpha-- \leftrightarrow --\alpha'--)$ follows from $T \cup T'$. Suppose that $\mathscr{I}$ and $\mathscr{J}$ are models of T that have the same domain and agree in what they assign to the β_is. Then $\mathscr{I}+\mathscr{J}$ is a model of $T \cup T'$, and therefore, since $\forall x_1 \ldots \forall x_k (--\alpha-- \leftrightarrow --\alpha'--)$ follows from $T \cup T'$, $\forall x_1 \ldots \forall x_k (--\alpha-- \leftrightarrow --\alpha'--)$ is true in $\mathscr{I}+\mathscr{J}$. $\mathscr{I}+\mathscr{J}$ therefore assigns the same thing to α as to α', and therefore, $\mathscr{I}$ and $\mathscr{J}$ agree in what they assign to α. α is thus implicitly definable from $\beta_1, \ldots, \beta_n$ in T.

One half of Beth's theorem is now quite easy. Suppose that α is explicitly definable from $\beta_1, \ldots, \beta_n$ in T. Then some definition

$$\forall x_1 \ldots \forall x_k (--\alpha-- \leftrightarrow \_\beta_1, \ldots, \beta_n\_)$$

of α from $\beta_1, \ldots, \beta_n$ is in T. Therefore

$$\forall x_1 \ldots \forall x_k (--\alpha'-- \leftrightarrow \_\beta_1, \ldots, \beta_n\_)$$

is in T'. (Recall that we do not replace the β_is by new symbols.) Therefore $\forall x_1 \ldots \forall x_k (--\alpha-- \leftrightarrow --\alpha'--)$, which follows from the conjunction of these two sentences, follows from $T \cup T'$, and therefore, by the lemma, α is implicitly definable from $\beta_1, \ldots, \beta_n$ in T. So explicit definability entails implicit definability.

Conversely, suppose that α is implicitly definable from $\beta_1, \ldots, \beta_n$ in T. By the lemma, $\forall x_1 \ldots \forall x_k (--\alpha-- \leftrightarrow --\alpha'--)$ follows from $T \cup T'$. By the compactness theorem, there is a finite subset of $T \cup T'$ from which it follows. By adding (finitely many) extra sentences to it, if necessary, we can regard this finite subset as $T_0 \cup T'_0$, where T_0 is a finite subset of T, and T'_0 comes from T_0 by replacing any non-logical symbol γ other than one of the β_is by γ'.

Let A (A') be the conjunction of the members of T_0 (T'_0). Then

$$\forall x_1 \ldots \forall x_k (--\alpha-- \leftrightarrow --\alpha'--)$$

follows from $(A \& A')$. Let $a_1, \ldots, a_k$ be k names that don't occur anywhere in sentences in T or T', and hence don't occur anywhere in A, A', $--\alpha--$, or $--\alpha'--$.

Let $--\alpha--(a_1, \ldots, a_k)$ be the result of replacing free occurrences of x_1 by occurrences of $a_1, \ldots,$ and free occurrences of x_k by occurrences of a_k in $--\alpha--$. Define $--\alpha'--(a_1, \ldots, a_k)$ similarly.

Then $(--\alpha--(a_1, \ldots, a_k) \leftrightarrow --\alpha'--(a_1, \ldots, a_k))$ follows from $(A \& A')$. So $(A \& A') \to (--\alpha--(a_1, \ldots, a_k) \leftrightarrow --\alpha'--(a_1, \ldots, a_k))$ is valid. Therefore $(A \& --\alpha--(a_1, \ldots, a_k)) \to (A' \to --\alpha'--(a_1, \ldots, a_k))$ is valid. By the Craig interpolation lemma, there is a sentence B implied by

$$(A \& --\alpha--(a_1, \ldots . a_k)),$$

implying $(A' \to --\alpha'--(a_1, \ldots, a_k))$, each of whose non-logical symbols occurs in both of these sentences, and hence is one of $\beta_1, \ldots, \beta_n, a_1, \ldots, a_k$.

Since $(A \& --\alpha--(a_1, \ldots, a_k))$ implies B, A implies

$$(--\alpha--(a_1, \ldots, a_k) \to B).$$

Since B implies $(A' \to --\alpha'--(a_1, \ldots, a_k))$, A' implies

$$(B \to --\alpha'--(a_1, \ldots, a_k)).$$

So $(A' \to (B \to --\alpha'--(a_1, \ldots, a_k)))$ is valid. By replacing each symbol γ' in $(A' \to (B \to --\alpha'--(a_1, \ldots, a_k)))$ by γ, we see that

$$(A \to (B \to --\alpha--(a_1, \ldots, a_k)))$$

is valid (as B contains no non-logical symbols other than $\beta_1, \ldots, \beta_n, a_1, \ldots, a_k$), and so A implies

$$(B \to --\alpha--(a_1, \ldots, a_k)).$$

A therefore implies $(--\alpha--(a_1, \ldots, a_k) \leftrightarrow B)$.

Let $\_\beta_1, \ldots, \beta_n\_$ be the result of replacing each a_i in B by x_i (after having first relettered the variables in B so that none of the x_is occurs bound in B). Since $a_1, \ldots, a_k$ do not occur in A, A implies

$$\forall x_1 \ldots \forall x_k (--\alpha-- \leftrightarrow \_\beta_1, \ldots, \beta_n\_).$$

All non-logical symbols occurring in $\_\beta_1, \ldots, \beta_n\_$ belong to $\{\beta_1, \ldots, \beta_n\}$, and all variables occurring free in $\_\beta_1, \ldots, \beta_n\_$ belong to $\{x_1, \ldots, x_k\}$. A therefore implies a definition of α from $\beta_1, \ldots, \beta_n$. Since A is in T, so is the definition, and therefore α is explicitly definable from $\beta_1, \ldots, \beta_n$ in T. So implicit definability entails explicit definability, and Beth's theorem is proved.

Exercise

Combine Beth's definability theorem, Tarski's indefinability theorem and Theorem 2 of Chapter 19 to obtain another proof of the existence of non-standard models of arithmetic. (*Hint*: consider the theory S consisting of the consequences in $L + G$ of $\{F\} \cup$ arithmetic (cf. Chapter 19), and show that G is implicitly definable in S from **0**, ′, +, and · if there are no nonstandard models of arithmetic.)

25
Monadic versus dyadic logic

A *monadic* formula is a formula of first-order logic, all of whose non-logical symbols are one-place predicate letters. A monadic formula may contain both '$=$' and some one-place predicate letters; it may contain '$=$' but no one-place predicate letters; and it may contain some one-place predicate letters, but not '$=$'. In the last case it is said to be a *pure* monadic formula.

One of the corollaries to the Skolem–Löwenheim theorem (Chapters 12 and 13) is that if S is a sentence of first-order logic which is satisfiable, then S is true in some interpretation whose domain is enumerable. In the present chapter we shall prove a related theorem about monadic sentences.

Theorem 1

If S is a monadic sentence which is satisfiable, then S is true in some interpretation whose domain contains at most $2^k \cdot r$ members, k being the number of (one-place) predicate letters and r being the number of variables in S.

In Chapter 22 we saw that there was no effective procedure for deciding whether or not pure dyadic sentences were valid (these are '$=$'-free sentences whose non-logical symbols are two-place predicate letters); at the end of this chapter we shall show that there isn't even a decision procedure for validity for pure dyadic sentences that contain only a single predicate letter. On the other hand, it follows from Theorem 1 that *there is an effective procedure for deciding whether or not a monadic sentence is valid.*

Thus the history of logic and the notion of decidability are related in the following curious way: a significant feature of the modern renaissance of logic that began with the work of Boole, Frege and others was a marked broadening of the concept of valid inference. Those inferences of whose validity 'classical', or pre-contemporary, logic gave an account are treated by contemporary logical theory as inferences of *monadic* logic, inferences whose premises and conclusions could be symbolized by formulas containing only one-place predicate letters, plus possibly, the equals-sign. Intuitively valid inferences such as

All horses are animals.
Therefore, all who ride all animals ride all horses;

or Everyone loves every lover
Therefore, either no one loves anyone or everyone loves everyone;

which could not be treated by classical logic, but which contemporary logical theory recognizes as valid as any inference of monadic logic, require two-place predicate letters in their symbolizations, however. Thus the price of the increased explanatory power of this theory is the consequent undecidability of the contemporary notion of valid inference, for undecidability, as we shall see, sets in precisely when two-place predicate letters (other than '$=$') are allowed in sentences used in symbolizations of inferences.

Let's begin by seeing how the decidability of monadic logic follows from Theorem 1. We are going to show how to effectively associate with each monadic sentence S a quantifier-free sentence S^*, which is satisfiable iff S is. So to tell whether a monadic sentence G is valid, we may apply the procedure described at the very end of Chapter 12 to test $(-G)^*$ for satisfiability: G is valid iff $(-G)^*$ is not satisfiable.

Let k be the number of predicate letters in S, and r the number of variables. Let $m = 2^k \cdot r$, and let $a_1, \ldots, a_m$ be m names that occur nowhere in S.

A *subformula* of S is a formula that is a (continuous) part of S. (S is considered to be a subformula of itself.) E.g. if $S =$ '$(\forall x\, Fx \vee \exists y\, Gy)$', then '$Fx$', '$\forall x\, Fx$', '$Gy$', '$\exists y\, Gy$', and '$(\forall x\, Fx \vee \exists y\, Gy)$' are all of its subformulas.

We inductively associate a quantifier-free formula H^* with each subformula H of S: if H is atomic, $H^* = H$; if H is a truth-functional compound, then H^* is the same compound of the formulas associated with the formulas of which H is compounded; and if $H = \exists v F$, then $H^* = F_v a_1 \vee \ldots \vee F_v a_m$ (and if $H = \forall v F$, $H^* = F_v a_1 \,\&\ldots\&\, F_v a_m$). S^* is then a quantifier-free sentence.

Observe that the terms occurring in S^* are precisely $a_1, \ldots, a_m$, and that S and S^* will have the same truth-value in any interpretation of both of them in which each object in the domain is denoted by at least one of $a_1, \ldots, a_m$. (Call these interpretations 'good'.)

Then S is satisfiable iff (by Theorem 1) S is true in some interpretation whose domain contains at most m members, iff S is true in some good interpretation, iff S^* is true in some good interpretation, iff (by Lemma III of Chapter 12) S^* is satisfiable. (S^* is satisfiable iff $\{S^*\}$ is O.K.)

Thus if Theorem 1 holds, monadic logic is decidable. We'll now prove Theorem 1. (The proof we give is due to Ronald Jensen.)

We suppose that $P_1, \ldots, P_k$ are the k one-place predicate letters (possibly $k = 0$) and $v_1, \ldots, v_r$ are the r variables occurring in a monadic sentence S, and that $\mathscr{I}$ is a model of S whose domain is D.

For each d in D, let $s(d) = \langle j_1, \ldots, j_k \rangle$, where, for each i between 1 and k, $j_i = 1$ or 0 according as $\mathscr{I}$ specifies that P_i is to be true or false of d. There are at most 2^k such sequences $s(d)$. (If $k = 0$, then for each d in D, $s(d)$ is $\langle\,\rangle$, the empty sequence.)

We shall say that c is *similar to* d iff $s(c) = s(d)$. Similarity is clearly an equivalence relation on D. Each member of D thus belongs to a unique equivalence class under the relation *similar to*. There are at most 2^k equivalence classes.

We can now construct our model $\mathscr{J}$ of S whose domain has at most $2^k \cdot r$ members. We form a set E by choosing from each equivalence class r members, if there are at least r members in the class, and *all* of the members of the class if there are fewer than r members in it. E then contains at most $2^k \cdot r$ members. $\mathscr{J}$ is the interpretation whose domain is E, and which specifies (for each i) that P_i is to be true of c iff $\mathscr{I}$ specifies that P_i is to be true of c (for any c in E). $\mathscr{J}$ specifies nothing further.

We know that $\mathscr{I}(S) = 1$; in order to see that Theorem 1 is true, we must see that $\mathscr{J}(S) = \mathscr{I}(S)$. There are two concepts we shall need for this end: that of *exact likeness* (of two finite sequences), and that of a *subsentence* of S.

Let $c_1, \ldots, c_n$ and $d_1, \ldots, d_n$ be finite sequences of elements of D. We shall say that $c_1, \ldots, c_n$ is *exactly like* $d_1, \ldots, d_n$ if for every $i\,(1 \leqslant i \leqslant n)$, c_i is similar to d_i, and for every i, j $(1 \leqslant i,j \leqslant n)$, $c_i = c_j$ iff $d_i = d_j$.

As for subsentences: let $a_1, a_2, \ldots, a_r$ be a sequence of r distinct names. A *subsentence* of S is a sentence that is either a subformula of S or that can be obtained from a subformula by substituting names for (free) variables: a_1 is always substituted for v_1, a_2 for v_2, etc. A subsentence of S need not, then, be a part of S, since the names substituted will not occur in S. E.g. if $a_1 =$ 'a', $a_2 =$ 'b', $v_1 =$ 'x', and $v_2 =$ 'y', then 'Fa', '$\forall x\, Fx$', 'Gb', '$\exists y\, Gy$', and '$(\forall x\, Fx \vee \exists y\, Gy)$' are all the subsentences of '$(\forall x\, Fx \vee \exists y\, Gy)$'. The only names that occur in subsentences of S, then, are $a_1, \ldots, a_r$.

The lemma that shows that $\mathscr{I}(S) = \mathscr{J}(S)$ is

Lemma 1

Suppose that G is a subsentence of S, that $b_1, \ldots, b_n$ are the n (distinct) names that occur in G ($0 \leqslant n \leqslant r$), that $d_1, \ldots, d_n$ is a sequence of elements of D, that $e_1, \ldots, e_n$ is a sequence of elements of E, and that $d_1, \ldots, d_n$ is exactly like $e_1, \ldots, e_n$. Then G is true in $\mathscr{I}^{b_1 \ldots b_n}_{d_1 \ldots d_n}$ iff G is true in $\mathscr{J}^{b_1 \ldots b_n}_{e_1 \ldots e_n}$.

Proof. The proof is an induction on the complexity of G, complexity being measured, as usual, by the number of occurrences of connectives and quantifiers in G.

If G is atomic, either $G = P_i b_1$ or $G = b_1 = b_1$ or $G = b_1 = b_2$. In the first case, G is true in $\mathscr{I}^{b_1}_{d_1}$ iff $\mathscr{I}$ specifies that P_i is to be true of d_1, iff – by the assumption of exact likeness – $\mathscr{J}$ specifies that P_i is to be true of e_1, iff G is true in $\mathscr{J}^{b_1}_{e_1}$. In the second case, G is true in both $\mathscr{I}^{b_1}_{d_1}$ and $\mathscr{J}^{b_1}_{e_1}$. In the third case G is true in $\mathscr{I}^{b_1 b_2}_{d_1 d_2}$ iff $d_1 = d_2$, iff – by the assumption of exact likeness – $e_1 = e_2$, iff G is true in $\mathscr{J}^{b_1 b_2}_{e_1 e_2}$.

If $G = -H$, H is of lower complexity than G, so we may assume that the lemma holds for H. So G is true in $\mathscr{I}^{b_1 \ldots b_n}_{d_1 \ldots d_n}$ iff H is not true in $\mathscr{I}^{b_1 \ldots b_n}_{d_1 \ldots d_n}$, iff H is not true in $\mathscr{J}^{b_1 \ldots b_n}_{e_1 \ldots e_n}$, iff G is true in $\mathscr{J}^{b_1 \ldots b_n}_{e_1 \ldots e_n}$.

The argument is similar if G is a conjunction or other truth-functional compound of simpler sentences or a *vacuous* quantification of a simpler sentence.

Suppose then that $G = \exists v_j H$, where v_j is free in H. (The argument is similar if $G = \forall v_j H$.) Then a_j does not occur in G. So if $b_1, \ldots, b_n$ are all of the n names that occur in G, they are not all of $a_1, \ldots, a_r$, and hence $n < r$. Let $b_{n+1} = a_j$.

Now if $\exists v_j H$ is true in $\mathscr{I}^{b_1 \ldots b_n}_{d_1 \ldots d_n}$, then for some d_{n+1} in D, $H_{v_j} a_j$ is true in $\mathscr{I}^{b_1 \ldots b_n b_{n+1}}_{d_1 \ldots d_n d_{n+1}}$. Let $X = d_{n+1}$'s equivalence class. If for some i between 1 and n, $d_{n+1} = d_i$, let $e_{n+1} = e_i$. If for no i between 1 and n, $d_{n+1} = d_i$, then let $d_{i_1}, \ldots, d_{i_m}$ be the m d_is among $d_1, \ldots, d_n$ that are in X ($0 \leqslant m \leqslant n < r$). There are then at least $m+1$, $\leqslant r$, members of X, and hence at least $m+1$ members of $X \cap E$, and hence there is a member of $X \cap E$ distinct from all of $e_{i_1}, \ldots, e_{i_m}$, and we take it for e_{n+1}. In either case, $e_1, \ldots, e_n, e_{n+1}$ is a sequence of elements of E that is exactly like $d_1, \ldots, d_n, d_{n+1}$, and therefore, by the hypothesis of the induction, $H_{r_j} a_j$ is true in $\mathscr{J}^{b_1 \ldots b_n b_{n+1}}_{e_1 \ldots e_n e_{n+1}}$, and therefore $\exists v_j H$ is true in $\mathscr{J}^{b_1 \ldots b_n}_{e_1 \ldots e_n}$.

If, on the other hand, $\exists v_j H$ is true in $\mathscr{J}^{b_1 \ldots b_n}_{e_1 \ldots e_n}$, then for some e_{n+1} in E, $H_{r_j} a_j$ is true in $\mathscr{J}^{b_1 \ldots b_n b_{n+1}}_{e_1 \ldots e_n e_{n+1}}$. By a similar argument, there is *a* d_{n+1} in D such that $d_1, \ldots, d_n, d_{n+1}$ is exactly like $e_1, \ldots, e_n, e_{n+1}$, and therefore, by the hypothesis of the induction, $H_{v_j} a_j$ is true in $\mathscr{I}^{b_1 \ldots b_n\, b_{n+1}}_{d_1 \ldots d_n\, d_{n+1}}$, whence $\exists v_j H$ is true in $\mathscr{I}^{b_1 \ldots b_n}_{d_1 \ldots d_n}$. This proves Lemma 1.

As S is a subsentence of itself that contains no names, it follows from Lemma 1 that $\mathscr{I}(S) = \mathscr{J}(S)$. Theorem 1 is therefore proved.

A Corollary of Theorem 1 is

Theorem 2

If S is a sentence containing *no* non-logical symbols (and thus only '$=$', variables, etc.), then if S is satisfiable, S is true in some interpretation whose domain contains no more members than there are variables in S.

Proof. S is then a monadic sentence, and as $k = 0$, $2^k \cdot r = 1 \cdot r = r$.

Another consequence of Theorem 1 is a result about pure monadic sentences:

Theorem 3

If S is a pure monadic sentence, then if S is satisfiable, S is true in some interpretation whose domain contains at most 2^k members, k being the number of predicate letters in S.

Theorem 3 is an immediate consequence of Theorem 1 and

Theorem 4

Any pure monadic sentence is equivalent to a pure monadic sentence containing exactly the same predicate letters and only one variable.

Proof. Theorem 4 follows from the fact that we can (inductively) associate with every pure monadic formula F a pure monadic formula, (a) which is equivalent to F, (b) which contains the same predicate letters and free variables as F, and (c) in which no variable v_i occurs in the scope of any quantifier, except possibly for $\exists v_i$ and $\forall v_i$. If F is a pure monadic *sentence*, then, the associated formula H will be a sentence equivalent to F that contains the same predicate letters, and in which no variable occurs in the scope of any quantifier containing a different variable. All variables in H can then be rewritten as (say) 'x'.

We associate with any pure monadic atomic formula that formula itself. We associate with any truth-functional compound of monadic formulas the same truth-functional compound of the associated formulas. Our only problem is to show what formula to associate with F, $= \exists v\, H$, where H is some monadic formula with which we have associated G. (Universal quantifications may be treated dually, in the obvious way.)

To find the formula associated with F, first use the ordinary propositional calculus rules to rewrite G as a disjunction $G_1 \vee \ldots \vee G_n$, where each G_i is a conjunction of atomic formulas, quantifications of truth-functional compounds of atomic formulas, and negations of atomic formulas and such quantifications. We may assume that each G_i has the form

$$B_1 \& \ldots \& B_j \& C_1 \& \ldots \& C_m,$$

where $B_1, \ldots, B_j$ are the conjuncts of G_i *in which v occurs free*, and $C_1, \ldots, C_m$ are the other conjuncts. The formula we associate with F is then

$$G'_1 \vee \ldots \vee G'_n,$$

where G'_i is

$$\exists v(B_1 \& \ldots \& B_j) \& C_1 \& \ldots \& C_m$$

if $j > 0$ (i.e. if v occurs free in at least one conjunct of G_i), and G'_i is G_i otherwise. The associated formula will then have properties (*a*), (*b*), and (*c*).

In the remainder of the chapter we shall sharpen the boundary between monadic and dyadic logic by using the main result of Chapter 21 to show that *there is no decision procedure for validity of pure dyadic sentences that contain only a single two-place predicate letter.*

The undecidability proof (due to Herbrand and Church) falls into two parts. The first part shows how to construct, for any pure dyadic sentence F, a pure sentence G containing only a single three-place predicate letter, which is valid if and only if F is. The second shows how to construct, for any pure sentence G containing only a single three-place predicate letter, a pure dyadic sentence H containing only a single two-place predicate letter, which is valid if and only if G is.

Going from F to G

Suppose that $A_1, \ldots, A_k$ are all the (two-place) predicate letters that occur in F. Let $v_1, \ldots, v_k$ be k (distinct) variables that occur nowhere in F. Let B be a three-place predicate letter that does not occur in F. And let s and t be variables. We obtain G from F by replacing each occurrence of each formula $A_i st$ in F by an occurrence of $Bv_i st$, and then prefixing $\forall v_1 \ldots \forall v_k$ to the result.

Example

If $F =$ '$\forall x \exists y (Jyx \,\&\, \forall z (Jyz \,\&\, Izz \,\&\, Kwx))$', then G might be

$$\text{'}\forall x_1 \forall x_2 \forall x_3 \forall x \exists y (Lx_2yx \,\&\, \forall z (Lx_2yz \,\&\, Lx_1zz \,\&\, Lx_3wx))\text{'}.$$

It is clear that if F is valid, then G is also valid, for G is obtained from F by prefixing universal quantifiers to a formula obtained from F by substitution, and substitution is an operation that preserves validity. More explicitly, suppose that there is an interpretation $\mathscr{I}$ in which G is false. Let D be the domain of $\mathscr{I}$. And let G^* be the result of deleting the prefix $\forall v_1 \ldots \forall v_k$ from G and replacing v_1 by $a_1, \ldots,$ and v_k by a_k. ($a_1, \ldots, a_k$ are k names.) Then there are $d_1, \ldots, d_k$ in D such that G^* is false in $\mathscr{I}^{a_1 \ldots a_k}_{d_1 \ldots d_k}$. Let $\mathscr{J}$ be the interpretation whose domain is also D and which specifies (for each i) that A_i is to be true of c, e iff $\mathscr{I}$ specifies that B is to be true of d_i, c, e. Then F has the same truth-value, *viz.*, false, in $\mathscr{J}$ that G^* has in $\mathscr{I}^{a_1 \ldots a_k}_{d_1 \ldots d_k}$. So F is not valid.

Suppose now that F is not valid. Then there is an interpretation $\mathscr{J}$ in which it is false. By Corollary 3 of Chapter 13 we may suppose that the domain D of $\mathscr{J}$ is the set of all natural numbers. We now let $\mathscr{I}$ be the following interpretation: the domain of $\mathscr{I}$ is also D, and if $1 \leqslant i \leqslant k$, then $\mathscr{I}$ specifies that B is to be true of i, c, e iff $\mathscr{J}$ specifies that A_i is to be true of c, e. (If $i = 0$ or $i > k$, then $\mathscr{I}$ specifies that B is to be true – it does not matter – of i, c, e.) Then G^* (as above) has the same truth-value, false, in $\mathscr{I}^{a_1 \ldots a_k}_{1 \ldots k}$ that F does in $\mathscr{J}$. And therefore G is false in $\mathscr{I}$ and thus not valid.

Going from G to H

We now want to see how to construct a formula H, containing only a single two-place predicate letter A, from a formula G containing only a single three-place predicate letter B, which is valid if and only if G is valid.

Well, let v_1, v_2, v_3, and v_4 be four distinct variables that occur nowhere in G. And let A be a predicate letter that does not occur in G. Let r, s, and t be variables. And let $Q(r, s, t)$ be the formula

$$\exists v_1 \exists v_2 \exists v_3 \exists v_4 (-Av_1v_1 \,\&\, Av_1v_2 \,\&\, Av_2v_3 \,\&\, Av_3v_4 \,\&\, Av_4v_1 \,\&\, Av_1r \,\&\, Av_2s \,\&\, Av_3t \,\&\, -Arv_2 \,\&\, -Asv_3 \,\&\, -Atv_4 \,\&\, Av_4r).$$

We obtain H from G by substituting an occurrence of $Q(r, s, t)$ for each occurrence of each formula $Brst$ in G.

It is clear that if G is valid, then H is valid too, for H is obtained from G by substitution. For the converse, we use

Lemma 2

Let R be a three-place relation on the natural numbers. Then there is a two-place relation S on the natural numbers such that for any natural numbers a, b, c, $R(a, b, c)$ iff for some natural numbers w, x, y, z,

$$-S(w,w) \mathbin{\&} S(w,x) \mathbin{\&} S(x,y) \mathbin{\&} S(y,z) \mathbin{\&} S(z,w) \mathbin{\&} S(w,a) \mathbin{\&} S(x,b) \\ \mathbin{\&} S(y,c) \mathbin{\&} -S(a,x) \mathbin{\&} -S(b,y) \mathbin{\&} -S(c,z) \mathbin{\&} S(z,a).$$

For suppose that Lemma 2 is true. Then if G is not valid, G is false in some interpretation $\mathscr{I}$ whose domain we may take to be the set of all natural numbers. Let $R(a, b, c)$ hold iff $\mathscr{I}$ specifies that B is to be true of a, b, c; let S be as in the conclusion of Lemma 2; and let $\mathscr{J}$ have the same domain as $\mathscr{I}$ and specify that A is to be true of a, b if $S(a, b)$ holds. Then $\mathscr{J}(H) = \mathscr{I}(G) = 0$, and H is not valid.

Proof of Lemma 2. Let R be a three-place relation on the natural numbers. In order to define our S we describe a certain repetition-free enumeration of all the triples $\langle x, y, z\rangle$ of natural numbers. One triple precedes another $\langle a, b, c\rangle$ in this enumeration if one of four conditions is met: (1) $x+y+z < a+b+c$; or (2) $x+y+z = a+b+c$ and $x < a$; or (3) $x+y+z = a+b+c$ and $x = a$ and $y < b$; or (4) $x+y+z = a+b+c$ and $x = a$ and $y = b$ and $z < c$. So the enumeration begins like this:

$\langle 0, 0, 0\rangle$
$\langle 0, 0, 1\rangle$
$\langle 0, 1, 0\rangle$
$\langle 1, 0, 0\rangle$
$\langle 0, 0, 2\rangle$
$\langle 0, 1, 1\rangle$
$\langle 0, 2, 0\rangle$
$\langle 1, 0, 1\rangle$
$\langle 1, 1, 0\rangle$
$\langle 2, 0, 0\rangle$
$\langle 0, 0, 3\rangle$
$\langle 0, 1, 2\rangle$
$\langle 0, 2, 1\rangle$
$\langle 0, 3, 0\rangle$
$\langle 1, 0, 2\rangle$
$\langle 1, 1, 1\rangle$
$\vdots$

We shall call $\langle 0,0,0\rangle$ the 1st triple, $\langle 0,0,1\rangle$ the 2nd triple, $\langle 0,1,0\rangle$ the 3rd triple, etc. There is no 0th triple. Notice that if the nth triple

$$= \langle a,b,c\rangle,$$

then all of a,b, and c are $< n$, and that if $a < n$ and $w = 4n+1$, then $a < w-4$ (for if $a < n$, then $a+1 \leqslant n$, and therefore $4a+4 < 4n+1$, whence $a \leqslant 4a < w-4$). Clearly, then, if all of a,b, and c are $< n$ and $w = 4n+1$, $x = 4n+2$, $y = 4n+3$, and $z = 4n+4$, then all of a,b, and c are less than $w-4$, $x-4$, $y-4$ and $z-4$.

We now define S: if $\langle a,b,c\rangle$ is the nth triple ($n \geqslant 1$), then

$$S(4n+1,u) \text{ iff } u = 4n+2 \text{ or } u = a$$

and $$S(4n+2,u) \text{ iff } u = 4n+2 \text{ or } u = 4n+3 \text{ or } u = b$$

and $$S(4n+3,u) \text{ iff } u = 4n+3 \text{ or } u = 4n+4 \text{ or } u = c$$

and $$S(4n+4,u) \text{ iff } u = 4n+4 \text{ or } u = 4n+1 \text{ or } (u = a \text{ and } R(a,b,c)).$$

'$S(t,u)$' is false in all other cases.

Observe that

$$\text{if } S(t,u), \text{ then } t+1 \geqslant u, \qquad (*)$$

and that

$$\text{there is at most one } s < r-4 \text{ such that } S(r,s). \qquad (**)$$

We must now show that $R(a,b,c)$ iff for some w,x,y,z,

$$-S(w,w) \,\&\, S(w,x) \,\&\, S(x,y) \,\&\, S(y,z) \,\&\, S(z,w) \,\&\, S(w,a) \,\&\, S(x,b) \\ \&\, S(y,c) \,\&\, -S(a,x) \,\&\, -S(b,y) \,\&\, -S(c,z) \,\&\, S(z,a).$$

Let's abbreviate '$-S(w,w)\,\&\ldots\&\,S(z,a)$' by '$Gw,x,y,z,a,b,c$'. It's clear that if $R(a,b,c)$, then for some w,x,y,z, Gw,x,y,z,a,b,c. For if $\langle a,b,c\rangle$ is the nth triple, and if $w = 4n+1$, $x = 4n+2$, $y = 4n+3$, and $z = 4n+4$, then, if $R(a,b,c)$, we have that Gw,x,y,z,a,b,c. ('$-S(a,x)$' holds because $a < x-4$, whence $a+1 \ngeqslant x$. Likewise for '$-S(b,y)$' and '$-S(c,z)$'. That the other clauses hold is evident from the choice of w,x,y, and z and from the definition of 'S'.)

So now suppose that Gw,x,y,z,a,b,c. We must show that $R(a,b,c)$. Since Gw,x,y,z,a,b,c, we have $S(w,x)$ and $-S(w,w)$. So, for some $n \geqslant 1$, $w = 4n+1$.

We have that $S(w,x)$, $S(x,y)$, $S(y,z)$, and $S(z,w)$. Therefore, by (*), $x+3 \geqslant y+2 \geqslant z+1 \geqslant w$, and so $x \geqslant w-3$. Similarly, $y \geqslant x-3$ and $z \geqslant y-3$. So neither $x < w-4$ nor $y < x-4$ nor $z < y-4$. Since

$S(w, x)$ and $x \nless w-4$, $x = w+1 = 4n+2$. Since $S(z, 4n+1)$, we have $z \neq w+1$ and $z \neq w+2$. As $S(x, y)$ and $y \nless x-4$, either $y = x$ or $y = x+1$. And, in either case, as $S(y, z)$ and $z \nless y-4$, either $z = y$ or $z = y+1$. But if either $y = x$ or $z = y$, then either $z = w+1$ or $z = w+2$, both of which are impossible. So $y = x+1 = 4n+3$ and $z = y+1 = 4n+4$.

If we can show that the nth triple is $\langle a, b, c\rangle$, then we can conclude that $R(a, b, c)$; for if $\langle a, b, c\rangle$ is the nth triple, then $S(z, a)$ iff $R(a, b, c)$; and we have that $S(z, a)$.

We have that $S(w, a)$ and $S(x, b)$ and $S(y, c)$, and we have that $-S(a, x)$ and $-S(b, y)$ and $-S(c, z)$. But since we know that $w = 4n+1$, $x = 4n+2$, $y = 4n+3$, and $z = 4n+4$, we also know that $S(x, x)$, $S(x, y)$, $S(y, y)$, $S(y, z)$, and $S(z, z)$. So $a \neq x$, $b \neq x$, $b \neq y$, $c \neq y$, and $c \neq z$. So $S(w, a)$ and $a < w-4$; $S(x, b)$ and $b < x-4$; and $S(y, c)$ and $c < y-4$. Let $\langle r, s, t\rangle$ be the nth triple. Then $S(w, r)$, $S(x, s)$, and $S(y, t)$, and $r < w-4$, $s < x-4$, and $t < y-4$. By (**) we have that $r = a$ and $s = b$ and $t = c$. So $\langle a, b, c\rangle$ *is* the nth triple, and so $R(a, b, c)$.

Exercises

25.1 Prove Theorem 3 directly, i.e. without deducing it from Theorem 1.

25.2 Show that the estimates $2^k \cdot r$, r, and 2^k in Theorems 1, 2, 3 cannot be reduced.

25.3 What happens if names are permitted to occur in the S of Theorem 1?

26
Ramsey's theorem

There is an old puzzle about a party, attended by six persons, at which any two of the six either like each other or dislike each other: the problem is to show that at the party either there are three persons, any two of whom like each other, or there are three persons, any two of whom dislike each other.

The solution: Let a be one of the six. Since there are five others, either there will be (at least) three others that a likes or there will be three others that a dislikes. Suppose that a likes them. (The argument is similar if a dislikes them.) Call the three b, c, d. Then if (case 1) b likes c, b likes d, or c likes d, then a, b, and c, a, b, and d, or a, c, and d, respectively, are three persons, any two of whom like each other; but if (case 2) b dislikes c, b dislikes d, and c dislikes d, then b, c, and d are three persons, any two of whom dislike each other. And either case 1 or case 2 must hold.

The number six cannot in general be reduced; if only five persons, a, b, c, d, e, are present, then the following situation can arise:

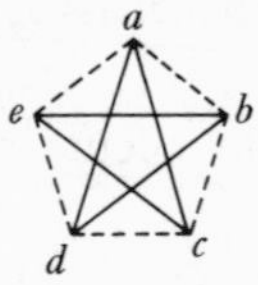

(A broken line means 'likes'; a straight, 'dislikes'.) In this situation there are no three of a, b, c, d, e, any two of whom like each other, and no three, any two of whom dislike each other.

A harder puzzle of the same type is to prove that at any party such as the previous one at which eighteen persons are present, either there are four persons, any two of whom like each other, or there are four persons, any two of whom dislike each other. (*Hint*: show that at any such party attended by nine persons, either there are four persons, any two of whom like each other, or three persons, any two of whom dislike each other.)

It is known that the number eighteen cannot be reduced.† But it is not

† See, e.g. F. Harary, *Graph Theory*, Addison-Wesley, Reading, Mass., 1969, pp. 16–17.

currently (1980) known what the minimum number m is such that if there are m persons present at a party like the previous ones, then it must be that either there are five persons, any two of whom like each other, or five persons, any two of whom dislike each other. It is known that m lies between 42 and 55,† but we do not yet have sufficient insight into the problem to be able to reduce the number of combinatorial possibilities to the point where a computer could feasibly be employed in surveying them in order to pinpoint the exact value of 'm'. And it is conceivable that because of such physical limitations as those imposed by the speed of light, the atomic character of matter, and the short amount of time before the next big bang, we shall never know what the value is, if not for five, then for six or seven or eight.‡

We are going to prove a theorem of F. P. Ramsey's that bears on these puzzles: *Let r, s, n be positive integers. Then there exists a positive integer m such that for any size-m set X, i.e. for any set X containing exactly m members, no matter how the size-r subsets of X are divided into s mutually exclusive classes, there will always be a size-n subset Y of X such that all size-r subsets of Y belong to the same one of the s classes.*§ In the puzzles, the size-2 subsets of the set of persons at the party were divided into two mutually exclusive classes, one consisting of the pairs of persons who like each other, and the other, of the pairs of persons who dislike each other. So in both problems $r = s = 2$. In the first, where $n = 3$, m may be taken to be 6; in the second, where $n = 4$, m may be taken to be 18. And despite the fact that the least value of 'm' that satisfies the conclusion of the theorem is unknown and possibly unknowable for all but a few non-trivial values of 'r', 's', 'n', Ramsey's theorem does give us a guarantee that for any r, s, n, a suitable m always *exists*, whatever it might be.

Our proof of Ramsey's theorem has two parts. In the first part, we prove a certain 'infinitary' version of the finitary version that was stated in the previous paragraph. In the second part, we use the compactness theorem and the Skolem–Löwenheim theorem to deduce the finitary version from the infinitary version. The infinitary version, which was also first proved by Ramsey,§ runs as follows: *Let r, s be positive integers. Then no matter how the size-r sets of natural numbers are divided into s*

† M. Gardner 'Mathematical Games', *Scientific American*, March 1978, p. 29.
‡ M. Gardner 'Mathematical Games', *Scientific American*, November 1977, p. 23.
§ In 'On a Problem of Formal Logic', in R. B. Braithwaite (ed.) *The Foundations of Mathematics*, Littlefield, Adams & Co., Paterson, N.J., 1960.

mutually exclusive classes, there will always be an infinite set Y of natural numbers such that all size-r subsets of Y belong to the same one of the s classes. (Thus if there are infinitely many angels and any two angels either like each other or dislike each other, then either there are infinitely many angels, any two of whom like each other, or infinitely many, any two of whom dislike each other.)

Here is the proof of the infinitary version:

A division of the set of size-r subsets of some set X into s mutually exclusive classes, $C_1, \ldots, C_s$, may be taken to be a function f from the set of size-r subsets x of X to the set of positive integers less than or equal to s such that $f(x) = j$ iff x is in C_j.

Let us denote by 'ω' the set of natural numbers; by '$[X]^r$', the set of size-r subsets of X; and let us write '$f\colon W \to Z$' to mean 'f is a function that assigns a member Z to each member of W'.

We thus desire to show that if $f\colon [\omega]^r \to \{1, \ldots, s\}$, then for some infinite set Y of natural numbers and some positive integer $j \leqslant s$, $f\colon [Y]^r \to \{j\}$.

We shall show this by defining an operation Φ such that if $f\colon [\omega]^r \to \{1, \ldots, s\}$ $(0 < r, s)$, then $\Phi(f)$ is an ordered pair $\langle j, Y \rangle$, consisting of a positive integer $j \leqslant s$ and an infinite set Y of natural numbers such that for all size-r subsets y of Y, $f(y) = j$.

We argue by induction on r ('s' denotes a fixed positive integer throughout what follows).

Basis step: $r = 1$. In this case the definition of $\Phi(f)$, $= \langle j, Y \rangle$, is easy. For if for each of the infinitely many size-1 sets $\{b\}$, $f(\{b\})$ is one of the finitely many positive integers $k \leqslant s$, then for at least one such k, $f(\{b\}) = k$ for infinitely many $\{b\}$. We can thus define j as the least positive integer $k \leqslant s$ such that $f(\{b\}) = k$ for infinitely many $\{b\}$, and define $Y = \{b \mid f(\{b\}) = j\}$.

Induction step: We assume as induction hypothesis that Φ has been suitably defined for all $g\colon [\omega]^r \to \{1, \ldots, s\}$. Suppose that $f\colon [\omega]^{r+1} \to \{1, \ldots, s\}$. In order to define $\Phi(f)$, $= \langle j, Y \rangle$, we define, for each natural number i, a natural number b_i, infinite sets Y_i, Z_i, W_i, a function $f_i\colon [\omega]^r \to \{1, \ldots, s\}$, and a positive integer $j_i \leqslant s$. Let $Y_0 = \omega$. We now suppose Y_i defined and show how to define b_i, Z_i, f_i, j_i, W_i, and Y_{i+1}.

Let b_i be the least member of Y_i.

Let $Z_i = Y_i - \{b_i\}$. Since Y_i is infinite, so is Z_i. Let the members of Z_i, in increasing order, be $a_{i0}, a_{i1}, \ldots$.

For any size-r set x of natural numbers, where $x = \{k_1, \ldots, k_r\}$, with

$k_1 < \ldots < k_r$, let $f_i(x) = f(\{b_i, a_{ik_1}, \ldots, a_{ik_r}\})$. Since b_i is not one of the a_{ik_m} and f is defined on all size-$(r+1)$ sets of natural numbers, f_i is well defined.

By the induction hypothesis, for some positive integer $j_i \leqslant s$ and some infinite set W_i, $\Phi(f_i) = \langle j_i, W_i \rangle$ and for every size-r subset x of W_i, $f_i(x) = j_i$. We have thus defined j_i and W_i, and we define $Y_{i+1} = \{a_{ik} \mid k \in W_i\}$.

Since W_i is infinite, Y_{i+1} is infinite. $Y_{i+1} \subseteq Z_i \subseteq Y_i$, and thus if $i_1 \leqslant i_2$, $Y_{i_1} \supseteq Y_{i_2}$. And since b_i is less than every member of Z_i and thus less than every member of Y_{i+1}, b_i is less than b_{i+1}, which is the least member of Y_{i+1}. Thus if $i_1 < i_2$, $b_{i_1} < b_{i_2}$.

For each positive integer $k \leqslant s$, let $E_k = \{i \mid j_i = k\}$. As in the basis step, some E_k is infinite, and we let j be the least k such that E_k is infinite and let $Y = \{b_i \mid i \in E_j\}$. This completes the definition of $\Phi(f)$.

Since $b_{i_1} < b_{i_2}$ if $i_1 < i_2$, Y is infinite. In order to complete the proof, we must show that if y is a size-$(r+1)$ subset of Y, then $f(y) = j$. So suppose that $y = \{b_i, b_{i_1}, \ldots, b_{i_r}\}$, with $i < i_1 < \ldots < i_r$ and $i, i_1, \ldots, i_r$ all in E_j. Since the Y_is are all nested, all of $b_{i_1}, \ldots, b_{i_r}$ are in Y_{i+1}. For each m, $1 \leqslant m \leqslant r$, let k_m be the unique member of W_i such that $b_{i_m} = a_{ik_m}$. And let $x = \{k_1, \ldots, k_r\}$. Then $x \subseteq W_i$, and since $i_1 < \ldots < i_r$, $b_{i_1} < \ldots < b_{i_r}$, $a_{ik_1} < \ldots < a_{ik_r}$, $k_1 < \ldots < k_r$, and x is thus a size-r subset of W_i. But $\Phi(f_i) = \langle j_i, W_i \rangle$ and thus $f_i(x) = j_i$. Since i is in E_j, $j_i = j$. Thus $f(y) = f(\{b_i, b_{i_1}, \ldots, b_{i_r}\}) = f(\{b_i, a_{ik_1}, \ldots, a_{ik_r}\}) = f_i(x) = j$.†

(Let us digress to discuss two possible strengthenings of the theorem we have just proved: *Let s be a positive integer.* [*Let k be a positive integer.*] *Then no matter how the finite sets of natural numbers are divided into s mutually exclusive classes, there will always be an infinite set Y of natural numbers such that for any positive integer r*[$\leqslant k$], *all size-r subsets of Y belong to the same one of the s classes.* With the material in brackets added, the proposed strengthening can easily be proved, by an induction on k that appeals to the theorem just proved. But without the bracketed material, the proposed strengthening is false for $s = 2$: Let $f(x) = 1$ if x contains the number that is the number of members in x; $f(x) = 2$ otherwise. Then there is no infinite set Y such that for every r, either

† This proof does not use the axiom of choice. In 'On a Problem of Formal Logic' Ramsey proved the stronger theorem that if X is an infinite set and $f\colon [X]^r \to \{1, \ldots, s\}$, then there exist a positive integer $j \leqslant s$ and an infinite subset Y of X such that $f\colon [Y]^r \to \{j\}$. Any proof of this stronger statement must appeal to (at least a weak form of) the axiom of choice.

$f\colon [Y]^r \to \{1\}$ or $f\colon [Y]^r \to \{2\}$: For if r is some positive integer that belongs to Y and $b_1, b_2, \ldots, b_r$ are r *other* members of Y, then

$$f(\{r, b_2, \ldots, b_r\}) = 1$$

and $f(\{b_1, b_2, \ldots, b_r\}) = 2$.)

Having proved the infinitary version of Ramsey's theorem, we can now deduce from it the finitary version, with the help of the compactness and Skolem–Löwenheim theorems. The finitary version has the form 'For any positive integers r, s, n, there exists a positive integer m such that ...'. We shall prove the theorem only for $r = 4$, $s = 3$, and $n = 7$. But as will be evident, the triple 4, 3, 7 is a sufficiently 'random' example for the proof for these particular values of 'r', 's', and 'n' to serve as a proof of the general case. And to prove the theorem for 4, 3, 7 it suffices to derive a contradiction from the supposition that for every positive integer m the following assertion holds:

> There exist a size-m set X and a function $f\colon [X]^4 \to \{1, 2, 3\}$ such that for each size-7 subset Y of X there are size-4 subsets Z_1, Z_2, Z_3 of Y such that $f(Z_1) \neq 1$, $f(Z_2) \neq 2$, and $f(Z_3) \neq 3$. (*m)

Let F_1, F_2, F_3 be three four-place predicate letters.

Let S be the following sentence:

$\forall w \forall x \forall y \forall z (w, x, y, z$ are all distinct $\leftrightarrow (F_1wxyz \vee F_2wxyz \vee F_3wxyz))$ &
$\forall w \forall x \forall y \forall z (F_1wxyz \to -(F_2wxyz \vee F_3wxyz))$ &
$\forall w \forall x \forall y \forall z (F_2wxyz \to -(F_1wxyz \vee F_3wxyz))$ &
$\forall w \forall x \forall y \forall z (F_3wxyz \to -(F_1wxyz \vee F_2wxyz))$ &
$\forall w \forall x \forall y \forall z (F_1wxyz \to (F_1wxzy$ &
the conjunction of all the 22, $= (4 \cdot 3 \cdot 2 \cdot 1) - 2$, other formulas $F_1 \ldots$ consisting of 'F_1' followed by all of 'w', 'x', 'y', 'z', in some order]))† &
$\forall w \forall x \forall y \forall z (F_2wxyz \to (F_2wxzy$ &
[the conjunction of the 22 other formulas $F_2 \ldots$])) &
$\forall w \forall x \forall y \forall z (F_3wxyz \to (F_3wxzy$ &
[the conjunction of the 22 other formulas $F_3 \ldots$])) &

† This conjunct is (logically) equivalent to:

$$\forall w \forall x \forall y \forall z (F_1wxyz \to (F_1wxzy \,\&\, F_1zwxy)).$$

$\forall w \forall x \forall y \forall z \forall a \forall b \forall c(w, x, y, z, a, b, c$ are all distinct $\rightarrow$

$(-F_1wxyz \vee -F_1wxya \vee$
[the disjunction of all the 838, $= (7\cdot6\cdot5\cdot4)-2$, other formulas $-F_1\ldots$, consisting of '$-F_1$' followed by four of 'w', 'x', 'y', 'z', 'a', 'b', 'c', in some order]) &
$(-F_2wxyz \vee$
[the disjunction of the 839 other formulas $-F_2\ldots$]) &
$(-F_3wxyz \vee$
[the disjunction of the 839 other formulas $-F_3\ldots$])).

For each m, let R_m be a sentence asserting the existence of at least m objects. (For example, if $m = 4$, R_m may be taken to be

$$\exists x_1 \exists x_2 \exists x_3 \exists x_4 \\ (x_1 \neq x_2 \;\&\; x_1 \neq x_3 \;\&\; x_1 \neq x_4 \;\&\; x_2 \neq x_3 \;\&\; x_2 \neq x_4 \;\&\; x_3 \neq x_4).$$

R_1 will be a valid sentence.)

Let S_m be the conjunction of S and R_m.

We want to see that (*m) holds if and only if Sm is satisfiable.

Suppose that (*m) holds. Let X and f be a set and function such as are asserted to exist in (*m). Let $\mathscr{I}$ be an interpretation whose domain is X and which specifies that F_i is to be true of an (ordered) quadruple a, b, c, d of elements of X if and only if a, b, c, d are all distinct and $f(\{a, b, c, d\}) = i$ $(i = 1, 2, 3)$. Since f is a function from the *un*ordered quadruples of X into $\{1, 2, 3\}$, if $f(\{a, b, c, d\}) = i$, $f(\{a, b, d, c\}) = i$, etc., and thus the first seven conjuncts of S hold in $\mathscr{I}$. The last conjunct of S also holds in $\mathscr{I}$, since for every seven objects in X, there are four of them that form a set Z_i such that $f(Z_i) \neq i$ $(i = 1, 2, 3)$. And since X contains m elements, R_m also holds in $\mathscr{I}$.

Suppose that S_m is satisfiable. Then for some $\mathscr{I}$, S and R_m hold in $\mathscr{I}$. Since R_m holds, there are at least m elements in the domain of $\mathscr{I}$. Let X be a subset of the domain containing exactly m elements. Define $f: [X]^4 \rightarrow \{1, 2, 3\}$ by: $f(\{a, b, c, d\}) = i$ $(a, b, c, d$ distinct members of $X)$ if and only if $\mathscr{I}$ specifies that F_i is to be true of the quadruple a, b, c, d. The first seven conjuncts of S insure that f is well defined. And the last insures that for every subset Y of X containing seven members, four of them will form a set Z_1 such that $f(Z_1) \neq 1$, four will form a set

Z_2 such that $f(Z_2) \neq 2$, and four will form a set Z_3 such that $f(Z_3) \neq 3$. Thus (*m) holds.

If $m_1 \leqslant m_2$, R_{m_2} implies R_{m_1} and therefore S_{m_2} implies S_{m_1}. Thus a finite non-empty subset $\{S_{m_1}, \ldots, S_{m_k}\}$ of $\{S_1, S_2, \ldots\}$ is satisfiable if and only if $S_{\max(m_1, \cdots, m_k)}$ is satisfiable.

We now suppose that for every positive integer m, (*m) holds. Then for every m, S_m is satisfiable. Therefore every finite subset of $\{S_1, S_2, \ldots\}$ is satisfiable, and by the compactness theorem $\{S_1, S_2, \ldots\}$ itself is satisfiable. There is therefore an interpretation $\mathscr{J}$ in which S and all sentences R_m hold. By the Skolem–Löwenheim theorem we may suppose that $\mathscr{J}$ has an enumerable domain. Since all the sentences R_m hold in $\mathscr{J}$, for every positive integer m, the domain of $\mathscr{J}$ contains at least m elements. Thus the domain of $\mathscr{J}$ is enumerably infinite. Let $\mathscr{I}$ be an interpretation isomorphic to $\mathscr{J}$ whose domain is ω. Since S holds in $\mathscr{J}$, S holds in $\mathscr{I}$. We define a function f: $[\omega]^4 \rightarrow \{1, 2, 3\}$ by $f(\{a, b, c, d\}) = i$ if and only if $\mathscr{I}$ specifies that F_i is to be true of the quadruple a, b, c, d. As before, the first seven conjuncts of S insure that f is well defined. And the last conjunct insures that for every subset Y of ω containing seven members, four of them will form a set Z_1 such that $f(Z_1) \neq 1$, four will form a set Z_2 such that $f(Z_2) \neq 2$, and four will form a set Z_3 such that $f(Z_3) \neq 3$.

But this is absurd. By the infinitary version of Ramsey's theorem, there is an infinite set X of ω such that either for every size-4 subset Z of X, $f(Z) = 1$, or for every size-4 subset Z of X, $f(Z) = 2$, or for every size-4 subset Z of X, $f(Z) = 3$. Let Y be a subset of X containing seven members. Then Y is a subset of ω, and either for every size-4 subset Z of Y, $f(Z) = 1$, or for every size-4 subset Z of Y, $f(Z) = 2$, or for every size-4 subset Z of Y, $f(Z) = 3$, and therefore there do *not* exist size-4 subsets Z_1, Z_2, Z_3 of Y such that $f(Z_1) \neq 1$, $f(Z_2) \neq 2$, and $f(Z_3) \neq 3$; contradiction.

27
Provability considered modal-logically

In Chapter 16 we defined the notion of a provability predicate for an extension T of Q and proved Löb's theorem, which states that if $B(y)$ is a provability predicate for T, then for any sentence A of the language of T, if $\vdash_T B(\ulcorner A \urcorner) \rightarrow A$, then $\vdash_T A$. In the present chapter we are going to make use of certain concepts and techniques of propositional modal logic in the investigation of provability predicates for theories extending Q. (We presuppose no prior knowledge of modal logic.) For the sake of definiteness, we shall suppose that $T = Z$ and $B(y) = \text{Prov}(y)$ (see Chapter 16), but in fact our treatment applies generally to theories that are extensions of Q and provability predicates for these theories.

We begin by describing a system of propositional modal logic called 'G'.† The syntax of G is quite standard; we suppose ourselves to have a denumerable stock of sentence letters and inductively define a *sentence* (*of* G) as follows:

> Every sentence letter is a sentence;
> the 0-place connective $\bot$ is a sentence;
> if B and C are sentences, so is $(B \rightarrow C)$; and
> if B is a sentence, so is $\Box B$.

$\bot$ is the 0-place truth-functional connective that is always evaluated as false; $-B$, the negation of B, can be defined as $(B \rightarrow \bot)$. Other connectives will be regarded as defined from $-$ and $\rightarrow$ in the usual manner. $\Box$ is the symbol ordinarily used in modal logic to mean 'necessarily'; $\Diamond$, which ordinarily means 'possibly', may be regarded as an abbreviation for $-\Box-$.

The notion of *subsentence* is defined in the usual way: any sentence is a subsentence of itself; if $(B \rightarrow C)$ is a subsentence of A, then B and C are also subsentences of A; and if $\Box B$ is a subsentence of A, then B is also a subsentence of A.

A sentence is an *axiom* (*of* G) if and only if it is either a tautology (possibly containing one or more occurrences of $\Box$), a sentence

† *The Unprovability of Consistency: An Essay in Modal Logic*, by George Boolos, Cambridge University Press, 1979, is an extended study of G. G is sometimes called 'L'.

$\Box(B \to C) \to (\Box B \to \Box C)$, a sentence $\Box B \to \Box\Box B$, or a sentence $\Box(\Box B \to B) \to \Box B$. The two *rules* (*of G*) are *modus ponens* (from A and $A \to B$, infer B) and *necessitation* (from A, infer $\Box A$). And, as usual, a sentence is a *theorem* (*of G*) if it is the last member of a finite sequence of sentences, each of which either is an axiom or is inferable by one of the rules from sentences that are earlier in the sequence.†

We write '$\vdash_G A$' to mean that A is a theorem of G.

The axioms and rules of G bear a strong and non-accidental resemblance to facts about Z and Prov(y) that were noted in Chapter 16, and the next definitions and theorem bring out one important connection between Z and G.

We use 'ϕ' as a variable for functions that assign to each sentence letter a sentence of Z.

If A is a sentence of G, then A^ϕ is inductively defined as follows:

$$\begin{aligned} p^\phi &= \phi(p) \ (p \text{ a sentence letter}); \\ \bot^\phi &= \mathbf{0} = \mathbf{1}; \\ (B \to C)^\phi &= (B^\phi \to C^\phi); \text{ and} \\ (\Box B)^\phi &= \mathrm{Prov}(\ulcorner B^\phi \urcorner). \end{aligned}$$

The arithmetical soundness theorem

If $\vdash_G A$, then for all ϕ, $\vdash_Z A^\phi$.‡

Proof. Prov(y) is a provability predicate for Z, and it is immediate from the definitions of provability predicate and A^ϕ that if A is a tautology or a sentence $\Box(B \to C) \to (\Box B \to \Box C)$ or $\Box B \to \Box\Box B$, then A^ϕ is a theorem of Z. Moreover, it is also immediate that B^ϕ is a theorem of Z if A^ϕ and $(A \to B)^\phi$ are and that $(\Box A)^\phi$ (= $\mathrm{Prov}(\ulcorner A^\phi \urcorner)$) is a theorem of Z if A^ϕ is. It thus remains to show that $\vdash_Z A^\phi$, where A is an axiom $\Box(\Box B \to B) \to \Box B$. By Löb's theorem it suffices to show that $\vdash_Z \mathrm{Prov}(\ulcorner A^\phi \urcorner) \to A^\phi$. Let $S = B^\phi$. Then

$$A^\phi = \mathrm{Prov}(\ulcorner(\mathrm{Prov}(\ulcorner S \urcorner) \to S)\urcorner) \to \mathrm{Prov}(\ulcorner S \urcorner).$$

† It is a technical curiosity that all sentences $\Box B \to \Box\Box B$ can be derived from the other axioms and rules: The sentence $B \to (\Box(\Box B \& B) \to \Box B \& B)$ can be derived from the tautology $B \to (\Box\Box B \& \Box B \to \Box B \& B)$, and then so can $\Box B \to \Box(\Box(\Box B \& B) \to \Box B \& B)$, $\Box B \to \Box(\Box B \& B)$, and $\Box B \to \Box\Box B$.

‡ The converse, *the arithmetical completeness theorem*, which states that if for all ϕ, $\vdash_Z A^\phi$, then $\vdash_G A$, was proved by Robert Solovay. See his 'Provability interpretations of modal logic', *Israel Journal of Mathematics* **25** (1976), 287–304 or Boolos, *Unprovability of Consistency*, chap. 12.

By (ii) of Chapter 16,

$$\text{Prov}(\ulcorner A^{\phi}\urcorner) \rightarrow [\text{Prov}(\ulcorner\text{Prov}(\ulcorner(\text{Prov}(\ulcorner S\urcorner) \rightarrow S)\urcorner)\urcorner) \rightarrow \text{Prov}(\ulcorner\text{Prov}(\ulcorner S\urcorner)\urcorner)]$$

and

$$\text{Prov}(\ulcorner(\text{Prov}(\ulcorner S\urcorner) \rightarrow S)\urcorner) \rightarrow [\text{Prov}(\ulcorner\text{Prov}(\ulcorner S\urcorner)\urcorner) \rightarrow \text{Prov}(\ulcorner S\urcorner)]$$

are theorems of Z and by (iii),

$$\text{Prov}(\ulcorner(\text{Prov}(\ulcorner S\urcorner) \rightarrow S)\urcorner) \rightarrow \text{Prov}(\ulcorner\text{Prov}(\ulcorner(\text{Prov}(\ulcorner S\urcorner) \rightarrow S)\urcorner)\urcorner)$$

is also a theorem of Z, and therefore

$$\text{Prov}(\ulcorner A^{\phi}\urcorner) \rightarrow [\text{Prov}(\ulcorner(\text{Prov}(\ulcorner S\urcorner) \rightarrow S)\urcorner) \rightarrow \text{Prov}(\ulcorner S\urcorner)],$$

i.e. $\text{Prov}(\ulcorner A^{\phi}\urcorner) \rightarrow A^{\phi}$, which is a truth-functional consequence of these three sentences, is a theorem of Z as well.

In order to state the theorems about G that we shall prove we make the following definitions:

S_p is the set of sentences containing no sentence letters other than p. (Sentences such as $\Box\bot \rightarrow \bot$ that contain no sentence letters at all are of course in S_p.)

We shall call a sentence A *modal in* the sentence letter p if $A \in S_p$ and every occurrence of p in A is in the scope of the necessity sign $\Box$. Thus $-\Box p$, $\Box p$, $\Box p \rightarrow \Box - p$, and $\Box(\Box\Box\bot \rightarrow \bot)$ are all modal in p, but neither p nor $\Box p \rightarrow p$ nor $\bot \rightarrow p$ is modal in p. If A is modal in p, then A is a truth-functional combination of sentences $\Box B$, where $B \in S_p$. ($\bot$ conventionally counts as a truth-functional combination of such sentences.)

For any sentence A, $\boxdot A$ is defined as the sentence $(\Box A \& A)$. (This definition has a point because $\Box A \rightarrow A$ is not always a theorem of G.)

$\Box^n\bot$ is defined by: $\Box^0\bot = \bot$; $\Box^{n+1}\bot = \Box\Box^n\bot$.

We are going to prove a fixed point theorem about G: if A is modal in p, then there is a truth-functional combination H of sentences $\Box^n\bot$ such that $\vdash_G \boxdot(p \leftrightarrow A) \rightarrow (p \leftrightarrow H)$.

We shall see that if A is taken as one of the sentences on the top line of Table 27-1, then H may be taken as the corresponding sentence on the bottom line.

TABLE 27-1

A	$\Box p$	$-\Box p$	$\Box - p$	$-\Box - p$	$-\Box\Box p$	$\Box p \to \Box - p$
H	$-\bot$	$-\Box\bot$	$\Box\bot$	$\bot$	$-\Box\Box\bot$	$\Box\Box\bot \to \Box\bot$

It follows, for example, that if $\vdash_Z S \leftrightarrow \text{Prov}(\ulcorner - S \urcorner)$, then $\vdash_Z S \leftrightarrow \text{Prov}(\ulcorner \mathbf{0} = \mathbf{1} \urcorner)$: Suppose that $\vdash_Z S \leftrightarrow \text{Prov}(\ulcorner - S \urcorner)$. Then

$$\vdash_Z \text{Prov}(\ulcorner (S \leftrightarrow \text{Prov}(\ulcorner - S \urcorner))\urcorner) \& (S \leftrightarrow \text{Prov}(\ulcorner - S \urcorner)).$$

Let ϕ be such that $\phi(p) = S$. Then $\vdash_Z (\boxdot(p \leftrightarrow \Box - p))^\phi$. By the fixed point theorem, $\vdash_G \boxdot(p \leftrightarrow \Box - p) \to (p \leftrightarrow \Box\bot)$, and therefore by the arithmetical soundness theorem, $\vdash_Z (\boxdot(p \leftrightarrow \Box - p) \to (p \leftrightarrow \Box\bot))^\phi$, i.e. $\vdash_Z (\boxdot(p \leftrightarrow \Box - p))^\phi \to (p \leftrightarrow \Box\bot)^\phi$. By modus ponens (in Z)

$$\vdash_Z (p \leftrightarrow \Box\bot)^\phi,$$

i.e. $\vdash_Z S \leftrightarrow \text{Prov}(\ulcorner \mathbf{0} = \mathbf{1} \urcorner)$.

Other similar theorems about the modal system G may be cashed as theorems about Z in similar ways. There is a general theorem about Z that follows from the fixed point theorem, which can be roughly put this way: if $P(y)$ is a predicate of Z that is 'built up from' $\text{Prov}(y)$ and truth-functional operations, then any fixed point of $P(y)$, i.e. any sentence S such that $\vdash_Z S \leftrightarrow P(\ulcorner S \urcorner)$, is equivalent in Z to some truth-functional combination of $\mathbf{0} = \mathbf{1}$, $\text{Prov}(\ulcorner \mathbf{0} = \mathbf{1} \urcorner)$, $\text{Prov}(\ulcorner \text{Prov}(\ulcorner \mathbf{0} = \mathbf{1} \urcorner) \urcorner)$, etc. These last sentences wear their truth-conditions on their face: since nothing false (in $\mathcal{N}$, the standard interpretation for the language L of arithmetic) is provable in Z, they are all *false*. Thus any sentence described 'self-referentially' as a fixed point of some predicate built up from truth-functions and $\text{Prov}(y)$ is in fact equivalent to a truth-functional combination of certain false sentences. The combination may be calculated from the predicate and, as we shall see, the calculation is quite simple.

The second theorem we shall prove is this: Let $p_0, \ldots, p_{n-1}$ be a sequence of sentence letters. Then

$$\vdash_G \Box\left(\bigwedge_{i<n} (\Box p_i \to p_i) \to - \Box^n \bot\right) \to$$
$$\left(- \Box^n \bot \to \bigwedge_{i<n} (\Box p_i \to p_i)\right).$$

We call this theorem *the reciprocation theorem*; we shall discuss its significance after we give its proof.

Here are some elementary facts about G:

Fact 1. The theorems of G are closed under truth-functional consequence. For suppose that $\vdash_G A_1, \ldots, \vdash_G A_n$ and B is a truth-functional consequence of $A_1, \ldots, A_n$. Then $(A_1 \rightarrow (A_2 \rightarrow \ldots (A_n \rightarrow B) \ldots))$ is a tautology and hence a theorem of G. Applying modus ponens n times yields that $\vdash_G B$.

Fact 2. $\vdash_G \Box A \& \Box B \leftrightarrow \Box(A \& B)$: $A \rightarrow (B \rightarrow (A \& B))$ is a tautology, and by necessitation, $\vdash_G \Box(A \rightarrow (B \rightarrow (A \& B)))$. Since

$$\Box(A \rightarrow (B \rightarrow (A \& B))) \rightarrow (\Box A \rightarrow \Box(B \rightarrow (A \& B)))$$

and $\Box(B \rightarrow (A \& B)) \rightarrow (\Box B \rightarrow \Box(A \& B))$ are axioms of G, fact 1 yields that $\vdash_G \Box A \& \Box B \rightarrow \Box(A \& B)$. Conversely, since $(A \& B) \rightarrow A$ and $(A \& B) \rightarrow B$ are tautologies, $\vdash_G \Box(A \& B) \rightarrow \Box A$ and

$$\vdash_G \Box(A \& B) \rightarrow \Box B,$$

and therefore $\vdash_G \Box A \& \Box B \leftrightarrow \Box(A \& B)$.

Fact 3. If $\vdash_G A_1 \& \ldots \& A_n \rightarrow B$, then $\vdash_G \Box A_1 \& \ldots \& \Box A_n \rightarrow \Box B$. The proof is by induction on n. If $n = 1$ and $\vdash_G A_1 \rightarrow B$, then by necessitation $\vdash_G \Box(A_1 \rightarrow B)$. And since $\vdash_G \Box(A_1 \rightarrow B) \rightarrow (\Box A_1 \rightarrow \Box B)$, we have that $\vdash_G \Box A_1 \rightarrow \Box B$. Suppose that $\vdash_G A_1 \& \ldots \& A_n \& A_{n+1} \rightarrow B$. Then $\vdash_G A_1 \& \ldots \& (A_n \& A_{n+1}) \rightarrow B$, whence by the hypothesis of the induction $\vdash_G \Box A_1 \& \ldots \& \Box(A_n \& A_{n+1}) \rightarrow \Box B$. Facts 1 and 2 then yield the conclusion that $\vdash_G \Box A_1 \& \ldots \& \Box A_n \& \Box A_{n+1} \rightarrow \Box B$.

Fact 4. If $\vdash_G \Box A_1 \& A_1 \& \ldots \& \Box A_n \& A_n \& \Box B \rightarrow B$, then

$$\vdash_G \Box A_1 \& \ldots \& \Box A_n \rightarrow \Box B.$$

For if the hypothesis holds, then

$$\vdash_G \Box A_1 \& A_1 \& \ldots \& \Box A_n \& A_n \rightarrow (\Box B \rightarrow B),$$

and thus by fact 3

$$\vdash_G \Box\Box A_1 \& \Box A_1 \& \ldots \& \Box\Box A_n \& \Box A_n \rightarrow \Box(\Box B \rightarrow B).$$

Since $\vdash_G \Box A_i \rightarrow \Box\Box A_i$, and $\vdash_G \Box(\Box B \rightarrow B) \rightarrow \Box B$, we have that $\vdash_G \Box A_1 \& \ldots \& \Box A_n \rightarrow \Box B$.

We shall now define the notion of a *model* (*for* G) and prove an adequacy (= soundness and completeness) theorem for this notion.† A model is a triple $\langle W, R, P\rangle$ consisting of a non-empty finite set W, a transitive (if $wRxRy$, then wRy) and irreflexive (not xRx) relation R on W ('on W' means: if wRx, then $w, x \in W$), and a function P that assigns a truth-value (1 or 0) to each pair consisting of a member of W and a sentence letter.

Let $\mathcal{M}$, $= \langle W, R, P\rangle$, be a model. (We shall always use '$\mathcal{M}$' to abbreviate '$\langle W, R, P\rangle$'.) For every w in W and sentence A we define the truth-value $\mathcal{M}(w, A)$ *of A at w in $\mathcal{M}$* as follows:

$$\mathcal{M}(w,p) = P(w,p) \quad (p \text{ a sentence letter}),$$
$$\mathcal{M}(w,\bot) = 0,$$
$$\mathcal{M}(w, B \to C) = \begin{cases} 1 & \text{if either } \mathcal{M}(w,B) = 0 \text{ or } \mathcal{M}(w,C) = 1, \\ 0 & \text{otherwise,} \end{cases}$$
$$\mathcal{M}(w, \Box B) = \begin{cases} 1 & \text{if for all } x \text{ such that } wRx, \mathcal{M}(x,B) = 1, \\ 0 & \text{otherwise.} \end{cases}$$

(Thus $\mathcal{M}(w, B\&C) = 1$ iff $\mathcal{M}(w,B) = \mathcal{M}(w,C) = 1$, and similarly for the other truth-functional connectives.)

We call the sentence A *valid in* the model $\mathcal{M}$ if $\mathcal{M}(w, A) = 1$ for all w in W, and we call A *valid* if A is valid in all models.

The adequacy theorem then states that A is valid if (soundness) and only if (completeness) A is a theorem.

The notion of *rank* will prove important in what follows.

If R is a transitive and irreflexive relation on W and $w_i R \ldots R w_1 R w_0$, then all of $w_0, \ldots, w_i$ are distinct: for if $i \geqslant j > k \geqslant 0$, by transitivity $w_j R w_k$ and then by irreflexivity $w_j \neq w_k$. Thus if W is also finite and contains (say) m members, then it is never the case that $w_i R \ldots w_1 R w_0$ if $i \geqslant m$. Thus if $\mathcal{M}$ is a model, then for each w in W there is a greatest natural number i $(< m)$ such that for some $w_i, \ldots, w_0$ in W,

$$w = w_i R \ldots R w_0.$$

We call this number the rank in $\mathcal{M}$ of w, and write 'rk(w)' whenever the context makes it safe to suppress the reference to $\mathcal{M}$. (If wRx for no x, then rk$(w) = 0$.) Clearly, if wRx, then rk$(w) >$ rk(x). And if $j < i =$ rk(w), then for some x such that wRx, rk$(x) = j$: for if

$$w = w_i R \ldots R w_j R \ldots R w_0,$$

† The adequacy theorem for G was first proved by Segerberg. Adequacy theorems of this type for systems of modal logic were first proved by Kripke.

then clearly $\text{rk}(w_j) \geqslant j$, but if $\text{rk}(w_j) > j$, then

$$\text{rk}(w) \geqslant (i-j)+\text{rk}(w_j) > i = \text{rk}(w),$$

a contradiction; thus $\text{rk}(w_j) = j$ and by transitivity wRw_j.

We show, by induction on i, that $\mathcal{M}(w, \Box^i\bot) = 1$ iff $\text{rk}(w) < i$. This is trivial for $i = 0$. Suppose that for all x in W,

$$\mathcal{M}(x, \Box^i\bot) = 1 \text{ iff } \text{rk}(x) < i.$$

Then $\mathcal{M}(w, \Box^{i+1}\bot) = 1$ iff for all x such that wRx, $\mathcal{M}(x, \Box^i\bot) = 1$; iff for all x such that wRx, $\text{rk}(x) < i$; iff $\text{rk}(w) \leqslant i$; iff $\text{rk}(w) < i+1$.

The Semantical soundness theorem†

If $\vdash_G A$, then A is valid.

Proof. We shall show that if $\vdash_G A$, then A is valid in an arbitrary model $\mathcal{M}$.

If A is a tautology, it follows from the clauses for $\bot$ and $\rightarrow$ of the definition of $\mathcal{M}(w, A)$ that $\mathcal{M}(w, A) = 1$. If $\mathcal{M}(w, \Box(B \rightarrow C)) = \mathcal{M}(w, \Box B) = 1$, then for every x such that wRx, $\mathcal{M}(x, B \rightarrow C) = \mathcal{M}(x, B) = 1$, and thus for every x such that wRx, $\mathcal{M}(w, C) = 1$, i.e. $\mathcal{M}(w, \Box C) = 1$. Thus $\mathcal{M}(w, \Box(B \rightarrow C) \rightarrow (\Box B \rightarrow \Box C)) = 1$. If $\mathcal{M}(w, \Box B) = 1$, then for every x such that wRx, $\mathcal{M}(x, \Box B) = 1$ and then $\mathcal{M}(w, \Box\Box B) = 1$; for if xRy, then wRy by the transitivity of R, and then $\mathcal{M}(y, B) = 1$. Thus $\mathcal{M}(w, \Box B \rightarrow \Box\Box B) = 1$.

We now consider axioms of the form $\Box(\Box B \rightarrow B) \rightarrow \Box B$. Suppose that $\mathcal{M}(w, \Box B) = 0$. Then for some x, wRx and $\mathcal{M}(x, B) = 0$. Let x be *of lowest rank* such that wRx and $\mathcal{M}(x, B) = 0$. Then if wRy and $\text{rk}(y) < \text{rk}(x)$, $\mathcal{M}(y, B) = 1$. Suppose that xRy. Then $\text{rk}(y) < \text{rk}(x)$. By transitivity wRy. Thus $\mathcal{M}(y, B) = 1$. So for all y such that xRy, $\mathcal{M}(y, B) = 1$ and therefore $\mathcal{M}(x, \Box B) = 1$. Since $\mathcal{M}(x, B) = 0$, $\mathcal{M}(x, \Box B \rightarrow B) = 0$, and since wRx, $\mathcal{M}(w, \Box(\Box B \rightarrow B)) = 0$. Thus for every w in W, $\mathcal{M}(w, \Box(\Box B \rightarrow B) \rightarrow \Box B) = 1$ and every axiom is valid in $\mathcal{M}$.

If $\mathcal{M}(w, A) = \mathcal{M}(w, A \rightarrow B) = 1$, then $\mathcal{M}(w, B) = 1$. And if A is valid in $\mathcal{M}$, then so is $\Box A$; for if wRx, then $\mathcal{M}(x, A) = 1$, and so $\mathcal{M}(w, \Box A) = 1$ for every w in W. Thus any sentence inferable from

† Unlike the arithmetical soundness theorem and the semantical completeness theorem, the semantical soundness theorem is not used in the rest of this chapter.

sentences valid in $\mathscr{M}$ by modus ponens or necessitation is also valid in $\mathscr{M}$. Every theorem of G is therefore valid in $\mathscr{M}$.

The semantical completeness theorem
If A is valid, then $\vdash_G A$.

Proof.† Suppose that A is not a theorem of G. We shall construct a model in which A is not valid. We shall call a sentence B a *formula* if B is either a subsentence of A or the negation of a subsentence of A. A set X of formulas will be called *consistent* if $-\bigwedge X$, which is the negation of the conjunction of all the members of X, is not a theorem of G. The set $\{-A\}$ is consistent; for if $\vdash_G --A$, then $\vdash_G A$, which contradicts our supposition. A consistent set X is called *maximal* if for every subsentence B of A, either B is in X or $-B$ is in X. Every consistent set X is a subset of some maximal set: $\bigwedge X$ is equivalent in the propositional calculus to some non-empty disjunction each of whose disjuncts is a conjunction of formulas that contains all members of X and contains exactly one of B and $-B$, for each subsentence B of A; since X is consistent, the conjuncts of at least one of these disjoined conjunctions will form a consistent set, and indeed a maximal set that contains all members of X. If X is maximal, with $B_1, \ldots, B_n \in X$, and C is a subsentence of A such that $\vdash_G B_1 \& \ldots \& B_n \rightarrow C$, then $C \in X$, since otherwise $-C \in X$ and $\vdash_G -\bigwedge X$, which contradicts the consistency of X.

Let W be the set of maximal sets. W is non-empty, since $\{-A\}$ is consistent and therefore a subset of some maximal set. W is finite: in fact, if there are k subsentences of A, there are at most 2^k maximal sets. Let wQx if and only if both $w, x \in W$ and whenever $\Box C$ is a subsentence of A such that $\Box C \in w$, then $\Box C \in x$ and $C \in x$. Q is clearly transitive: if $wQxQy$, and $\Box C$ is a subsentence of A such that $\Box C \in w$, then $\Box C \in x$, and so $\Box C \in y$ and $C \in y$. Let wRx if and only if both wQx and not xQx. R is evidently irreflexive. And R is transitive: if $wRxRy$, then wQx, xQy, and not yQy; and since wQy by the transitivity of Q, we have that wRy. Let $P(w,p) = 1$ if $p \in w$, and $= 0$ otherwise (for all w in W and sentence letters p). We shall now show by induction on their complexity that for all subsentences B of A and all w in W, $B \in w$ if and only if $\mathscr{M}(w, B) = 1$. (As ever, $\mathscr{M} = \langle W, R, P \rangle$.)

† Our proof incorporates a number of simplifications due to Solovay and Warren Goldfarb.

If B is a sentence letter p, then $p \in w$ iff $P(w, p) = 1$ iff $\mathcal{M}(w, p) = 1$. If $B = \bot$, then $\bot \notin w$, which is consistent. But also $\mathcal{M}(w, \bot) = 0$. If $B = (C \to D)$, then C and D are subsentences of A and $\vdash_G -B \leftrightarrow (C \& -D)$. Thus $B \notin w$ iff (by maximality) $-B \in w$, iff $C \in W$ and $-D \in w$, iff $C \in w$ and $D \notin w$, iff (by the induction hypothesis)

$$\mathcal{M}(w, C) = 1 \quad \text{and} \quad \mathcal{M}(w, D) = 0,$$

iff $\mathcal{M}(w, B) = 0$. So $B \in w$ iff $\mathcal{M}(w, B) = 1$.

Assume now that $B = \Box C$. Suppose $\Box C \in w$. If wRx, then wQx and $C \in x$, whence by the induction hypothesis $\mathcal{M}(x, C) = 1$; thus for all x such that wRx, $\mathcal{M}(x, C) = 1$ and $\mathcal{M}(w, \Box C) = 1$. For the converse suppose that $\mathcal{M}(w, \Box C) = 1$. Let $\Box D_1, \ldots, \Box D_n$ be all the sentences $\Box D$ that are in w. Let $X = \{\Box D_1, D_1, \ldots, \Box D_n, D_n, \Box C, -C\}$. Is X consistent? If it is, then for some maximal set x, $X \subseteq x$. But since all of $\Box D_1, D_1, \ldots, \Box D_n, D_n$ are in x, wQx, and since $\Box C$ and $-C$ are in x, not xQx. Thus wRx, $\mathcal{M}(x, C) = 1$, and by the induction hypothesis, $C \in x$. But $-C \in X \subseteq x$. Thus X is *not* consistent and so $\vdash_G \Box D_1 \& D_1 \& \ldots \& \Box D_n \& D_n \& \Box C \to C$. By fact 4, $\vdash_G \Box D_1 \& \ldots \& \Box D_n \to \Box C$. Since $\Box D_1, \ldots, \Box D_n \in w$ and $\Box C$ is a subsentence of A, $\Box C \in w$.

We conclude that if B is any subsentence of A and $w \in W$, then $B \in w$ iff $\mathcal{M}(w, B) = 1$. As $\{-A\}$ is consistent, for some w in W, $-A \in w$, and $A \notin w$. Thus $\mathcal{M}(w, A) \neq 1$, and A is not valid in the model $\mathcal{M}$.

We now turn to the fixed point theorem for G. Suppose that A is modal in p. To each pair consisting of a natural number i and a sentence B in S_p we assign a truth-value $\theta(i, p)$ as follows:

$$\theta(i, \bot) = 0;$$

$$\theta(i, B \to C) = \begin{cases} 1 & \text{if either } \theta(i, B) = 0 \text{ or } \theta(i, C) = 1, \\ 0 & \text{otherwise;} \end{cases}$$

$$\theta(i, p) = \theta(i, A); \text{ and}$$

$$\theta(i, \Box B) = \begin{cases} 1 & \text{if for all } j < i \ \theta(j, B) = 1, \\ 0 & \text{otherwise.} \end{cases}$$

(Thus $\theta(0, \Box B) = 1$ for all B in S_p.)

θ is well defined: Suppose that $\theta(j, B)$ is defined for all $j < i$ and all B in S_p. Then $\theta(i, \Box B)$ is defined for all $\Box B$ in S_p, and hence $\theta(i, C)$ is defined for all truth-functional combinations C of sentences $\Box B$ in S_p. This means that $\theta(i, A)$ is defined, since A, being modal in p, is such a truth-functional combination. Hence $\theta(i,p)$ is defined, and therefore $\theta(i, C)$ is defined for all sentences C in S_p, since every C in S_p is a truth-functional combination of p and sentences $\Box B$ in S_p. Thus $\theta(i, C)$ is defined for all i and all C in S_p.

Let n be the number of subsentences of A that are of the form $\Box B$.

If $\theta(i, \Box B) > \theta(i, B)$, then $\theta(i, B) = 0$, and thus if $h > i$, $\theta(h, \Box B) = 0 \leqslant \theta(h, B)$. There is therefore at most one i such that $\theta(i, \Box B) > \theta(i, B)$. There are $n+1$ natural numbers $i \leqslant n$ and n subsentences $\Box B$ of A. It follows that for some $i \leqslant n$ there is no subsentence $\Box B$ of A such that $\theta(i, \Box B) > \theta(i, B)$. Thus for every subsentence $\Box B$ of A, $\theta(i, \Box B) \leqslant \theta(i, B)$. If $\theta(i, \Box B) = 1$, then $\theta(i, B) = 1$ and therefore $\theta(i+1, \Box B) = 1$; but if $\theta(i, \Box B) = 0$, then $\theta(i+1, \Box B) = 0$. In either case $\theta(i+1, \Box B) = \theta(i, \Box B)$, and thus for every sentence C that is either a subsentence of A or the sentence letter p, $\theta(i+1, C) = \theta(i, C)$, since C will either be a subsentence $\Box B$ of A, a truth-functional combination of these, A (which is such a combination), p, or a truth-functional combination of p and subsentences $\Box B$ of A. And then for every subsentence $\Box B$ of A, we have that $\theta(i+1, \Box B) = \theta(i, \Box B) \leqslant \theta(i, B) = \theta(i+1, B)$. Iterating, we see that for every $h > i$ and every such C, $\theta(h, C) = \theta(i, C)$. So for every $h \geqslant n$, $\theta(h,p) = \theta(n,p)$.

There are therefore only finitely many h for which $\theta(h,p) \neq \theta(n,p)$. If $\theta(n,p) = 0$, we let $H_A = \bigvee \{\Box^{h+1} \bot \,\&\, - \Box^h \bot \mid \theta(h,p) = 1\}$; if $\theta(n,p) = 1$, we let $H_A = \bigwedge \{-(\Box^{h+1} \bot \,\&\, - \Box^h \bot) \mid \theta(h,p) = 0\}$. (An empty disjunction is truth-functionally equivalent to $\bot$; an empty conjunction, to $-\bot$.)

The fixed point theorem

Suppose that A is modal in p. Then $\vdash_G \boxdot(p \leftrightarrow A) \rightarrow (p \leftrightarrow H_A)$.†

† Claudio Bernardi and C. Smoryński independently proved that for every A modal in p there exists H such that $\vdash_G \boxdot(p \leftrightarrow A) \rightarrow (p \leftrightarrow H)$. D. de Jongh and Giovanni Sambin have independently proved that if every occurrence of p in A lies in the scope of $\Box$, then there exists a sentence H such that $\vdash_G \boxdot(p \leftrightarrow A) \rightarrow (p \leftrightarrow H)$, H does not contain p, and all sentence letters contained in H are contained in A. For two very different proofs, see C. Smoryński, 'Calculating Self-referential Statements, I: Explicit Calculations', *Studia Logica* **38** (1979), 17–36, and Boolos, *Unprovability of Consistency*, chap. 11.

Proof. We apply the semantical completeness theorem. We suppose that $\mathscr{M}$ is a model and $\mathscr{M}(w, \boxdot(p \leftrightarrow A)) = 1$ and show that $\mathscr{M}(w,p) = \mathscr{M}(w, H_A)$ and thus that $\mathscr{M}(w, p \leftrightarrow H_A) = 1$. We first show by induction on i, where $i = \text{rk}(x)$, that if $B \in S_p$, and either wRx or $w = x$, then $\mathscr{M}(x, B) = \theta(i, B)$, and we may therefore suppose that for all $j < i$, where $j = \text{rk}(y)$, if $C \in S_p$, and either wRy or $w = y$, then $\mathscr{M}(y, C) = \theta(j, C)$. If $B \in S_p$, then B is either a sentence $\Box C$, the sentence A, which is a truth-functional combination of sentences $\Box C$ (since A is modal in p), p, or a truth-functional combination of p and sentences $\Box C$. If B is a truth-functional combination of sentences D for which $\mathscr{M}(x, D) = \theta(i, D)$, then $\mathscr{M}(x, B) = \theta(i, B)$. And if $\mathscr{M}(x, A) = \theta(i, A)$, then by the definition of θ, $\theta(i, p) = \theta(i, A)$, and since either wRx or $w = x$, $\mathscr{M}(x, p \leftrightarrow A) = 1$, $\mathscr{M}(x, p) = \mathscr{M}(x, A)$, and therefore $\mathscr{M}(x, p) = \theta(i, p)$. We may thus suppose that $B = \Box C$. $C \in S_p$, as $B \in S_p$.

Assume that $j < i$. Then for some y, xRy and $\text{rk}(y) = j$. wRy, since xRy and either wRx or $w = x$. The induction hypothesis then yields that $\mathscr{M}(y, C) = \theta(j, C)$. Thus if $\mathscr{M}(x, B) = 1$, $\mathscr{M}(y, C) = 1$ (as xRy) and so $\theta(j, C) = 1$. Thus if $\mathscr{M}(x, B) = 1$, then for all $j < i$, $\theta(j, C) = 1$ and so $\theta(i, B) = 1$. Conversely, assume that xRy. Let $j = \text{rk}(y)$. Then $j < i$ and wRy. The induction hypothesis yields that $\mathscr{M}(y, C) = \theta(j, C)$. Thus if $\theta(i, B) = 1, \theta(j, C) = 1$ and so $\mathscr{M}(y, C) = 1$. Thus if $\theta(i, B) = 1$, then for all y such that xRy, $\mathscr{M}(y, C) = 1$ and so $\mathscr{M}(x, B) = 1$. For all B in S_p, then, $\mathscr{M}(x, B) = \theta(i, B)$.

Now let $i = \text{rk}(w)$. Then $\mathscr{M}(w, p) = \theta(i, p)$. $\mathscr{M}(w, \Box^h \bot) = 1$ iff $i = \text{rk}(w) < h$, whence $\mathscr{M}(w, \Box^{h+1} \bot \,\&\, - \Box^h \bot) = 1$ iff $i = h$. Thus $\mathscr{M}(w, H_A) = 1$ iff i is an h such that $\theta(h, p) = 1$, iff $\theta(i, p) = 1$. So $\mathscr{M}(w, H_A) = \theta(i, p)$, and $\mathscr{M}(w, p) = \mathscr{M}(w, H_A)$.

The correctness of the entries in Table 27-1 can now be verified.

A corollary to the fixed point theorem: If A contains no sentence letters (and is therefore modal in p), then $\vdash_G A \leftrightarrow H_A$. For neither A nor H_A contains any sentence letters, the result of substituting a sentence for a sentence letter in a theorem of G is also a theorem of G, and $\vdash_G \boxdot(p \leftrightarrow A) \to (p \leftrightarrow H_A)$. Substituting A for p thus gives us that $\vdash_G \boxdot(A \leftrightarrow A) \to (A \leftrightarrow H_A)$. Since $\vdash_G \boxdot(A \leftrightarrow A)$, $\vdash_G A \leftrightarrow H_A$.

For any sentence S of Z built up from $\mathbf{0} = \mathbf{1}$ by repeated applications of Prov(⌜ ⌝) and truth-functional connectives, e.g.

$$\text{Prov}(\ulcorner(\text{Prov}(\ulcorner \mathbf{0} = \mathbf{1}\urcorner) \to -\text{Prov}(\ulcorner\text{Prov}(\ulcorner \mathbf{0} = \mathbf{1}\urcorner)\urcorner))\urcorner),$$

we can therefore find a truth-functional combination S' of the false sentences $\mathbf{0} = \mathbf{1}$, Prov(⌜$\mathbf{0} = \mathbf{1}$⌝), Prov(⌜Prov(⌜$\mathbf{0} = \mathbf{1}$⌝)⌝), etc. such that $\vdash_Z S \leftrightarrow S'$. Given S we find an A containing no sentence letters and such that $A^\phi = S$. (ϕ may be arbitrarily chosen; as A contains no sentence letters, all A^ϕ are identical.) Then $\vdash_Z A^\phi \leftrightarrow H_A{}^\phi$, and we may take $S' = H_A{}^\phi$, which is a truth-functional combination of $\mathbf{0} = \mathbf{1}$, Prov(⌜$\mathbf{0} = \mathbf{1}$⌝), etc. Thus the truth-value of any such S can be calculated, as can its provability-value, for S is provable if and only if Prov(⌜S⌝) is true.

The reciprocation theorem

Let $p_0, \ldots, p_{n-1}$ be a sequence of sentence letters. Then

$$\vdash_G \Box\left(\bigwedge_{i<n} (\Box p_i \to p_i) \to -\Box^n \bot\right) \to \left(-\Box^n \bot \to \bigwedge_{i<n} (\Box p_i \to p_i)\right).$$

Proof. We again apply the semantical completeness theorem. Suppose that $\mathcal{M}$ is a model,

$$\mathcal{M}\left(w, \Box\left(\bigwedge_{i<n} (\Box p_i \to p_i) \to -\Box^n \bot\right)\right) = \mathcal{M}(w, -\Box^n \bot) = 1,$$

but that

$$\mathcal{M}\left(w, \bigwedge_{i<n} (\Box p_i \to p_i)\right) = 0.$$

We shall obtain a contradiction and thereby establish the theorem.

Since $\mathcal{M}(w, -\Box^n \bot) = 1$, $\mathrm{rk}(w) \geqslant n$, and therefore for some $w_{n-1}, \ldots, w_0$ in $W, wRw_{n-1}R \ldots Rw_0$, where $\mathrm{rk}(w_j) = j$ for all $j < n$. Since $\mathrm{rk}(w_j) < n$, $\mathcal{M}(w_j, -\Box^n \bot) = 0$. And since wRw_j,

$$\mathcal{M}\left(w_j, \bigwedge_{i<n} (\Box p_i \to p_i) \to -\Box^n \bot\right) = 1.$$

So

$$\mathcal{M}\left(w_j, \bigwedge_{i<n} (\Box p_i \to p_i)\right) = 0.$$

Lemma

There are at least j i such that $\mathcal{M}(w_{n-j}, \Box p_i \& p_i) = 1 (1 \leqslant j \leqslant n)$.

Proof. Induction on j. Suppose that $j = 1$. Since

$$\mathcal{M}\left(w, \bigwedge_{i<n} (\Box p_i \rightarrow p_i)\right) = 0,$$

for some i $\mathcal{M}(w, \Box p_i \rightarrow p_i) = 0$, $\mathcal{M}(w, \Box p_i) = 1$, and, as wRw_{n-1}, $\mathcal{M}(w_{n-1}, \Box p_i) = \mathcal{M}(w_{n-1}, p_i) = \mathcal{M}(w_{n-1}, \Box p_i \& p_i) = 1$. Suppose that $j = k+1 \leqslant n$. Let $I = \{i | \mathcal{M}(w_{n-k}, \Box p_i \& p_i) = 1\}$. Then by the induction hypothesis, there are at least k members of I. If $i \in I$, then

$$\mathcal{M}(w_{n-j}, \Box p_i \& p_i) = 1$$

(since $w_{n-k} R w_{n-j}$). And since

$$\mathcal{M}\left(w_{n-k}, \bigwedge_{i<n} (\Box p_i \rightarrow p_i)\right) = 0,$$

for some $h < n$, $\mathcal{M}(w_{n-k}, \Box p_h \rightarrow p_h) = 0$, $\mathcal{M}(w_{n-k}, p_h) = 0$, whence $h \notin I$, and $\mathcal{M}(w_{n-k}, \Box p_h) = 1$. Since $w_{n-k} R w_{n-j}$, $\mathcal{M}(w_{n-j}, \Box p_h \& p_h) = 1$. Thus there are at least $k + 1$ i such that $\mathcal{M}(w_{n-j}, \Box p_i \& p_i) = 1$. This completes the proof of the lemma.

By the lemma there are at least n i such that $\mathcal{M}(w_0, p_i) = 1$, and thus for every $i < n$, $\mathcal{M}(w_0, p_i) = 1$ and therefore $\mathcal{M}(w_0, \Box p_i \rightarrow p_i) = 1$. Thus

$$\mathcal{M}\left(w_0, \bigwedge_{i<n} (\Box p_i \rightarrow p_i)\right) = 1,$$

which contradicts the assertion that for all $j < n$,

$$\mathcal{M}\left(w_j, \bigwedge_{i<n} (\Box p_i \rightarrow p_i)\right) = 0.$$

Let $Z_1 = Z$ and let Z_{n+1} be the theory obtained from Z_n by adding to the axioms of Z_n the sentence Con_n of the language of Z that expresses the consistency of Z_n (i.e. the unprovability in Z_n of $\mathbf{0} = \mathbf{1}$). Con_n is equivalent in Z (identical if $n = 1$) to the sentence

$$C_n, = -\mathrm{Prov}(\ulcorner \ldots \mathrm{Prov}(\ulcorner \mathbf{0} = \mathbf{1} \urcorner) \ldots \urcorner)$$

(n nested $\mathrm{Prov}(\ulcorner\ \urcorner)$s). Every theorem of Z is true (in $\mathcal{N}$), and therefore no false sentence is a theorem of Z_n, Z_n is consistent, and C_n is true.

$\vdash_G \Box B \rightarrow \Box\Box B$, and thus if $n \leqslant m$, $\vdash_G \Box^n \bot \rightarrow \Box^m \bot$, $\vdash_G -\Box^m \bot \rightarrow -\Box^n \bot$, and by the arithmetical soundness theorem, $\vdash_Z C_m \rightarrow C_n$. But if $m < n$ and $\vdash_Z C_m \rightarrow C_n$, then $m+1 \leqslant n$ and, as we have just seen,

$\vdash_Z C_n \to C_{m+1}$, whence $\vdash_Z C_m \to C_{m+1}$. Contraposing and applying Löb's theorem, we infer that $\vdash_Z -C_m$, which is impossible, as no false sentence is a theorem of Z. Thus $\vdash_Z C_m \to C_n$ if and only if $n \leqslant m$.

A *reflection principle* is a sentence $\text{Prov}(\ulcorner S \urcorner) \to S$. Löb's theorem asserts that if a reflection principle is provable, then its consequent is. No sentence consistent with Z implies all reflection principles: if for all S', $\vdash_Z S \to (\text{Prov}(\ulcorner S' \urcorner) \to S')$, then in particular,

$$\vdash_Z S \to (\text{Prov}(\ulcorner - S \urcorner) \to -S);$$

by the propositional calculus, $\vdash_Z \text{Prov}(\ulcorner - S \urcorner) \to -S$, and then by Löb's theorem, $\vdash_Z -S$, and S is inconsistent with Z.

C_1, i.e. $-\text{Prov}(\ulcorner \mathbf{0} = \mathbf{1} \urcorner)$, is equivalent in Z to the reflection principle $\text{Prov}(\ulcorner \mathbf{0} = \mathbf{1} \urcorner) \to \mathbf{0} = \mathbf{1}$. C_2, i.e. $-\text{Prov}(\ulcorner \text{Prov}(\ulcorner \mathbf{0} = \mathbf{1} \urcorner) \urcorner)$, is equivalent to the conjunction of the two reflection principles

$$\text{Prov}(\ulcorner \text{Prov}(\ulcorner \mathbf{0} = \mathbf{1} \urcorner) \urcorner) \to \text{Prov}(\ulcorner \mathbf{0} = \mathbf{1} \urcorner)$$

and $\text{Prov}(\ulcorner \mathbf{0} = \mathbf{1} \urcorner) \to \mathbf{0} = \mathbf{1}$. And in like manner, C_n is equivalent to a conjunction of n reflection principles.

It follows from the reciprocation theorem that C_n implies any conjunction of n reflection principles that implies it: Suppose that

$$\vdash_Z \bigwedge_{i<n} (\text{Prov}(\ulcorner S_i \urcorner) \to S_i) \to C_n.$$

Then

$$\vdash_Z \text{Prov}\left(\ulcorner \left(\bigwedge_{i<n} (\text{Prov}(\ulcorner S_i \urcorner) \to S_i) \to C_n\right) \urcorner\right).$$

Let $p_0, \ldots, p_{n-1}$ be a sequence of n distinct sentence letters and let $\phi(p_i) = S_i$ for each $i < n$. Then

$$\vdash_Z \left(\Box\left(\bigwedge_{i<n} (\Box p_i \to p_i) \to -\Box^n \bot\right)\right)^\phi.$$

By the reciprocation theorem and the arithmetical soundness theorem,

$$\vdash_Z \left(\Box\left(\bigwedge_{i<n} (\Box p_i \to p_i) \to -\Box^n \bot\right) \to \left(-\Box^n \bot \to \bigwedge_{i<n} (\Box p_i \to p_i)\right)\right)^\phi.$$

Thus

$$\vdash_Z \left(-\Box^n \bot \rightarrow \bigwedge_{i<n} (\Box p_i \rightarrow p_i) \right)^\phi,$$

i.e.

$$\vdash_Z C_n \rightarrow \bigwedge_{i<n} (\mathrm{Prov}(\ulcorner S_i \urcorner) \rightarrow S_i).$$

It follows further that no conjunction of fewer than n reflection principles implies C_n. For if B is a conjunction of m reflection principles, $m < n$, and $\vdash_Z B \rightarrow C_n$, then since $\vdash_Z C_n \rightarrow C_m, \vdash_Z B \rightarrow C_m$, and therefore $\vdash_Z C_m \rightarrow B$ and $\vdash_Z C_m \rightarrow C_n$, which we have seen to be impossible in virtue of Löb's theorem.

Exercise

Let $A = \Box(-p \rightarrow \Box \bot) \rightarrow \Box(p \rightarrow \Box \bot)$. Find H_A.

Index